Mechanical Behavior of Materials
Deformation and Design

Mechanical Behavior of Materials
Deformation and Design

Rajiv S. Mishra,
University of North Texas, Denton, Texas

Indrajit Charit,
University of Idaho, Idaho Falls, Idaho

Ravi Sankar Haridas,
University of North Texas, Denton, Texas

ELSEVIER

Academic Press is an imprint of Elsevier
125 London Wall, London EC2Y 5AS, United Kingdom
525 B Street, Suite 1650, San Diego, CA 92101, United States
50 Hampshire Street, 5th Floor, Cambridge, MA 02139, United States
The Boulevard, Langford Lane, Kidlington, Oxford OX5 1GB, United Kingdom

Copyright © 2026 Elsevier Ltd. All rights are reserved, including those for text and data mining, AI training, and similar technologies.

Publisher's note: Elsevier takes a neutral position with respect to territorial disputes or jurisdictional claims in its published content, including in maps and institutional affiliations.

No part of this publication may be reproduced or transmitted in any form or by any means, electronic or mechanical, including photocopying, recording, or any information storage and retrieval system, without permission in writing from the publisher. Details on how to seek permission, further information about the Publisher's permissions policies and our arrangements with organizations such as the Copyright Clearance Center and the Copyright Licensing Agency, can be found at our website: www.elsevier.com/permissions.

This book and the individual contributions contained in it are protected under copyright by the Publisher (other than as may be noted herein).

Notices

Knowledge and best practice in this field are constantly changing. As new research and experience broaden our understanding, changes in research methods, professional practices, or medical treatment may become necessary.

Practitioners and researchers must always rely on their own experience and knowledge in evaluating and using any information, methods, compounds, or experiments described herein. In using such information or methods they should be mindful of their own safety and the safety of others, including parties for whom they have a professional responsibility.

To the fullest extent of the law, neither the Publisher nor the authors, contributors, or editors, assume any liability for any injury and/or damage to persons or property as a matter of products liability, negligence or otherwise, or from any use or operation of any methods, products, instructions, or ideas contained in the material herein.

ISBN: 978-0-12-804554-1

For information on all Elsevier publications visit our website at
https://www.elsevier.com/books-and-journals

Publisher: Andrea Gallego Ortiz
Acquisition Editor: Stephen R. Merken
Editorial Project Manager: Shilpa Kumar
Production Project Manager: Nandhini Thanga Alagu
Cover Designer: Matthew Limbert

Typeset by GW Tech

Printed in India

Last digit is the print number: 9 8 7 6 5 4 3 2 1

To our gurus and families

The guiding and loving forces that shaped our "matter" and "energy" to function!

Contents

Preface and acknowledgement ix

CHAPTER 1	Mechanical response of materials 1
CHAPTER 2	Framework of five basic design approaches for structural components 19
CHAPTER 3	Introduction to systems approach to materials 33
CHAPTER 4	A brief survey of microstructural elements in engineering structural materials 47
CHAPTER 5	Simple mechanical tests 63
CHAPTER 6	Elastic response of materials: stiffness limiting design 91
CHAPTER 7	General dislocation theory for crystalline materials 127
CHAPTER 8	Yielding and work hardening: strength limiting design 149
CHAPTER 9	Toughness of materials: toughness limiting design 213
CHAPTER 10	Fatigue behavior of materials: fatigue limiting design 251
CHAPTER 11	High-temperature deformation of materials: creep limiting design 301

Index 347

Preface and acknowledgement

Mechanical behavior of materials is a core subject for students of materials science and engineering and an optional elective for mechanical engineering students. Therefore a large number of textbooks are available, and they can be typically divided into two groups: materials centric and mechanics centric. The materials-centric books cater primarily to the materials science and engineering students. These textbooks follow a linear presentation of information, that is, they start with atoms, crystal structure, defects in crystals—dislocations—and then properties. A typical undergraduate student takes the core course in the junior year of a 4-year degree program. The mechanics-centric books catering to mechanical engineering students tend to focus more on stress concentration and life prediction. Typically, those books dwell less on microstructural aspects and more on continuum mechanics approaches. The primary author of this book (RSM) has researched the mechanical behavior of materials for 40+ years at the time of this writing and has been teaching various aspects of the mechanical behavior of materials for more than 25 years. This intersection of teaching and research on advanced metallic materials led to a shift in his thinking and the development of a conviction that the real answers in this field lie at the crossover between microstructural aspects and mechanics. Also, an observation was that most undergraduate students struggled to appreciate the "real" importance of this field. After all, all the structural components and engineered systems are designed based on the mechanical properties of materials. Also, a critical question that students need to start pondering is, "Why do fail-safe designed components fail in service?" They need to be able to see the connections between the fundamentals of deformation mechanisms and the performance of engineered systems!

That became the motivation for putting this textbook together, which is written to serve as an initial book for students at the undergraduate level or graduate level. Engineers and researchers getting into the field of design and performance of structural materials may find the book useful as a resource for continuing education. *We have purposely kept the book short and avoided the urge to make it a comprehensive review. The entire content of the book can be easily covered in a 15-week semester.*

The presentation of this book differs significantly from that of other books. We start with the first four generic chapters to provide students with a broader perspective of "why they should be interested in mechanical behavior," other than the fact that it is a required course for many! Ashby's framework of five design approaches, namely, stiffness, strength, toughness, fatigue, and creep, guides the entire presentation, starting with the introduction of this framework in Chapter 2. Intertwined is Olson's framework of materials system approach linking materials science and the engineering chain of processing → microstructure → properties → performance, presented in Chapter 3 along with the concepts of design allowables. Chapter 4 very briefly presents the microstructure of engineering materials so that the students develop an understanding of the length scales of various features. With the initial broader perspective, hopefully students will be motivated to understand the fundamentals behind various mechanical properties and learn to question the assumptions behind certain theories. Keeping an alert mind as they pose the question, how does it apply to a particular microstructure? The first part of the book concludes with a generic survey of techniques used to evaluate mechanical properties in Chapter 5.

Then the book pivots to finally getting into the five mechanical property-based design approaches. Chapter 6 delves into stiffness-limited design where only the elastic properties are important. The components will not go through any appreciable plastic deformation. Note that we have crossed half of the book without discussing "dislocations" that form the basis for the remaining four plasticity-based design approaches of strength, toughness, fatigue, and creep. The dislocations are formally introduced in Chapter 7, and then Chapters 8 to 11 are plasticity-based chapters. As is quite common, there is always a gap between desire, intention, and the reality of the final product. The book uses a "concept" builder and text boxes to highlight certain aspects. The desire and intention at the start of the book writing phase was to have "many" such examples in each chapter. The reality is that we only put in a few in this first edition. Our hope is to revise the chapters based on the feedback from instructors and students to make these more impactful. Nevertheless, we think that students who are "global learners" will enjoy this presentation more, whereas in our opinion the previous textbooks on mechanical behavior cater to "sequential learners." We would love to hear your feedback on the impact of our presentation and book content. Does this approach and keeping the content short work for you?

At the time of completion of this book (December 2024), one of the authors (RSM) had spent over 40 years researching deformation mechanisms in metallic materials. He would like to acknowledge the start he got as a Master's student working with Prof. G.S. Murty on superplasticity and serving as a teaching assistant for the mechanical behavior of materials class at Indian Institute of Technology Kanpur (IIT Kanpur) in 1983. RSM went on to learn creep from Prof. G.W. Greenwood and Prof. H. Jones at the University of Sheffield. The learning continued with Prof. Amiya Mukherjee at UC-Davis with the first experience of co-teaching the course. After starting as a faculty at the University of Missouri-Rolla (UMR), inheriting the teaching material from Prof. Ron Kosher and Prof. David van Aken built the knowledge further. The generosity of Prof. Greg Olson in providing his class notes on the systems approach became another formative step. Finally, the series of books by Prof. Ashby shaped and refined the thinking that has gone into the structure of this book. Over the years, many students and postdoctoral researchers contributed to RSM's understanding of mechanical behavior as they completed research projects and wrote their papers. The authors have learned much of what they know from these interactions in the research group, where Indrajit and Ravi Sankar at one point were researchers.

The support from NSF, DARPA, ARL, ONR, Boeing, General Motors, Pacific Northwest National Laboratory, Pratt & Whitney, Friction Stir Link, AFRL, Magnesium Elektron North America, NASA, Naval Surface Warfare Center, Hitachi North America, and several others over the years allowed continuous growth of our work and understanding. Without sustained support, it would not have been possible to reach this stage where we put this book together!

One of the authors (IC) would like to acknowledge his major professors who left a deep impression on him and from whom he has learned a lot in the area of the mechanical behavior of materials. IC has been serving on the faculty of the University of Idaho, spanning over 18 years, and many of the courses he has taught there are more or less related to this area only. During his Master's degree at the Indian Institute of Science in 1999–2000, IC worked under the supervision of Prof. Atul H. Chokshi on the high-temperature deformation characteristics of 3 mol% yttria-stabilized tetragonal zirconia. Later, while completing his Ph.D. degree at the University of Missouri-Rolla (2000–2004) under Prof. Rajiv Mishra's supervision, IC worked on the investigation of superplasticity in friction stir–processed aluminum alloys. Subsequently, IC worked on his postdoctoral research (2005–2007) under Prof. K. Linga Murty's guidance on the creep behavior of zirconium and titanium alloys. IC considers himself very fortunate to have had the opportunity to work with these foremost experts in the field of mechanical behavior and has developed lifelong research and teaching interests therein. The collaborations with Profs. Mishra and Murty continue to this day. While it is not possible to mention all the professional colleagues by name here, IC would like to gratefully acknowledge their support and encouragement here as well. Over several years, IC has been fortunate to obtain funding support from different agencies, such as the Department of Energy—Office of Nuclear Energy, Idaho National Laboratory/Center for Advanced Energy Studies, National Technology Laboratory, Pacific Northwest National Laboratory, Nuclear Regulatory Commission, Idaho Department of Commerce, and National Science Foundation, which helped him to sustain his efforts in this very important area.

One of the authors, RSH, embarked on his exploration of the mechanics of materials during his Master's program at the IIT Kanpur, nearly 13 years ago. The foundational knowledge of the mechanics of materials acquired through extensive coursework at IIT Kanpur served as the bedrock for his research. Special gratitude is owed to Prof. P. Venkitanarayanan, Prof. N. N. Kishore, Prof. P. M. Dixit, Prof. Sumit Basu, Prof. S. Mahesh, Prof. Sovan Das, and Prof. Ishan Sharma for illuminating his path through their teachings. He extends his heartfelt appreciation to Dr. P. Venkitanarayanan, his Ph.D. advisor, for his unwavering support, particularly during challenging moments at IIT Kanpur. His visionary guidance, knowledge, patience, and adept handling of intricate experiments consistently served as a motivating force for a budding researcher like RSH. RSH proceeded to serve as a postdoctoral research associate under Prof. Rajiv Mishra at the University of North Texas (UNT). The outstanding facilities in Prof. Mishra's lab provided the ideal environment to broaden his research into the assessment of material-centric mechanical behavior. The expertise and vision of Prof. Mishra played a pivotal role in shaping and advancing his research endeavors. He expresses sincere gratitude to Prof. Mishra for extending an invitation to contribute to the present textbook. He expresses deep gratitude for the dedicated efforts of his colleagues, graduate students, and undergraduate students in achieving the research objectives at UNT. Currently, RSH is serving as a research assistant professor in the department of mechanical engineering at UNT and his research merges at the interface of mechanics, materials, and manufacturing.

Although we have not listed students and postdoctoral researchers by name, their tireless efforts are very much behind the body of this work. Many of their figures have been cited and sometimes without full acknowledgement. Apologies to all those whose work we could not properly cite.

We are very thankful to the Elsevier team, in particular Steve Merken, for providing the opportunity to put this textbook together and for his extraordinary patience. Shilpa Kumar consistently moved this project along, and we are grateful for that. We are sure we tested the patience of the Elsevier team by missing numerous deadlines!

Finally, none of this would be possible without the support of our families and their sacrifices. So, we close with a big THANKS to our family members.

Rajiv S. Mishra
Indrajit Charit
Ravi Sankar Haridas

Resources that accompany this book

Resources are available to adopting instructors who register on the Elsevier textbook website, https://educate.elsevier.com/book/details/9780128045541
   Teaching ancillaries for this book, including a solutions manual
   PowerPoint lecture slides
   Image bank

CHAPTER 1

Mechanical response of materials

Chapter outline

1.1 Mechanical response of materials – introduction 3
    1.1.1 Ashby's property charts 5
1.2 Broad categories of materials... 6
    1.2.1 Metals/alloys 7
    1.2.2 Ceramics... 8
    1.2.3 Polymers 10
    1.2.4 Composites 11
1.3 Engineered systems and materials 12
    1.3.1 Automobile... 13
    1.3.2 Advanced aircraft 14
    1.3.3 Power plant.. 14
    1.3.4 Miscellaneous examples of engineering systems 15
1.4 Exercises 17

Learning objectives

- A broad understanding of mechanical response of materials.
- Classification of materials.
- Understanding of what constitutes engineered systems.

The setup and layout of this book are very different from those of other books on the mechanical behavior of materials. First, we will set up the context of mechanical response of materials under loading in different types of systems. The initial chapters provide a broader framework and context of why we are concerned with mechanical behavior and what the critical first steps to learn are. Use of the systems approach for materials guides us to fundamentals that are important to master and that explain how they are relevant in engineering applications. Similarly, the types of design approaches used for engineering applications bring certain clarity to fundamentals that need to be mastered. With this approach, the introductory chapter provides a broader framework of mechanical response of materials, a broad classification of materials, and some examples of engineered systems.

    The process of developing engineered systems involves (1) design, (2) materials, and (3) processes. The process starts with a conceptual design; further progress depends on the choice/availability of materials and processes. The importance of mechanical behavior of materials and availability of the

Chapter 1 Mechanical response of materials

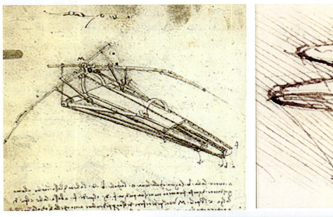

Figure 1.1

Examples of da Vinci's sketches (top) and current concepts (bottom) that are limited by the availability of engineered materials and processes. While da Vinci was able to conceptualize flying machines, materials and processes did not exist at that time to support engineering development. Similar constraints apply today to the concepts of hyperplanes. An artist's conception of the X-43A Hypersonic Experimental Vehicle, or "Hyper-X", in flight.[1]

right combination of materials and processes in the design of engineered systems can be appreciated by considering Fig. 1.1. The top figures show sketches by Leonardo da Vinci. If we consider the concepts and systems that he drew, the first part of conceptualization was done! What limited the progress or building of these systems? The availability of engineered materials and processes! After about 500 years, we live in very exciting times, where engineered spacecraft are traversing through the vast universe. Still, the conceptual designs are not fully attained even with the availability of current materials and processes. As an example, the concept of hyperplanes, which are supposed to fly at Mach

---

[1]Hyperplane figure adapted from https://www.nasa.gov/centers/dryden/multimedia/imagegallery/X-43A/ED99-45243-01.html. Credits: NASA Illustration.

5 to 7, is still undeveloped. The hyperplanes have severe requirements of lightweight for structural systems and extremely high temperatures associated with Mach 5 to 7 speeds.

1.1 Mechanical response of materials – introduction

To begin with, an application of load or displacement may be considered as the starting point. This is because both are linked by a cause-effect relationship. At a very basic level, if load is applied, the material will undergo displacement. The engineering terms to discuss normalized load and displacement are *stress* and *strain*, respectively. The assumption here is that the readers of this book have completed at least an introductory course on materials science and that they have at least a basic familiarity with the keywords used to describe mechanical response of materials. The primary approach in this book is "materials science" based, and the concepts considered to describe deformation may differ at times from the purely "mechanics" approach. A simple way to understand this difference is to consider the origin of theories for plastic deformation. While Tresca (1864) and von Mises (1913) described deformation of metals in continuum mechanics framework, Orowan (1934) and Taylor (1934) introduced the concept of dislocations. Both approaches describe the transition from elastic to inelastic states, but the fundamental steps are quite different. It is important to capture the appropriate physics behind an observation. This aspect is emphasized throughout this textbook, focus being on concepts rather than mere mathematical expressions.

Fig. 1.2 lists some basic mechanical responses of materials to loading as a function of temperature and time. The rate of loading and the type of loading (monotonic or cyclic) can also vary. The initial response of the material is elastic or fully recoverable, and above a critical value of load the material undergoes plastic or permanent, nonrecoverable, deformation. The plastic deformation is also referred to as inelastic response. Incorporation of time dependence of the response changes the terms to viscoelasticity and viscoplasticity. The rate of loading is another important factor, and the response is described by the terms *quasistatic* (behavior at low rates of deformation) or *dynamic* (high deformation rates). Very low rates of deformation are referred to as *creep*. The common types of testing are discussed in **Chapter 5**. The simplest tests, in terms of ease of testing, are hardness measurement and tensile or compression tests. The rate of loading is low during these tests, and material response is considered static or quasistatic. Useful information about the elastic response of material, strength, and ductility is easily obtained from these tests.

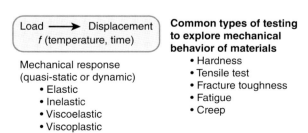

Figure 1.2

A simple depiction of common mechanical responses of materials to different applications of loading. Note that the key interest is to link different types of mechanical loading with properties of interest for mechanical design.

Adding the nature of loading and duration of test cycle expands the response space. In typical hardness testing or tensile testing, the load or displacement is applied monotonically. Note that while uniaxial tension or compression tests involve elementary loading that can be easily quantified, an indentation test for hardness measurement has complex loading leading to a multiaxial stress state. If the material is subjected to cyclic loading, it undergoes fatigue deformation. Fatigue testing also brings in the effect of test duration. A number of engineering structures experience fatigue deformation, and in fact, fatigue is the most common cause of failure of engineered materials. The number of cycles to failure of a component then brings in another line of demarcation between low cycle fatigue and high cycle fatigue (**Chapter 10**). A typical high cycle fatigue test is done below the yield strength of the material. Still, by undergoing long-term cycle loading, the material fails below the yield strength of the material. A key point to ponder is why the material fails in the so-called elastic deformation regime! Understanding the response of such loading on localized internal deformation and propagation of deformation through the volume of material being tested is important for linking material performance to microstructural features.

Metallic materials are typically considered "tough" and are used in applications where damage tolerance is required. For engineered materials, toughness may be quantified by different measures or concepts related to performance. One is damage tolerance. Damage tolerance can be described as the ability of a structure to sustain defects safely until inspection and/or repair can be performed. The basic resistance of material to failure is defined as toughness. In other words, how much energy is absorbed during the process of deformation and rupture? A simple tensile test provides the first glimpse of toughness of the material. The area under the tensile curve gives a measure of energy absorbed during the process of deformation and failure. However, this is considered static loading. The failure of a material under dynamic loading conditions, such as a crash event, is quite different. **Chapter 9** presents a good comparison of various toughness evaluation approaches and what we can glean from these tests. It should be noted that fracture toughness is not the same as toughness, but they are related. Fracture toughness (or critical energy release rate) is a measure of strain energy consumed in growing a crack (mostly by inelastic deformation).

Fig. 1.3 shows a general depiction of the application of load to determine (1) hardness, (2) strength, (3) fatigue data, and (4) fracture toughness values. All testing techniques are presented in detail in **Chapter 5**. The hardness tests are very simple to perform and involve pushing an indenter into the material. Instrumented hardness testing also leads to a measurement of the stiffness of the material being tested, which are material property values that are needed for a **stiffness-limiting design approach**. Similarly, pulling the specimen during tensile testing or squeezing it during compression testing provides strength values needed for a **strength-limiting design approach**. Application of cyclic loading results in fatigue data, which can be obtained under stress-amplitude testing or strain-amplitude testing. Stress-amplitude testing is relatively easy to perform and provides stress *(S)*–number of cycles to failure *(N)* data. The *S-N* data is used for the **fatigue-limiting design approach**, whereas the **toughness-limiting design approach** is used when the existence of flaws is an important consideration. To determine fracture toughness value, resistance to crack growth is measured under certain loading conditions and geometrical constraints. The **creep-limited design approach** is the last of the five approaches considered in this book. Creep tests are conducted on deadweight machines where either rupture life or time to a particular creep strain value is determined. The type of approach depends on the particular application. **Chapter 11** presents the creep-limiting design approaches along with creep deformation mechanisms, which dominate deformation above 0.4–0.5 T_m, where T_m is the melting point of the material in K. Here the ratio of temperature (also in K) of interest and the melting point is called homologous temperature; so by definition the fraction is the homologous temperature.

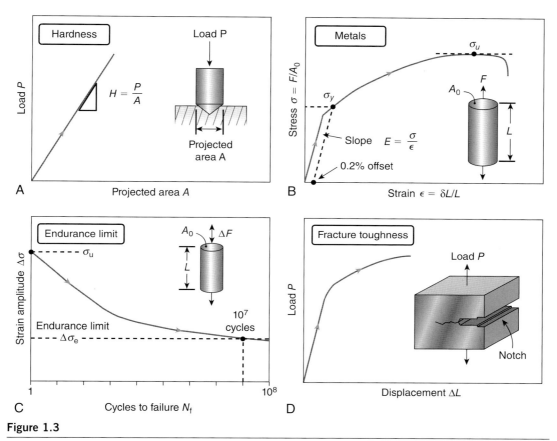

Figure 1.3

Typical types of loading during mechanical tests that result in obtaining (a) hardness, (b) strength, (c) fatigue, and (d) fracture toughness data.

1.1.1 Ashby's property charts

Professor Michael Ashby of Cambridge University in the UK started property charts as a way to depict material properties in a space to aid selection of materials for engineering applications. Previously we mentioned the five design approaches. The entire book is concerned with these approaches, and Ashby's approach is developed further in **Chapter 2**. Fig. 1.4 shows four charts to just highlight typical values of four of the five properties and to begin to get a feel for the property space. An important feature of these charts is the relative ease with which different materials can be compared visually. The charts of a specific property against the density of material allow for comparison on the basis of normalized properties. When a particular property is divided by the density, it is referred to as "specific property." For example, considering specific stiffness and specific strength for aerospace materials is very common. Additional design considerations, such as space or physical size, would mean that the low-density material may become too large for a particular physical space. Most engineering applications must

Chapter 1 Mechanical response of materials

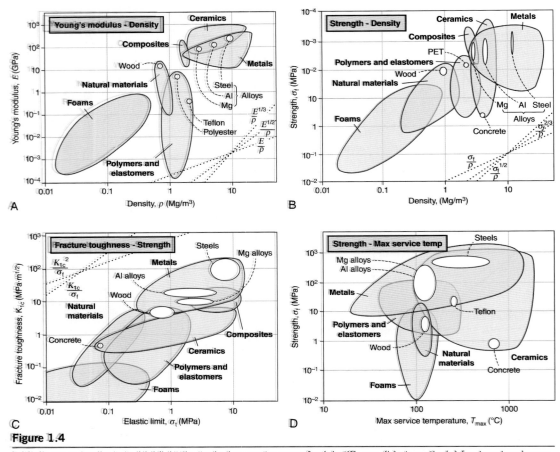

Figure 1.4

Ashby's property charts to highlight the typical property space for (a) stiffness, (b) strength, (c) fracture toughness, and (d) maximum service temperature.[2]

consider and meet multiple design considerations, and typically an index approach is pursued to find a balance among various constraints.

1.2 Broad categories of materials

We live in a "Materials World" that consists of both natural and manmade/engineered materials. Given the materials' unique status, and based on the importance of materials, progress of civilization has been divided into distinct chronological periods such as Stone Age, Copper Age, Bronze Age, Iron Age, Plastic Age, and Silicon Age. Even though some of the materials used today were in use in ancient

---

[2] Adapted from Ashby MF. *Materials Selection in Mechanical Design*. 2nd ed. Butterworth-Heinemann; 1999.

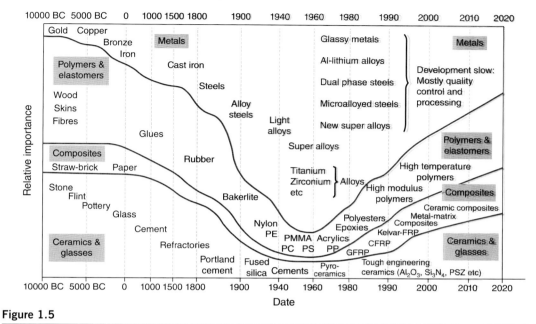

Figure 1.5

The evolution of engineering materials with time. Note that the horizontal scale is highly nonlinear.[3]

times, since the Industrial Revolution, materials have become even more ubiquitous, more numerous, and more utilitarian (Fig. 1.5). For instance, the relative importance of metals increased in the last century, but now it is diminishing with the increased availability of high-performance polymers and composites. We will discuss in this section various materials classes including metals, ceramics (including glasses), polymers, and composites. We will not discuss semiconductors since they are seldom used as load-bearing components. Similarly, a whole category of advanced materials, including nanomaterials, biomaterials, intermetallics, smart materials, bulk amorphous alloys, and metallic foams, also will not be discussed in this particular section. However, they will be discussed throughout the book when they are relevant, and their importance will be highlighted as appropriate.

1.2.1 Metals/alloys

Metals tend to be naturally strong, ductile/malleable, electrically and thermally conductive, and electropositive. Approximately three-fourths of the elements in the Periodic Table are metals. The different ways by which metals are categorized include but are not limited to alkali metals, alkaline earth metals, rare earth metals, transition metals, and noble metals.

The bonding between metal atoms is different in the sense that bonding occurs as a result of the formation of a common "electron sea" that is contributed by the valence shell electrons of each metal atom and results in attractive forces. Approximately two-thirds of metals have simple crystal structures,

---

[3]From Ashby MF. *Materials Selection in Mechanical Design*. 2nd ed. Butterworth-Heinemann; 1999.

Figure 1.6

The Golden Gate Bridge, north of San Francisco, California, is one of the most famous steel bridges in the world.[4]

mainly body-centered cubic (BCC), face-centered cubic (FCC), and hexagonal close-packed (HCP). Examples of FCC metals are copper (Cu), aluminum (Al), and lead (Pb); BCC metals are alpha-iron (α-Fe), chromium (Cr), and vanadium (V); and HCP metals are alpha-zirconium (α-Zr), alpha-titanium (α-Ti), and cadmium (Cd). Properties of metals vary widely. For example, tungsten (W) has a high melting point, whereas mercury is liquid at room temperature. Pure metals are rarely used; instead, most often, they are used as alloys – to be qualified for an alloy designation, at least one element of the alloy needs to be a metal. Alloying may serve many purposes, such as improving corrosion resistance or increasing strength. One common example are Fe-C alloys, which are commonly known as steel.

We can broadly classify engineering alloys into two main types of alloy systems: ferrous and nonferrous. Ferrous metals and alloys typically contain iron as the base metal and generally contain some carbon. In terms of tonnage use of metals/alloys, ferrous alloys (steel and cast iron) rank first, followed by nonferrous aluminum–based alloys. Steels have become invaluable materials for driving civilization forward since the Industrial Revolution. Fig. 1.6 shows a shining example of steels in the form of the Golden Gate Bridge. Steel is a versatile material in terms of its intrinsic structure and properties. There are plain carbon steels, low, medium, and high alloy steels, and a variety of specialty steels. For instance, a US car may contain ferrous alloys up to 60% of its weight. Some of the nonferrous metals/alloys are aluminum, magnesium, and titanium (also called light metals); nickel-based superalloys and refractory metals/alloys (W, Mo, Ta, Nb, Re – melting point >2000°C); and precious metals (Au, Ag).

1.2.2 Ceramics

Ceramics are crystalline and generally are composed of both metallic and nonmetallic components (C, O, N, S, P). Examples of oxide ceramics are Al_2O_3 (aluminum oxide or alumina), MgO (magnesium oxide or magnesia), and CaO (calcium oxide or calcia). Also, think about silicon nitride and silicon carbide – they are also ceramics! Fig. 1.7 illustrates some ceramic products. Often crystal structures of ceramics are complicated compared to those of metals. Bonding in ceramics is generally covalent and/or

---

[4]Courtesy: http://goldengatebridge.org

Figure 1.7

Some ceramics for conventional engineering applications. These various components have characteristic that make them resistant to damage by high temperatures and aggressive environments and are therefore used in furnaces and chemical processing plants.[5]

ionic in nature. In general, ceramics are strong but brittle (i.e., not malleable like metals), as well as thermally and electrically insulating. Indeed, ceramics based on compounds like $YBa_2Cu_3O_7$ and $Ba_2Sr_2CaCu_2O_x$ are superconducting at temperatures to about 95 K! Ceramics can resist very high temperatures and are superior in chemical/oxidation resistance compared to metals and polymers in general.

While the compositions of glasses are close to those of ceramics, they are different in structure. Atoms in the structure of glasses are amorphous, being arranged irregularly and randomly. Fig. 1.8 demonstrates that point. A noncrystalline silica glass shows schematically the random arrangement of atoms with only short-range order, whereas the crystalline cristobalite shows truly crystalline structure. Glasses are generally transparent, brittle, and chemically inert. Ordinary window glass contains approximately 72% of SiO_2 along with Na_2O and CaO. Some common silicate glasses combine the important qualities of transmitting clear images and resisting chemically aggressive environments. A major revolution in the telecommunications industry happened when traditional metal cables were replaced with optical glass fibers (photonic materials)! Optical glass fibers carry data in laser light pulses rather than electrical signals. Thus glasses are used in structural materials (such as in large buildings) as well as high-tech products.

---

[5]Juste, E., Petit, F., Lardot, V. et al. Shaping of ceramic parts by selective laser melting of powder bed. Journal of Materials Research (Springer Nature Link) 29, 2086–2094 (2014). https://doi.org/10.1557/jmr.2014.127

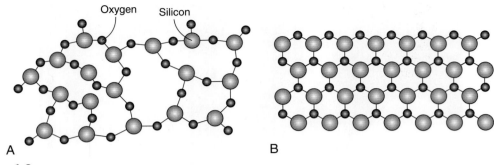

Figure 1.8

Schematic comparison of the atomic-scale structure of (a) silica glass (noncrystalline) and (b) crystalline silica (such as cristobalite).

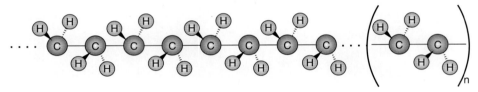

Figure 1.9

Molecular structure of polyethylene (n is the number of mers in a long polymer chain).

1.2.3 Polymers

Polymers (generally known as plastics) are quite different from metals. Polymers contain *macromolecules* instead of simple atoms/molecules in their structures. These molecules are generally giant chains of atoms. The unit "mer" is repeated several times and creates the giant molecule. The molecular structure of polyethylene (or polythene) is shown in Fig. 1.9. Polymers may contain a few elements such as C, H, O, N, Si, Cl, S, and other occasional elements. Although polymers have existed in nature for several millions of years in the form of wool, silk, DNA, cellulose, and proteins, their manmade (synthetic) versions are relatively new. A few synthetic polymers include polypropylene, polyvinyl chloride, nylon, polyethylene, polyester, Teflon, epoxy, and polymethylmethacrylate. Despite being a relative newcomer among the engineering materials classes, polymers have an array of applications in such various products as computers, houseware, toys, appliances, and automobile components. In many applications, polymers have even replaced metals, for example, polymer matrix composites (PMCs) in place of aluminum alloys in aircraft structures. Some of the advantages of polymers are low cost, low density, corrosion resistance, and low electrical and thermal conductivity. However, low strength and elastic modulus, low thermal stability, low creep resistance, and high thermal expansion coefficient may cause issues in certain applications that require high performance. Kevlar fibers developed by Du Pont are used in making body armors!

Properties of polymers depend strongly on the structure of the individual molecules, their shape and size, and molecular arrangement. The structure of polymers is varied – crystalline, amorphous, or a mixture of crystalline and amorphous, i.e., semicrystalline. Increasing crystallinity also affects various properties of polymers. Polymers are formed by linking monomers through a polymerization process. Monomers are linked via covalent bonding to form giant molecular chains. However, these chains are held together by secondary bonds (such as hydrogen bonding, van der Waals bonding). Polymers can be of different types depending on the polymer chains, such as linear polymers, branched polymers, cross-linked polymers, network polymers, copolymers, and terpolymers. Note that thermoset polymers, once given a final curing or hardening processing step, remain in solid state and cannot be softened by heating. They would char if heated above a critical temperature. On the other hand, thermoplastics can be remelted. A very common example is metallographic sample preparation. During cold or hot mounting procedure, thermoset or thermoplastic polymers are used. For example, the class of thermosets are cross-linked polymers that include epoxies, phenolics, and silicones. That is why the molecular weight of the polymers becomes quite large and can range from 10,000 to 10,000,000 g/mol for most commercial polymers. Many properties such as tensile strength, impact strength, elastic modulus, and viscosity (in molten condition) strongly depend on the molecular weight.

1.2.4 Composites

Composites are widely applied in modern industrial applications (aerospace, automotive, sports, electronics, and many more). Indeed, their use may be more common than we think! Consider the many examples of natural composites in nature, e.g., wood, bone, and seashells. Nature took millions to billions of years to perfect these biological structures. Thus emerged the field of *biomimetics,* which deals with learning the fundamental principles of how nature works and trying to emulate those principles to make engineered materials. Humans made various forms of engineering composites in the early stages of civilization to help gain advantage over natural materials. Since midtwentieth century, manmade composites have assumed increasing significance. Most recently, the use of PMCs in Boeing's Dreamliner (787) airplane is particularly noteworthy. Composites are generally characterized as strong, stiff, and lightweight but expensive.

Composites can be defined as the combination of two or more materials that result in a better combination of properties than the properties of individual constituents. So, basically, the analogy word could be *"The Best of Both Worlds!"* Basically, the principle of combined action applies in composite design. Different ways to group composites include particle-reinforced (large-particle and dispersion strengthened), fiber-reinforced (continuous and discontinuous), and structural composites (sandwich and laminates). Depending on the matrix materials, there could be three types of composites: metal matrix composites (MMCs), ceramic matrix composites (CMCs), and PMCs. Examples of PMC materials include glass fiber-reinforced polymer (GFRP) composites, carbon fiber-reinforced polymer (CFRP) composites, and aramid fiber-reinforced polymer (AFRP) composites. Sports products such as tennis rackets, bicycle frames, and skis are made of CFRP composites. MMCs include Al-SiC composites, which are noted for being lightweight and stiff and having high strength-to-weight ratios. An example of CMCs is the transformation toughening system of Al_2O_3-ZrO_2, wherein partially stabilized ZrO_2 particles are distributed in alumina matrix. Carbon-carbon composites are used for high-performance conditions. Hybrid composites also have applications in the mix.

Table 1.1 summarizes relative properties of metals, ceramics, and polymers with some examples.

Table 1.1 Comparison of metals, polymers and ceramics properties

Properties	Metals	Polymers	Ceramics
Melting point	Low to High Gallium (22.76°C) Tungsten (3,422°C) Fun fact: Mercury is a metal, but its melting point is -38.83°C, which means it is essentially liquid at room temperature.	Low for most polymers Exception: Certain Kevlars can melt at much above 500°C.	High Maximum 4,215°C (Ta_4HfC_5)
Density at room temperature	Low to High, 1 to 22 Lithium: 0.534 g/cm^3 Osmium: 22.6 g/cm^3	Low 1 to 2 g/cm^3	Most intermediate 2 to 19 g/cm^3
Thermal expansion coefficient	Medium to high	Very high	Low to medium
Thermal conductivity	Medium to High Silver (highest in metal): 429 W/m.K at 20°C	Quite low	Medium
Electrical conductivity	Good Conductors	Insulators	Electrical Insulators One exception: Super-conducting ceramics
Chemical resistance	Low to medium	Excellent	Good
Oxidation resistance	Generally poor	-	Oxides excellent; Certain carbides and nitrides are good
Hardness	Medium	Low	High
Young's modulus	15 to 400 GPa	0.001 to 10 GPa	150 to 450 GPa
Compressive strength	Up to 2,500 MPa	Up to 350 MPa	Up to 5,000 MPa Ceramics are strong under compression
Tensile strength	Up to 2,500 MPa	Up to 140 MPa	Up to 400 MPa
Elevated temperature creep resistance	*Poor to medium* High for metal-based systems like nickel base superalloys and oxide dispersion strengthened alloys	*Generally poor*	*Good*

1.3 Engineered systems and materials

An engineered system is comprised of many components and subsystems. Usually these are made of many engineered materials that are put together by various forms of fastening and joining techniques. The earlier Ashby charts featured the property space of specific materials. Manufacturing processes that are used to put the system together usually impact the properties. It is important to keep the impact of secondary processes in mind. For example, if a joining process involves local melting of material, its impact on microstructure and properties must be considered. We wrap up this section with illustrations of different types of systems. For readers who do not have an in-depth knowledge of the details of engineered systems, consider these as you read the next subsections:

- Type of component
- Nature of loading – static, cyclic, dynamic
- Type of loading – tension, compression, bending, shear, mixed
- Which material it may be made of and what other materials may be appropriate?
- Temperature range of application
- Which mechanical design approach would be appropriate (elastic, yield or ultimate strength, fatigue, toughness, creep)?
- Importance of environmental effects – corrosion, erosion?

1.3.1 Automobile

Fig. 1.10 shows an automobile cutaway picture. As a student, when you look at this picture, what comes to mind? How many subsystems and components do you see? As you look at a particular subsystem or component, if you start making a set of questions, what is important? Consider only components that are undergoing mechanical loading. At this stage, normal curiosity should lead to questions such as:

- Which type of material would be best?
- What type of loading is experienced?
- Which design approach would be most appropriate?

As you navigate through this, certain fundamentals that are key to making the best engineered system in this category would become obvious. How would the thought process change if the key design goal changes from performance to cost?

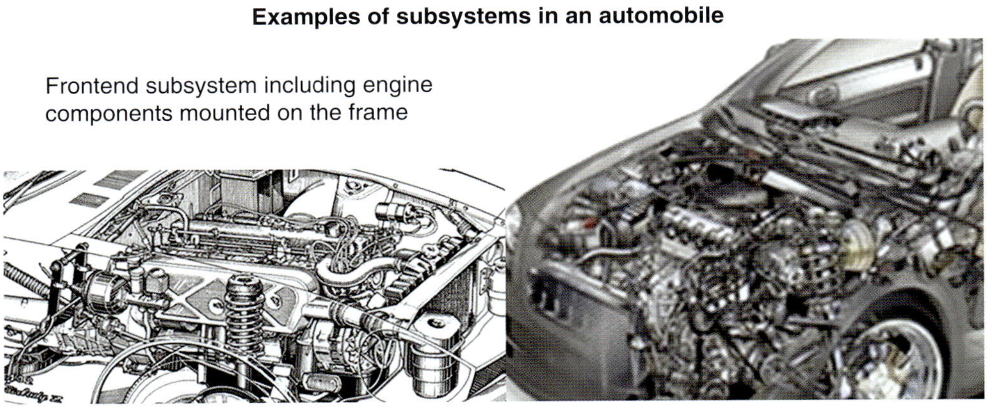

Examples of subsystems in an automobile

Frontend subsystem including engine components mounted on the frame

Figure 1.10
A cutaway of an automobile showing various mechanical subsystems.

Figure 1.11
Picture of Boeing 787 plane with some design considerations for various sections of the plane. Note that many properties are considered for selection of a particular part of the aircraft.

1.3.2 Advanced aircraft

Fig. 1.11 shows the latest Boeing 787 airplane that makes extensive use of composites. Superimposed in the picture are various properties that are considered, as well as the complexity in the advanced engineering system that performs at the highest structural efficiency. The key takeaway point here is that as you go through the various sections of the book and learn the microstructural dependence of a particular property, also consider how it affects other properties. For example, certain microstructural features may lead to higher strength. Then ask, how do these microstructural features impact fatigue, fracture toughness, and other properties? By the time you finish this book, ideally you will have a good understanding of microstructural control for a balanced set of properties for a mechanical component or system.

1.3.3 Power plant

Now consider a very large land-based engineered system like a power plant (Fig. 1.12). The scale obviously is quite different from the previous examples, and also the cost components for various parts have different considerations than the previous examples of mobile systems. Even the turbine in a power plant has different considerations than the aircraft turbine engine. Extensive use of low-cost welding to bring together various steel parts is quite different from automobile and aircraft joining solutions. Think of how the materials for such large systems would be driven by cost.

1.3 Engineered systems and materials   15

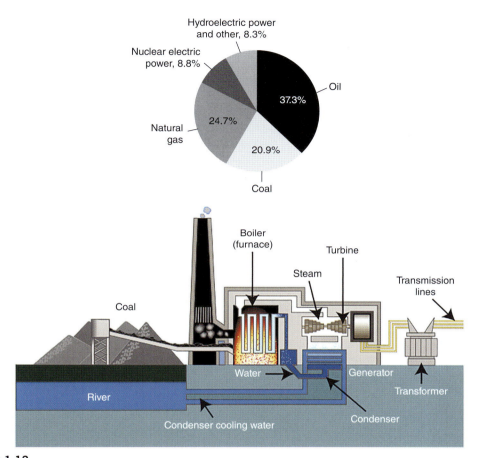

Figure 1.12
A photograph of a thermal power plant with a cutout depiction.[6]

1.3.4 Miscellaneous examples of engineering systems

Fig. 1.13 shows a few examples of sports equipment. We do not always think of the level of materials-based advantage that sports equipment provides to athletes. The subfigure of the pole vault record confirms the advantage of a stiffer pole. It blurs the discussion of the ethics of athletes using advanced sports equipment. Golf and tennis provide other examples. Golf driver and iron designs and choice of materials have significantly increased players' ability to hit longer shots and with greater control of the

---

[6]From United States Tennessee Valley Authority, https://commons.wikimedia.org/wiki/File:Coal_fired_power_plant_diagram.svg

16 Chapter 1 Mechanical response of materials

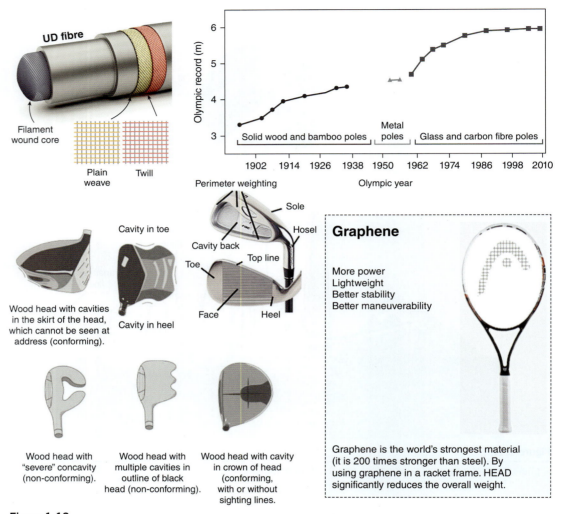

Figure 1.13

A few examples of sports equipment. The upper row shows the impact of improved pole material on the progression of pole vault Olympic record. Similarly, the combination of materials and design has transformed the sports of golf and tennis.

ball trajectory. The sport's governing body has the unenviable task of regulating equipment to restrict any unfair advantage.

Microelectronics is another area where intuitively the importance of mechanical property may be discounted. The interconnects are very small, and they go through thermal cycles that lead to thermal stresses. The most common cause of failure relates to temperature (Fig. 1.14). Creep deformation and thermomechanical fatigue of solders are quite important. In general, the scale of components in

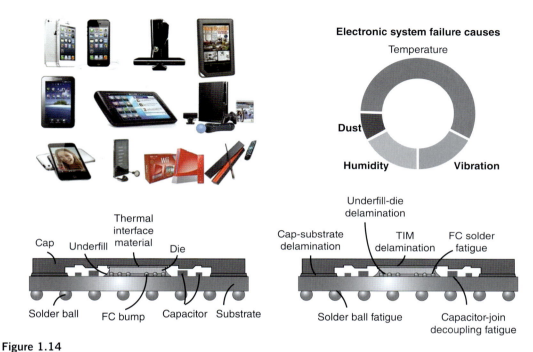

Figure 1.14

Examples of electronic instruments and issues of component failure.

microelectronic devices is miniature, and one needs to think about how various deformation mechanisms and failure mechanisms change with length scale.

1.4 Exercises

1. Complete the following worksheet for Figs 1.10–1.13. You can also take any engineering system picture and fill in this type of worksheet.

    Worksheet

    Component _____

    Nature of loading _____

    Type of loading _____

    Which material _____

    Temperature range _____

    Design approach _____

    Environmental effects _____

2. What is your primary motivation for taking this course? List five things you would like to learn by the end of the semester. (CHECK – at the end of the semester you can check this list to see if it fulfilled your interests.)

3. What are the five materials that interest you most?

4. Which engineered system most fascinates you? Can you list the subsystems and components? For each component, try to write down the manufacturing process used.

Further readings

This list presents the most popular introductory materials science and engineering books. Students should have followed one of these and be comfortable with the basic concepts of microstructure and properties before continuing with the current book. A number of figures throughout the book are taken from Ashby's *Materials Selection in Mechanical Design*; however, familiarity with this book is not a prerequisite for this textbook.

Ashby MF. *Materials Selection in Mechanical Design*. 5th ed. Butterworth-Heinemann; 2016.

Ashby MF, Shercliff H, Cebon D. *Materials: Engineering, Science, Processing and Design*. 3rd ed. Butterworth-Heinemann; 2013.

Askeland DR, Wright WJ. *The Science and Engineering of Materials*. 7th ed. Cengage Learning; 2015.

Callister Jr WD, Rethwisch DG. *Materials Science and Engineering: An Introduction*. 8th ed. John Wiley and Sons; 2009.

Henri T. "Mémoire sur l'Écoulement des Corps Solides soumis à des Fortes Pressions," (extrait par l'auteur). *Compt Rend Acad Sci Paris*. 1864;59:754-758.

Orowan E. Zeitschrift für Physik. 1934;89:605-613, 614-633, 634-659.

Shackelford JF. *Introduction to Materials Science for Engineers*. 8th ed. Pearson; 2014.

Taylor GI. The mechanism of plastic deformation of crystals. part I-Theoretical, *Proc R Soc Lond A*. 1934;145:362-387.

Von Mises R. Mechanik der festen korper im plastisch deformablen zustand. *Nachr Ges Wiss Gottingen*. 1913;582.

CHAPTER 2

Framework of five basic design approaches for structural components

Chapter outline

2.1 Concept of design allowables	19
2.2 Ashby's basic mechanical design framework	21
2.2.1 Stiffness-limited design	22
2.2.2 Strength-limited design	23
2.2.3 Toughness-limited design	25
2.2.4 Fatigue-limited design	28
2.2.5 Creep-limiting design	29
2.3 Exercises	30

Learning objectives

- Brief understanding of statistical approaches for design allowables.
- Ashby's basic mechanical design framework focused on five mechanical properties: stiffness, strength, toughness, fatigue, and creep.

This book is geared toward increasing knowledge and understanding of students in materials science and engineering and mechanical engineering programs. It begins with a quick review of a simple framework for structural applications. The primary objective of mechanical design of structural materials is to maximize the efficient use of engineered materials. The two elements in this chapter are, (1) concept of design allowables and (2) mechanical design framework. The basic design allowable approaches are built on statistical analysis of data and will use the very well-established MIL Handbook 5 approach that is easily accessible on the internet (see links in references). The mechanical design framework takes key elements from Ashby's *Materials Selection in Mechanical Design*. In succeeding chapters, after the fundamental understanding of various properties has been established, students will have opportunity to confirm links between statistical property distribution and microstructural distribution.

2.1 Concept of design allowables

In introductory materials science and engineering, students were introduced to the basic mechanical behavior terms used thus far in this book. The broader context will be repeated many times; we want not only to understand the fundamental underpinning of the mechanical behavior but also to understand the path to application. The focus on linking fundamentals to design and application is also a key differentiator between the approaches taken in this textbook as compared to all other mechanical behavior of materials textbooks.

Of first note is that properties of any engineering material exhibit statistical variations. All mechanical design approaches start from these statistical variations and build a path to property values that can be used for "design." In a simple primer, we look merely at a few keywords that are core to the design process. What is the key difference between scientific documents reporting values of mechanical properties and the property value for design? Designers use *guaranteed minimum value*, whereas the scientific community reports *average values*. This is an important concept to keep in mind as we move forward and build the basic mechanistic understanding. For design purposes, we are interested in reducing variation while maximizing the minimum value of a specific property. So, an understanding of what leads to property variation is as important as understanding of the fundamental mechanisms of deformation and the key stress values that trigger a particular deformation or failure mechanism.

We begin with the most familiar statistical approach that is presented in most foundational courses. Fig. 2.1 shows a graph of the most widely used normal distribution of properties. The other two statistical distributions that we will consider are (1) log-normal distribution and (2) Weibull distribution. Both the normal distribution and the three-parameter Weibull distribution are used in the Metallic Materials Properties Development and Standardization (MMPDS) document, a design guideline used extensively for aerospace applications, and can serve as an illustrative example for students going through this basic course on mechanical behavior of materials. The MMPDS emerged from the MIL Handbook 5, which is widely available on the internet from a variety of sources, and can be used in such a course. The Society of Automotive Engineers (SAE) maintains the other popular design document called aerospace materials specifications (AMSs). Both these documents provide extensive materials property data. There are four types of room-temperature mechanical properties included in the handbook. The one with the least statistical confidence but is most often quoted is the "Typical Basis"; a typical property value is an average value and has no statistical assurance aasociated with it. The

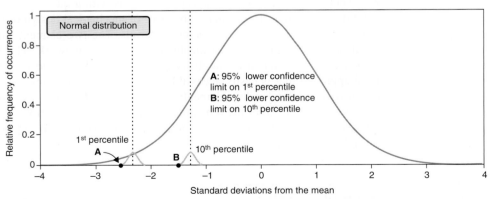

Figure 2.1

Normal distribution of properties and definition of design basis described in Metallic Materials Properties Development and Standardization (MMPDS) document.

following text is reproduced directly from the MMPDS presentation about other three properties basis definitions, which are based on statistical assurance:

- "A-basis: The lower of either a statistically calculated T99 or specification minimum. T99 is the value at which at least 99% of population is expected to equal or exceed with 95% confidence.
- B-basis: Same as T90; at least 90% of population is expected to equal or exceed with 95% confidence.
  NOTE: A- and B-basis for static design properties are based on minimum population of 100.
- S-basis: Specification minimum, or value based on specification minimum."

Designers use this approach, and the specification minimum can be defined based on a certain population. For example, for certain applications the AMS specifications can define the minimum lot requirement of 30 for S-basis. Note that the discussion in this chapter is used to define the need for fundamental understanding of mechanical behavior of materials. The goal is to determine how to connect the fundamentals to the issues that concern designers. In other words, as a practicing engineer, your primary goal is to design engineered components and structures that are safe. On the other hand, as a materials scientist or researcher, you want to understand the fundamentals behind these numbers and the origin of scatter in the data.

Of the many other resources for design allowables, some are available on the web without charge. As this course progresses, students are encouraged to use these resources to do "food for thought" exercises as well as to complete some exercises at the end of chapters. One example of material-specific resource is data available from the Aluminum Association. This resource is used and highlighted in the book *Aluminum Structures* by Kissell and Ferry.[1] Note that the "typical values" given in most of the databases are average values and not the "minimum values" used by designers. The minimum values are determined on the A-basis, B-basis, and S-basis as per the design requirements and approach. Although we will use the typical values throughout the book, students should keep in mind the difference. Every time a calculation for a specific property is considered, students should look at the microstructure-based parameters and ponder how that would lead to property variation.

FOOD FOR THOUGHT—What leads to distribution in properties?

Hint: Metallic materials are ordered by specifying chemical composition and minimum acceptable properties. When bulk material is produced, what type of variations can occur in composition? Think of the maximum compositional variation possible and compounding effect. Also, note that microstructure is typically not specified while ordering the material, and during this course you will learn about the microstrcture-property linkages.

2.2 Ashby's basic mechanical design framework

The layout of this book is based on five design approaches. These are introduced here one by one, and each property will be treated extensively in subsequent chapters.

---

[1] Kissell JR, Ferry RL. *Aluminum Structures: A Guide to Their Specifications and Design*. 2nd ed. Wiley; 2002.

2.2.1 Stiffness-limited design

This approach is for applications where the modulus of material is the most important property. In these applications, the deflection of structure is critical. We are interested in the intrinsic stiffness of materials; here intrinsic stiffness refers to elastic properties of the material without any other approach to enhance the stiffness of the structure. Although Ashby's book on *Materials Selection in Mechanical Design* covers the additional aspect of geometrical shape selection to increase effective stiffness, that approach is not considered in this book. Design of composites is an important part of stiffness of materials. Consider two basic aspects covered in the introductory materials science and engineering book: (1) Materials are either crystalline or amorphous, and (2) atomic spacing is an indication of bond strength. Most of the designs consider either Young's modulus *(E)* or shear modulus *(G)*. At this stage of the book, consider a crystalline material and think of the atomic spacing in different directions. It is intuitive that bond strength in a specific crystal direction will change depending on the corresponding atomic spacing. Therefore the elastic response of the crystal will be direction dependent. For a single crystal component, like turbine blades used in aircraft engines, orientation of the blade will govern elastic response. However, for a polycrystalline material, elastic response will average out the responses of individual grains. This will be addressed in greater detail in **Chapter 6**. For design of a component or a system, the specification will limit the overall acceptable deflection, and elastic properties will be critical.

Fig. 2.2 shows a plot of elastic modulus and density. Note the use of design guide lines. Table 2.1 gives a list of materials indices for certain applications. Ashby's approach of materials index to down-select a few materials for a particular application is an efficient method. This discussion focuses on just the type

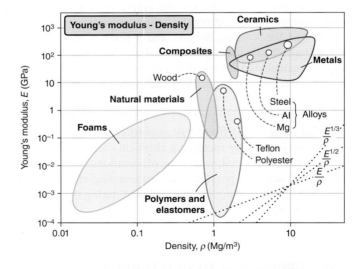

Figure 2.2

A plot of elastic modulus and density of material. Note the use of design guide lines on the plot. Ashby's approach of using a material index for a particular application is an efficient way to down-select materials for a particular application.[2]

---

[2] Ashby MF. *Materials Selection in Mechanical Design*. 5th ed. Butterworth-Heinemann; 2016.

Table 2.1 Materials indices for stiffness-limited design at minimum mass taken from Ashby's material selection approach (Ashby MF. Materials Selection in Mechanical Design. 5th ed. Butterworth-Heinemann; 2016.). Note that the highlighted materials indices are used as guidelines in Fig. 2.2 as an example.

Function and constraints	Maximize
Tie (tensile strut)	
Stiffness, length specified; section area free	E/ρ
Shaft (loaded in torsion)	
Stiffness, length, shape specified; section area free	$G^{1/2}/\rho$
Stiffness, length, outer radius specified; wall thickness free	G/ρ
Stiffness, length, wall thickness specified; outer radius free	$G^{1/3}/\rho$
Beam (loaded in bending)	
Stiffness, length, shape specified; section area free	$E^{1/2}/\rho$
Stiffness, length, height specified; width free	E/ρ
Stiffness, length, width specified; height free	$E^{1/3}/\rho$
Column (compression strut, failure by elastic buckling)	
Buckling load, length, shape specified; section area free	$E^{1/2}/\rho$
Panel (flat plate, loaded in bending)	
Stiffness, length, width specified; thickness free	$E^{1/3}/\rho$
Cylinder with internal pressure	
Elastic distortion, pressure and radius specified; wall thickness free	E/ρ
Spherical shell with internal pressure	
Elastic distortion, pressure and radius specified; wall thickness free	$E/(1-\nu)\rho$

of indices for various loading conditions. Consider the beam in the bending condition with different geometrical constraints as listed in Table 2.1. Three different indices can result in minimizing the weight of the beam based on the geometrical constraints of section area or width or height: (1) E/ρ, (2) $E^{1/2}/\rho$, and (3) $E^{1/3}/\rho$. Note that these materials indices have been used in Fig. 2.2 as an example. What should be taken forward from this is the relative importance of Young's modulus and density changes in material selection. However, in all cases maximizing modulus value while lowering density will result in higher performance. Also, note that the vast unoccupied regions in Fig. 2.2 indicate the lack of material with that combination of properties. Composites are excellent examples of combining materials to fill the unoccupied region that may have design opportunities. Principles of composite design are presented in **Chapter 6**.

2.2.2 Strength-limited design

This approach is used for applications where yield strength or ultimate tensile strength of the material is used as the design stress. Obviously, enhancing the strength of material is desirable. The Aluminum Association's aluminum design manual gives the following strength values:

- Tensile yield strength (F_{ty})
- Tensile ultimate strength (F_{tu})
- Compression yield strength (F_{cy})
- Shear yield strength (F_{sy})
- Shear ultimate strength (F_{su})

Recall that the stress-strain curves can be very different for different classes of materials. While the typical stress-strain curves for metallic materials have well-defined yield strength, for brittle ceramics, glasses, and rigid plastics, there will be just a failure stress or fracture stress. Additionally, very flexible elastomers will not have any yielding either. Strength of a metallic material is very sensitive to the microstructure. Fig. 2.3 shows plots of strength-density and modulus-strength. Only for metals and polymers, the σ_f represents yield strength of material. In each of these plots, the spread of strength

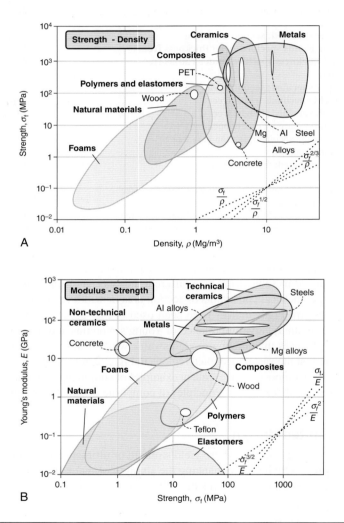

Figure 2.3

Plots showing variation of strength with (a) density and (b) elastic modulus of material. Note the use of design guide lines on the plot. Ashby's approach of using a material index for a particular application is an efficient way to down-select materials for a particular application.[3]

---

[3] Ashby MF. *Materials Selection in Mechanical Design*. 5th ed. Butterworth-Heinemann; 2016.

values for a given class of metallic material is higher than the other properties. Consider aluminum alloys; while density changes marginally with alloying content, strength variation is greater than one order of magnitude. This requires understanding of individual strengthening mechanisms and then the ways of calculating superposition of strengthening mechanisms. **Chapter 8** points out that while the individual strengthening mechanisms are well established, the addition or superposition of multiple mechanisms is not straightforward. Similarly, the impact of loading type can also be important, depending on the details of microstructure. We are tasked with understanding the relationship between microstructure and level of strength during different types of loading. Postyield behavior for metallic materials is also very important and represents change in the flow stress as a function of strain. The ultimate strength represents a point on stress-strain curve where failure mechanisms start. For a simple tensile test, the onset of failure mechanism can be external like necking or internal like void/microcrack formation. These aspects will be covered in **Chapters 8 to 10** under fracture mechanisms.

Table 2.2 gives a list of materials indices for certain applications. Again, note that these can be very useful design guide lines, and it is given here just to introduce the topic. Throughout the book, we will emphasize that for any engineering application, a range of properties needs to be optimized. In this particular case, two different goals have been used. For designs that emphasize weight, strength is normalized by density, and different indices emerge based on the type of application: (1) σ_f/ρ, (2) $\sigma_f^{1/2}/\rho$, and (3) $\sigma_f^{2/3}/\rho$. The relative importance of enhancing strength or the payoff depends on the type of loading. On the other hand, if the focus is on performance only, let us consider two of the indices listed in Table 2.2, (1) σ_f/E and (2) σ_f^2/E. When the indices have a power-law dependence, the payoff in increasing strength is very high. From this limited list we can note that springs, knife edge, and pivots are excellent examples of components that can benefit from strength increase. At this stage, no emphasis has been given to cost. A continuing emphasis throughout the book will be that for any materials implementation, cost considerations are always important, i.e., performance enhancement at what cost? So, when in **Chapter 8** various strengthening fundamentals and approaches are considered, this question should be remembered, to shape up the thinking of performance enhancement.

2.2.3 Toughness-limited design

For applications where damage tolerance or impact resistance is key, fracture toughness of material is used for design. Note that "toughness" and "fracture toughness" are not the same, though related. Toughness is the ability of a material to absorb (or dissipate) energy in a volume, while fracture toughness is specific to this energy resulting in increasing the size (surface) of a crack. Throughout the book, Ashby's materials selection nomenclature of toughness-limited design is used. Note that the word "toughness" in this refers to fracture toughness. Microstructure is key to enhancing fracture toughness. Usually large constituent particles, embedded pores, brittle intermetallic particles, or large inclusions knock down or lower the fracture toughness values. The fracture mechanics approach to design differs from the strength of material approach. The fracture mechanics approach introduces the flaw size as an additional structural variable. This approach is critical for the toughness-limiting design, and it uses the relationships among crack size, applied stress, and toughness, which for tensile opening of a preexisting crack is expressed as

$$K_{Ic} = \sigma \sqrt{\pi a_c} \tag{2.1}$$

where K_{Ic} is the critical stress intensity factor referred to as fracture toughness of material, σ is the remotely applied stress normal to the crack plane, and "$2a_c$" is the critical crack length. This will be discussed in more detail later in **Chapter 9**. But at this stage, a grasp of the approach designers take is sufficient.

Table 2.2 Materials indices for strength-limited design (a) at minimum mass and (b) for maximum performance taken from Ashby's material selection approach (Ashby MF. Materials Selection in Mechanical Design. 5th ed. Butterworth-Heinemann; 2016.). Note that the highlighted materials indices are used as guide lines in Fig. 2.3 (a) as an example.

Function and constraints	Maximize
(a)	
Tie (tensile strut)	
Stiffness, length specified; section area free	σ_f/ρ
Shaft (loaded in torsion)	
Stiffness, length, shape specified; section area free	$\sigma_f^{2/3}/\rho$
Stiffness, length, outer radius specified; wall thickness free	σ_f/ρ
Stiffness, length, wall thickness specified; outer radius free	$\sigma_f^{1/2}/\rho$
Beam (loaded in bending)	
Stiffness, length, shape specified; section area free	$\sigma_f^{2/3}/\rho$
Stiffness, length, height specified; width free	σ_f/ρ
Stiffness, length, width specified; height free	$\sigma_f^{1/2}/\rho$
Column (compression strut)	
Load, length, shape specified; section area free	σ_f/ρ
Panel (flat plate, loaded in bending)	
Stiffness, length, width specified; thickness free	$\sigma_f^{1/2}/\rho$
Plate (flat plate, compressed in-plane, buckling failure)	
Collapse load, length, and width specified; thickness free	$\sigma_f^{1/2}/\rho$
Cylinder with internal pressure	
Elastic distortion, pressure and radius specified; wall thickness free	σ_f/ρ
Spherical shell with internal pressure	
Elastic distortion, pressure and radius specified; wall thickness free	σ_f/ρ
Flywheels, rotating disks	
Maximum energy storage per unit volume; given velocity	σ
Maximum energy storage per unit mass; no failure	σ_f/ρ
(b)	
Springs	
Maximum stored elastic energy per unit volume; no failure	σ_f^2/E
Maximum stored elastic energy per unit mass; no failure	$\sigma_f^2/E\rho$
Elastic hinges	
Radius of bend to be minimized (max. flexibility without failure)	σ_f/E
Knife edges, pivots	
Minimum contact area, maximum bearing load	σ_f^3/E^2 and H
Compression seals and gaskets	
Maximum conformability; limit on contact pressure	$\sigma_f^{3/2}/E$ and I/E
Diaphragms	
Maximum deflection under specified pressure or force	$\sigma_f^{1/2}/E$
Rotating drums and centrifuges	
Maximum angular velocity; radius fixed; wall thickness free	σ_f/E

For damage-tolerant design, designers consider that every material contains a flaw. In the event that all inspection techniques return negative results, meaning no detectable flaw, the minimum resolution of the most sensitive technique can be used as the flaw size. For a given crack size $2a$ and a given fracture toughness, Eq. (2.1) provides the required failure stress. Note that in this approach, to maintain the same level of damage tolerance for a given flaw size, if the failure stress of a material is increased, the fracture toughness must be increased proportionately! Fig. 2.4 presents the variation of fracture toughness with strength and elastic modulus. One should use caution in using this approach for nonmetallic materials. Plastic deformation of material is important for higher fracture toughness of metallic material, as it is a reflection of energy required for crack growth. **Chapters 8 and 9** point out that conventional approaches of increasing strength lead to decrease in fracture toughness values in metallic materials. This can be noted from Table 2.3, and when energy control is the primary consideration, the value of fracture toughness is more important. Note that the highlighted materials indices are used as guide lines in Fig. 2.4 (a) as an example. The relative importance of fracture toughness

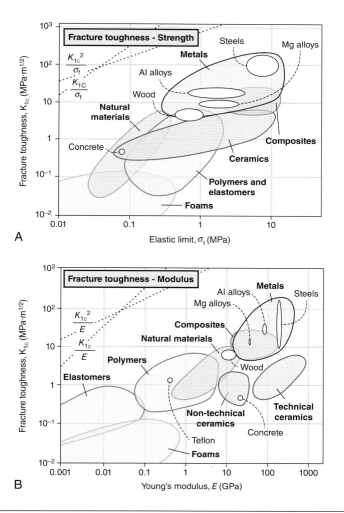

Figure 2.4

Plots showing variation of fracture toughness with (a) strength and (b) elastic modulus of material. Note the use of design guide lines on the plot. Ashby's approach of using a material index for a particular application is an efficient way to down-select materials for a particular application.[4]

---

[4]Ashby MF. *Materials Selection in Mechanical Design*. 5th ed. Butterworth-Heinemann; 2016.

Table 2.3 Materials indices for toughness-limited design taken from Ashby's material selection approach (Ashby, 2016). Note that the highlighted materials indices are used as guidelines in Fig. 2.4 (a) as an example.

Function and constraints	Maximize
Ties (tensile member)	
Maximum flaw tolerance and strength, load-controlled design	K_{IC} and σ_f
Maximum flaw tolerance and strength, displacement-control	K_{IC}/E and σ_f
Maximum flaw tolerance and strength, energy control	K^2_{IC}/E and σ_f
Shafts (loaded in torsion)	
Maximum flaw tolerance and strength, load-controlled design	K_{IC} and σ_f
Maximum flaw tolerance and strength, displacement-control	K_{IC}/E and σ_f
Maximum flaw tolerance and strength, energy-control	K^2_{IC}/E and σ_f
Beams (loaded in bending)	
Maximum flaw tolerance and strength, load-controlled design	K_{IC} and σ_f
Maximum flaw tolerance and strength, displacement-control	K_{IC}/E and σ_f
Maximum flaw tolerance and strength, energy-control	K^2_{IC}/E and σ_f
Pressure vessel	
Yield-before-break	K_{IC}/σ_f
Leak-before-break	K^2_{IC}/σ_f

changes when the design approach shifts from "yield-before-break" to "leak-before-break." The microstructural aspects and relationship with deformation mechanisms are considered in **Chapter 9**.

2.2.4 Fatigue-limited design

For applications where low cycle or high cycle fatigue strength is important, the inherent microstructure can be important. For example, high cycle fatigue is sensitive to pre-existing porosity, large constituent particles, or large intermetallic particles that act as sites for crack nucleation. At the same time, low cycle fatigue can be quite different because it is driven more by plasticity, and the element of time is not so important. In terms of materials selection chart, capturing this difference is not easy.

The major data used in engineering applications for fatigue life is in the form of *S-N* (stress-number of cycles to failure) curves. While details are discussed in **Chapter 10**, readers are expected to remember the basic concepts of *S-N* curve from the introductory materials science and engineering course. Four more commonly used design philosophies are discussed below.

2.2.4.1 *Infinite life approach*

This approach depends on the basic concept of fatigue endurance limit. A fundamental difference among various metallic materials is the shape of *S-N* curve at lower stresses. Steels exhibit very well-defined fatigue endurance limit, and the *S-N* becomes perfectly horizontal at lower stresses. However, aluminum alloys do not exhibit true endurance limit, and designers assume a design life of 10^7 or 10^8 cycles to determine the stress value that is then considered as the pseudo endurance limit. If the design stresses are below the fatigue endurance limit, infinite life would be expected for structural parts like rotating components. **Chapter 10** details empirical correlations between ultimate tensile strength and fatigue endurance limit for various classes of metallic materials.

2.2.4.2 Safe-life approach
The safe-life design approach is based on the required lifetime of a structure or part. The infinite life approach leads to lower design stresses, which can make the structure heavy or inefficient. Therefore if a component is designed for only a known life, the *S-N* curve data can be used to determine the stress corresponding to desired life. The allowable stress would be higher than the endurance limit. An example of such an approach would be systems that operate for a short time, systems such as rockets.

2.2.4.3 Fail-safe design approach
This design philosophy was developed by the aircraft industry and assumes that the structure is constructed such that cracking or even failure of a component would not lead to catastrophic failure of the entire structure. Incorporation of redundancy or having multiple loading paths in the structure is the main feature of this approach. An example of this approach is the fuselage or wings of the aircraft.

2.2.4.4 Damage-tolerant approach
The aircraft industry is the primary user of this approach. As mentioned in the toughness section, the damage-tolerant approach assumes the presence of a preexisting flaw. This approach ties in crack growth and the inspection regime. The initial flaw size is used to calculate the inspection period during which the crack will not become critical.

2.2.5 Creep-limiting design
For high-temperature applications, creep deformation and fracture are critical. This is the only design approach where fine grain size is detrimental. Various factors can set the upper service temperature for a material. Fig. 2.5 shows the variation of strength with maximum service temperature. Loss of strength is an intrinsic reason to set the upper service temperature, whereas excessive oxidation is an extrinsic reason for the upper service temperature. Titanium alloys are good examples of a metallic material that loses environmental resistance before intrinsic strength drops.

The intrinsic creep strength-limited design considers two distinct approaches. The creep life approach is for components that have no geometrical constraints, and as detailed in **Chapter 11**, creep life at a constant stress provides the design data. Components in power plants are good examples of this approach. On the other hand, rotating components in the aircraft engine represent an application where the change in blade length is important. In such cases, creep strain at a given stress and specified time is the important design data. Again, there are no common materials indices for creep in Ashby's materials selection approach. Contrast Fig. 2.5 with Figs. 2.2 to 2.4, where guide lines were plotted. As mentioned before, these materials index based quantitative approach for materials selection is quite attractive.

Chapter 2 Framework of five basic design approaches for structural components

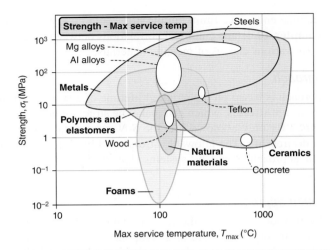

Figure 2.5
Plot showing variation of strength with temperature.[5] Maximum service temperature can be set by loss of strength or severe oxidation.

2.3 Exercises

1. How do designers define the minimum guaranteed property for design? If you are a materials developer in charge of enhancing the strength of material, what type of distribution in property would you target?

2. Perform a web search for specified composition of a high-strength aluminum alloy (2XXX or 7XXX alloy). List the range of composition for each alloying element. Note that elements that have a "maximum" specified are not desired alloying elements. Comment on what happens to the total alloying element if your alloy has all minimum values as compared to all maximum values. What would be your expectation of mechanical property variability within this "acceptable composition range"?

3. List the five design approaches covered in this chapter and make a list of materials you are interested in. From the materials selection chart, write down the corresponding property values for these materials.

4. Perform a web search for typical values of Young's modulus (E) and density (ρ) for Mg, Al, Ti, and Fe. What are the values of E/ρ for these elements? Comment on the pattern observed.

5. Connect your answer in Chapter 1 regarding the engineering system and components. Pick three components and create a list to include type of loading and your guess of design approach used.

---

[5]Ashby MF. *Materials Selection in Mechanical Design*. 5th ed. Butterworth-Heinemann; 2016.

Further readings

This list gives some introductory resources for design-related topics. Students are encouraged to browse these references to become familiar with the terms that designers consider. While mechanical engineering students get formal exposure to design principles from books like Shigley's *Mechanical Engineering Design*, a typical materials science and engineering student does not get such exposure. We maintain that cross-knowledge of design approaches and basic mechanical behavior of materials is foundationally valuable. A number of figures throughout the book are taken from Ashby's *Materials Selection in Mechanical Design*; however, familiarity with this book is not a prerequisite for this textbook.

Access to an older version of MIL Handbook 5 at https://archive.org/details/milhdbk-5-j; the Metallic Materials Properties Development and Standardization (MMPDS) website has additional resources: https://www.mmpds.org/

Ashby MF. *Materials Selection in Mechanical Design*. 5th ed. Butterworth-Heinemann; 2016.

Budynas R, Nisbett K. *Shigley's Mechanical Engineering Design*. 10th ed. McGraw-Hill Education; 2014.

Kissell JR, Ferry RL. *Aluminum Structures: A Guide to Their Specifications and Design*. 2nd ed. Wiley; 2002.

Sharp ML, Nordmark GE, Menzemer CC. *Fatigue Design of Aluminum Components and Structures*. McGraw-Hill Professional; 1996.

CHAPTER 3

Introduction to systems approach to materials

Chapter outline

3.1 Olson's systems approach to materials by microstructural design 35
    3.1.1 Specific examples of the success of the approach: Olson's flying cybersteel 39
3.2 Importance of processing-microstructure-properties correlations 43
3.3 Exercises 45

Learning objective

- Brief understanding of systems approach to materials.

This short chapter introduces the systems approach to materials. The classic materials chain is path for preparing, examining, and documenting materials and their properties. It is expressed as: processing → microstructure → properties → performance. This sequence has been recognized from as early as the 1950s (Zener (1948) and Smith (1981)). Yet, a formal nomenclature of "systems approach" became popular after the work of Olson (1990, 1997). An understanding of this systems approach will enhance appreciation of the broader context as we examine the various mechanical behavior concepts. Note that the mechanical behavior of materials forms the structural properties group. The materials link chain implies that mechanical behavior originates from the microstructure, which in turn evolves during the processing of material.

While a materials engineer or scientist builds understanding from the atomic level up, the designer starts from the required performance of the component or system (Fig. 3.1). As Fig. 3.1 (a) illustrates, the Mechanical Behavior of Materials course relates the components of material's structure with properties. While the aforementioned course is a core for a degree in Materials Science and Engineering, clearly mechanical engineers who become designers or deal with materials and process selection would benefit from such an understanding. The layout of Fig. 3.1 (b) is inverse, as it links the concepts from Chapter 2 to microstructural features that influence specific properties related to any particular design goal. At this stage a new key phrase is introduced, "*unintended microstructural feature*." An unintended microstructural feature can be defined as anything that is not a part of the original design of the material. This is a very important distinction, as the unexpected failure of material is often linked to such features. A philosophical question that can be mulled over becomes, "Why do fail-safe designed components fail?" The answer can lie in the probabilistic nature of material microstructure and component loading, or it can be pinned to an unintended microstructural feature that was incorporated in the material during the processing stage. An easy example of unintended microstructural feature is inclusions in metals, particularly introduced during the casting stages. Control of impurity elements in

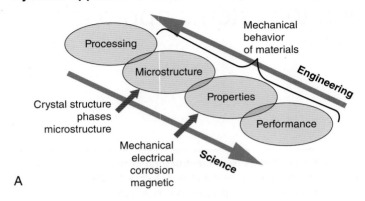

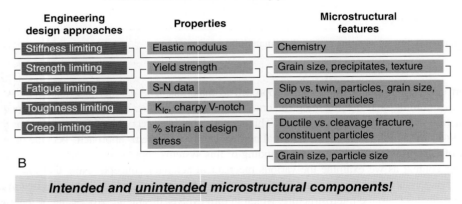

Figure 3.1

(a) The depiction of processing-microstructure-properties-performance link for materials. Note that while the mechanical behavior of materials books focuses primarily on the microstructure-property linkage, understanding the impact of processing on microstructure is important. (b) An illustration of the dependence of design approaches on key mechanical properties that in turn depend on key microstructural features. Note that the constituent particles are highlighted here as "unintended" microstructural features and differ from other microstructural features that are "intended."

the metallic materials is a key to the control of the "unintended microstructural features." Later in the chapter we consider an example flow chart of steel production.

Olson's pioneering work to establish the "systems approach" predates the onset of a formal materials field of "Integrated Computational Materials Engineering (ICME)" (NRC, 2008). Students are referred to two key papers that give background to the thinking behind the systems approach framework (Olson, 1997, 2000).

3.1 Olson's systems approach to materials by microstructural design

We begin with the cause-effect sequence of processing → microstructure → properties → performance, which in the reverse form is treated as goals-means sequence. A design engineer specifies a certain level of performance of material for the component based on the likely service conditions, which is translated into the required material properties. Material selection is done on the basis of these required properties. Fundamentally, these properties derive from the microstructural features. And these microstructural features originate from the processing steps. The processing steps can be classified as primary or secondary. The primary processing steps are the ones used in the production of material. In the illustrative sequence in Fig. 3.2, the origin of material is depicted by the boxes labeled casting and powder consolidation. Typical manufacturing of a material starts with these steps. If the component is manufactured by casting, the next steps would be heat treatment and machining. Note that the cast material will have most of the microstructural features listed under the "microstructure" box. This is where two core concepts and the connector lines are introduced. The first core concept is that *the origin of microstructure from the processing steps is <u>probabilistic</u>*. For example, a casting will have pores, which are small internal voids. The size and distribution of pores in castings are probabilistic in nature and are influenced by casting practices, gating design, and other process variables. Similarly, the size, distribution, and orientation of grains and second-phase particles will be probabilistic. Once a microstructure is established, the second core concept is that *the properties are <u>deterministic</u>*. This means that for a given grain size and orientation, the strength will be fixed. As the book progresses, a connection between the influence of processing on the microstructure and the subsequent influence of

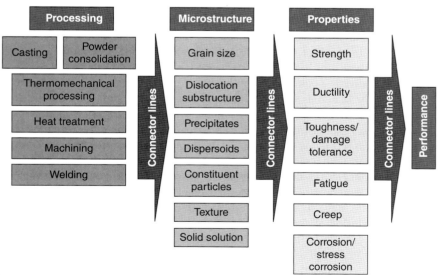

Figure 3.2

A generic illustration of Olson's systems approach. The connector lines are drawn among the specific boxes.

the microstructure on properties will become increasingly important. While we will focus mainly on the microstructure-properties linkage, to keep linking back to processing through thought experiments is important. If the engineering system requires bringing various components together, the joining is achieved by mechanical fastening, in which case there is no metallurgical change to the microstructure, or it can be welded together, in which case the microstructure will change at the joint and surrounding regions. Welding is considered a secondary manufacturing process. Let us reflect on the engineering approach in the context of a secondary manufacturing process. Consider two types of materials. In the first example, let us assume that a wrought material was picked during the material selection stage. Its properties are derived from precipitates from a previous heat treatment step. Now for a particular application, the material needs to be welded by a fusion welding technique, to change the microstructure from wrought to cast, and thereby to change the properties completely. The systems chart allows for laying out the connections and thereby helping to make the proper decision. For the second example, consider a powder-processed metal matrix composite. A powder metallurgy processed metal matrix composite contains a second phase, which is not processable by liquid route (casting) or in which the liquid processing route is not desirable. For example, consider the Al-SiC metal matrix composite. It can be processed through the liquid processing route, but the powder metallurgy processing route gives much better microstructure and avoids undesirable interaction between liquid Al and SiC. So, in the systems chart, fusion welding would not be a viable option for powder metallurgy processed Al-SiC metal matrix composite, as the microstructure would be destroyed by such a process selection. The connector lines connect individual boxes to depict dependence. Multiple lines can originate from a box and connect to a number of boxes that depend on this box. For example, grain size can impact strength, ductility, toughness, fatigue, creep, and corrosion properties. So, connector lines need to be drawn from grain size to all the properties that are affected by it. Similarly, grain size depends on the thermo-mechanical processing step, heat treatment, and any secondary process like welding.

While we started with a generic use of Olson's system chart, it is a very effective tool for "materials by design." It is interesting to reflect on the paradigm shift with this approach and how the computational tools are now central to engineered materials. DARPA's Accelerated Insertion of Materials (AIM) program is one of the best examples of this paradigm shift. Conventional discovery-based materials development typically took 15 to 20 years from initial development to implementation. Materials Genome Initiative is a culmination of the emergence of computational tools and their pervasive use. The website for Materials Genome Initiative (https://www.mgi.gov/) states, "*The Materials Genome Initiative is a multi-agency initiative designed to create a new era of policy, resources, and infrastructure that support U.S. institutions in the effort to discover, manufacture, and deploy advanced materials twice as fast, at a fraction of the cost.*" A brief review of the progress should guide students to learn the basic concepts with a view toward engineering implementation. Most of the material presented in this chapter is taken from the writings of Olson and coworkers. Olson and Kuehmann (2014) summarized Materials Genome progression over the years and the tools that are essential to progress (Fig. 3.3). Although at this stage students are not likely to have enough background to understand this figure in its entirety, inclusion of such a complex figure gives a broad overview. Fig. 3.3 (a) reviews early efforts, which progressed to the Materials Genome. At a simpler level, we can note that early efforts were limited to calculation of a particulate phase on a binary phase diagram. Integration of atomistic simulations and continuum simulations with thermodynamic calculations leads to a holistic approach (Fig. 3.3 [b]). Note that the goal of developing materials faster at lower cost requires fundamental understanding of microstructural evolution during processing and quantitative databases. The

3.1 Olson's systems approach to materials by microstructural design

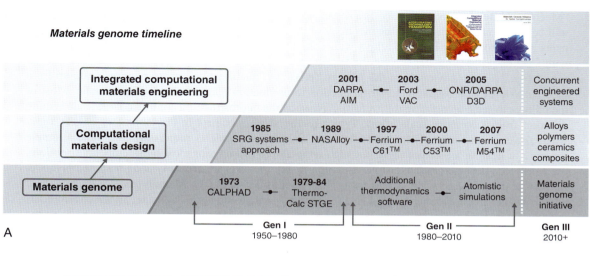

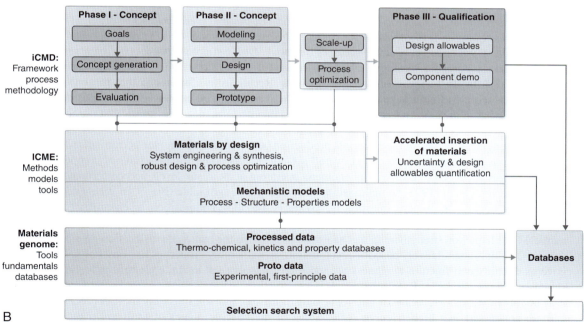

Figure 3.3

(a) Materials Genome technology timeline. (b) Hierarchy of present and future Materials Genome methods, tools, and databases.[1]

---

[1]Olson GB, Kuehmann CJ. Materials genomics: from CALPHAD to flight. *Scr Mater*. 2014;70:25-30.

"Mechanical Behavior of Materials" is a key to development of advanced structural materials. Students will benefit from thinking about the hierarchy of the Materials Genome methods and their subcomponents. The theoretical design of alloys with complex microstructural features needs to be validated with specific processing steps. Note the various project phases of alloy development depicted at the top of Fig. 3.3 (b). Moving from the "design" phase to "qualification" phase requires scale-up from laboratory-level small-scale material production to production quantities at industrial scale. Most of the failures to scale up occur in this transition phase as the microstructural control at large scale is quite different from laboratory scale effort. Typically, the process control in the laboratory is very tight, whereas very tight control of processing steps at industrial scale is rather difficult. Take a simple example of heating a piece of metal for high-temperature rolling. For most high-temperature rolling, the workpiece is heated to the desired temperature, but the rolls are typically at ambient temperature. In the research laboratory, the furnace is typically controlled to $\pm 1°C$. A small piece of experimental alloy is heated and then rolled using a small laboratory-scale rolling mill. Consider all the uncertainties involved in this simple experiment. The workpiece will cool during transfer from the furnace to the rolling mill. The surface of the workpiece will be quenched by the cold rolls, and a temperature gradient will develop from the surface to the interior of the workpiece. This gradient will depend on the size of material. If we now consider these variations in the industrial setup, first the temperature control of large furnaces has wider fluctuations. The time for handling from the furnace to the rolling mill will have larger variation depending on the skill of operator. The thermal gradients will be different depending on the size of ingot. It is important for students to develop an understanding of an acceptable variability, details of which are beyond the scope of this book. Doing simple thought experiments is a good way to link these potential variations during processing to the extent of microstructural variability in workpiece and finally to the mechanical properties. Recall from **Chapter 2** that the engineering specifications require minimum guaranteed values. So, if an industrial process leads to large variations in lot-to-lot performance of materials, then the acceptable performance level for design purposes will be lower. A good understanding of systems approach and defining the critical connector paths can help a materials engineer to control the materials production better, leading to tighter performance window.

Discussions in previous chapters and overall systems charts confirm that a suite of computational modeling tools is needed to deal with phenomena operating at atomistic levels and performance at the continuum level. Olson (1997) summarized this according to length scale (Fig. 3.4). Whenever talk of a new alloy begins, an obvious question is, "What is the alloy composition or chemistry?" In this figure, modeling at the electronic level quantum-mechanical calculations predicts surface thermodynamics. The chapter on strengthening mechanisms clarifies that the nanoscale dislocation/precipitate (or particles) models quantitatively describe precipitation strengthening. Properties like onset of failure that governs ductility and fracture toughness are modeled by tools such as finite element modeling at the submicron to micron scale. Recall that many microstructural features emanate from solidification and heat treatment stages. Modeling of allotropic transformation and solidification address processability and are conducted at a higher length scale. Some of the specific names of tools are mentioned in Fig. 3.3 (b). Specific chapters that focus on various properties can enhance learning by connecting concepts with realistic microstructures (to be discussed in Chapter 4). Engineered structural materials are selected based on a combination of attributes (properties). Achieving a balanced set of attributes is critical for successful performance, and the systems approach-based thinking becomes increasingly important. A simple example would be material selection for a fatigue-critical application. Even though

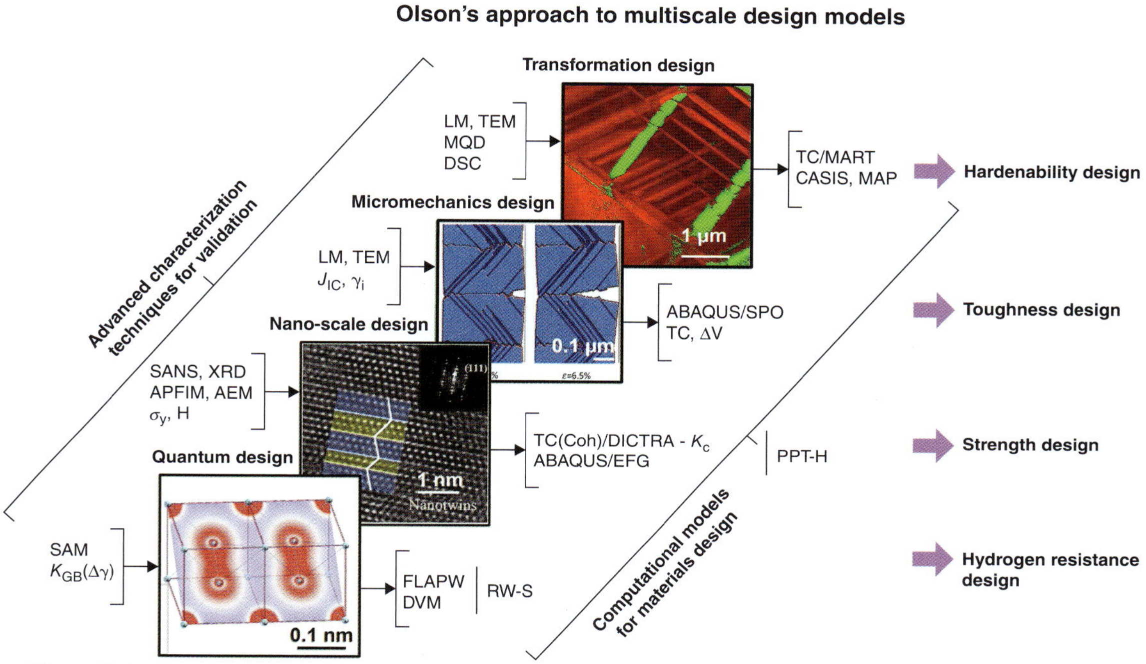

3.1 Olson's systems approach to materials by microstructural design

Figure 3.4

Multiscale hierarchy of interdisciplinary mechanistic models supporting computational materials design[2].

the main property would be fatigue life or fatigue strength, minimum acceptable static strength and ductility will be specified as well. Additionally, if the structure is going to be built using a secondary processing technique, then the impact of that process on all the specified properties will be needed. This is the challenging part in incorporating process modeling that links with microstructural evolution and resultant mechanical properties.

3.1.1 Specific examples of the success of the approach: Olson's flying cybersteel

The examples taken here are from Olson (2013) and Olson and Kuehmann (2014), where the efforts of QuesTek Innovations, a small business specializing in materials by computational design, are summarized. The example shown in Fig. 3.5 gives a very good list of desired properties for a corrosion-resistant

---

[2]Modified from G.B. Olson, "Computational Design of Hierarchically Structured Materials", Science 277, (1997) 1237.

Chapter 3 Introduction to systems approach to materials

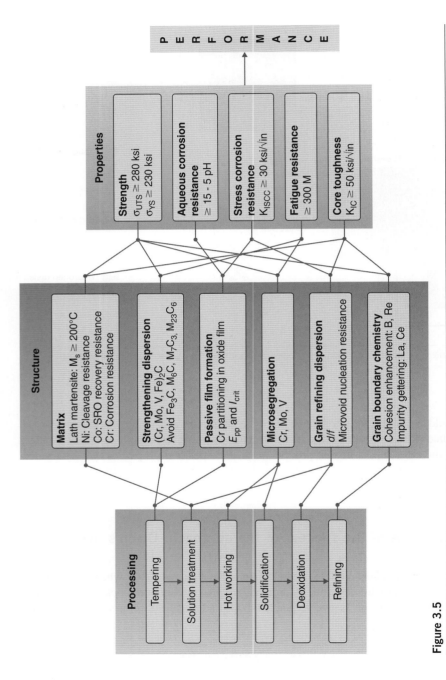

Figure 3.5

System chart representing corrosion-resistant martensitic landing-gear steel and indicating the desired property objectives, the microstructural subsystems, and the required sequence of processing steps. Links between system blocks denote quantitative models needed to effect the design via science-based computation.[3]

[3]Olson GB, Kuehmann CJ. Materials genomics: from CALPHAD to flight. *Scr Mater*. 2014;70:25-30.

martensitic landing-gear steel. It highlights the complexity of balancing properties for high-performance applications. Note the microstructural features that need to be controlled to obtain the right properties. This is where quantitative computational prediction is a key to *Materials by Design*. It can also involve resolving conflicting microstructural requirements. The first designer alloy to demonstrate the full cycle of computational design and Accelerated Insertion of Materials (AIM) qualification is the Ferrium® S53 (AMS 5922) corrosion-resistant landing gear steel, which allows a drop-in replacement for current non-stainless landing-gear steels, and thus eliminates the need for cadmium plating. S53 was developed by QuesTek Innovations and is a secondary hardening martensitic steel efficiently strengthened by M_2C carbide precipitates and contains sufficient Cr content to provide passivation against general corrosion.

The computationally derived S53 steel was designed, developed, and implemented in 10 years. Fig. 3.6 shows the summary of progress through various steps. QuesTek Innovations also developed another variant called M54 with higher fracture toughness. There are several keywords and phrases that students should pick from this figure that have industrial context. The use of technology readiness levels (TRL) is an important phrase that is used for both materials and processes. The higher levels track with maturity. The corresponding presentation of material development milestones shows the early step of design goals to completion of MMPDS handbook data that is based on statistical data analysis. The transition from laboratory-level results to production scale is always a major challenge, as the industrial setup has higher production variability. Fig. 3.7 further shows the interconnected nature of funding and

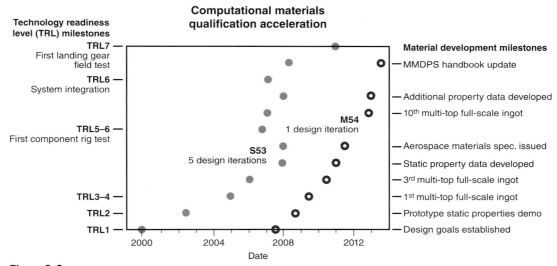

Figure 3.6

Timeline of landing gear level technology readiness levels (TRL) and the corresponding materials development milestones for computationally designed S53 and M54 alloys.[4]

---

[4]Olson GB, Kuehmann CJ. Materials genomics: from CALPHAD to flight. *Scr Mater.* 2014;70:25-30.

42 Chapter 3 Introduction to systems approach to materials

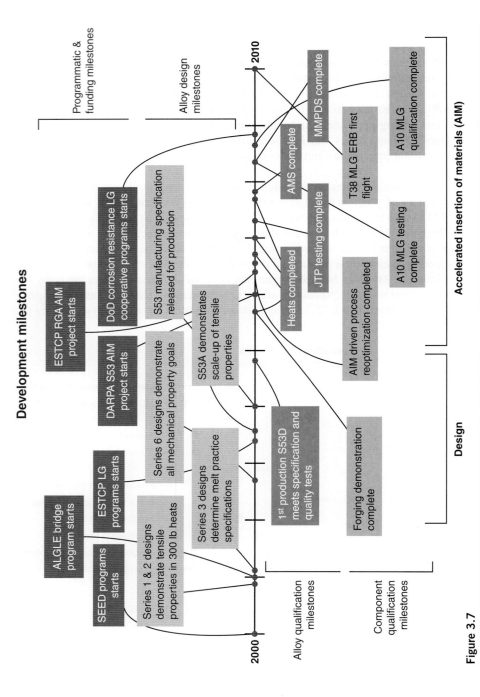

Figure 3.7

Case study summary chart of the computational design and qualification of Ferrium S53 landing gear steel. The top level represents the programmatic and funding milestones of the multiagency effort. The second level represents iterative alloy design milestones leading to the composition and processing specifications. The bottom levels represent alloy and component level AIM-based qualification, which led to the first flight in December 2010.[5]

[5]Olson GB. Genomic materials design: the ferrous frontier. *Acta Mater.* 2013;61(3):771-781.

technology transition to various levels. Very important is developing a broader understanding of various aspects involved in design and qualification of new materials. As emphasized earlier, this steel was developed and implemented in less than 10 years. For clarity, the overall figure is divided into four broad categories. The uppermost row lists various funding sources that were instrumental in making it happen. The funding was triggered by milestones under "alloy design," "alloy qualification," and "component qualification." Note that this is similar conceptually to the layout in Fig. 3.3. With each progression, the TRL of the overall effort increases as outlined in Fig. 3.6. The mechanical behavior of materials course is sequenced such that juniors take this course along with thermodynamics, phase transformation, and introduction to manufacturing processes. These courses build the foundation for the Senior Design course. The layout of this book is different from other textbooks on this topic, as the intention is not only to master the fundamentals of mechanical behavior of materials but also to get an understanding of the connections that ultimately drive the effective use of structural materials.

3.2 Importance of processing-microstructure-properties correlations

Before we wrap up this chapter, let us zoom out to consider the entire material flow, from production to end use. The context of mechanical behavior of materials can be appreciated for two separate stages of materials life. For wrought products, the processing of materials involved thermomechanical steps, which depend on the mechanical properties in different microstructural states. The microstructure evolves during the processing steps, and the final microstructure determines the material properties for end use. Fig. 3.8 is an overview of the steel production process as well as the processing of end use products. Let us examine the direct or indirect impact mechanical properties of materials can have at many steps. Generation of such a list can provide context and motivation for why learning of fundamentals is important. Particularly, which of the steps can impact microstructure and how that can impact properties. Note that the entire Materials by Design approach requires quantification. Which of these steps can add uncertainties or may not be quantifiable? If we start with the initial stages of hot metal being poured in casting mold, improper conditions can lead to hot tear. Hot tear is a defect that forms during the final stages of solidification. What is the origin of such a defect? The solidification shrinkage leads to residual stresses, and in turn the residual stresses can lead to cracking under certain conditions. So, the importance of understanding the high-temperature property of material as well as nature of chemical segregation can be appreciated.

For wrought products, the material is deformation processed. During thermomechanical processing, the cast microstructure evolves or converts to a wrought microstructure. Again, understanding the material property as a function of microstructure is needed to model the entire process and to be able to predict the final resultant microstructure. How much deformation can be imparted in a step? What governs the change in grain size as a function of deformation strain and temperature? How does deformation alter the texture of the wrought product? Does texture depend on strain path? Looking at the overall process chart and asking such questions can be key to answering many of these questions. The expectation is that by the time students finish reading this book, most of these questions will become easy to answer. Indeed, successful navigation through the materials by design approach will require knowing the relationship between each step. To predict microstructural changes and the resultant properties, students must be able to express the relationships quantitatively. Fig. 3.9 shows the length scales of various microstructural features and their relationship with various properties. At this stage, we should look at the length scale

44 Chapter 3 Introduction to systems approach to materials

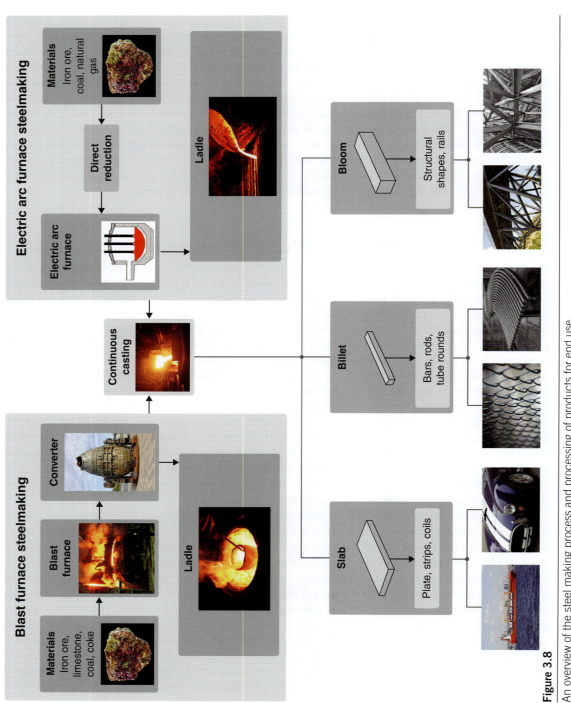

Figure 3.8
An overview of the steel making process and processing of products for end use.

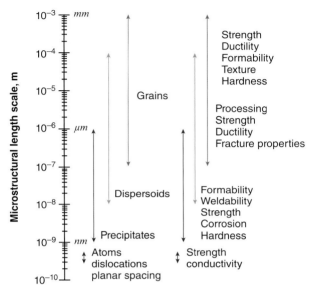

Figure 3.9

An overview of relationships between microstructural features and properties.

range of a particular feature and consider how that changes during processing. For example, cast material is likely to have large grain size. As the material is deformed during thermomechanical processing, the grain size becomes finer. What should be the relationship between processing strain and grain size? What is the most effective way to achieve grain refinement? Similar to Fig. 3.5, each microstructural feature influences a number of properties. Each manufacturing step influences the microstructure and therefore impacts final properties. Manufacturing processes are termed as "primary" or "secondary." In Fig. 3.8, the steel making steps and casting of large ingots are considered primary processes. The thermomechanical processing steps to give finished shapes are considered secondary processes. In examples of products at the bottom of Fig. 3.8, several use joining or welding. Again such secondary manufacturing steps alter the microstructure and impact different microstructural features differently. Quantification of these cause-effect relationships is part of the overall effort in Figs. 3.6 and 3.7.

3.3 Exercises

1. Materials Science and Engineering departments use the systems chain link in their logos. Write a small description of the processing-microstructure-properties-performance chain.

2. Practice drawing a systems chart as shown in Fig. 3.2 based on your current knowledge of any engineering material. Write down the unanswered questions that come to your mind about the linkages that you need to understand.

3. Draw the systems charts for different processing methods.
4. Draw the systems charts for ceramics, metals, and composites.
5. List all the microstructural features mentioned in this chapter.
6. What is unintended microstructural feature?
7. **Thought experiment**: Take the overview chart for steel manufacturing in Fig. 3.8. For production of a steel sheet, what steps will be involved? For each step, can you think of what process parameters an engineer will need to control? From your limited knowledge at this stage, do you think these process parameters will have an impact on the microstructure? List those and organize your thoughts on what knowledge you need to fully answer these linkages.
8. Using Fig. 3.7, draw a chart that shows key steps in design and development of a new alloy.

Further readings

This list gives some resources for Olson's systems approach and related key publications before his papers. Students should browse some of these references to familiarize themselves with the details of the systems approach.

Jenkins GM. The systems approach. *J Syst Eng.* 1969;1:1.

National Research Council. *Integrated Computational Materials Engineering: A Transformational Discipline for Improved Competitiveness and National Security*. The National Academies Press; 2008. https://doi.org/doi:10.17226/12199.

Olson GB, Azrin M, Wright ES. Science of steel. In: *Innovations in Ultrahigh-Strength Steel Technology. Sagamore Army Materials Research Conference Proceedings: 34th.* 1990:3-66.

Olson GB. Computational design of hierarchically structured materials. *Science.* 1997;277:1237-1242.

Olson GB. Genomic materials design: the ferrous frontier. *Acta Mater.* 2013;61(3):771-781.

Olson GB, Kuehmann CJ. Materials genomics: from CALPHAD to flight. *Scr Mater.* 2014;70:25-30.

Olson GB. Designing a New Material World. *Science.* 2000;288:993-998.

Smith CS. *A Search for Structure*. MIT Press; 1981.

Zener C. *Elasticity and Anelasticity of Metals*. University of Chicago Press; 1948.

CHAPTER 4

A brief survey of microstructural elements in engineering structural materials

Chapter outline
4.1 Metallic materials 49
4.2 Nonmetallic materials 55
4.3 Exercises 60

Learning objective
- An understanding of microstructural features in metallic and nonmetallic materials.

This last introductory chapter is the culmination of our efforts to establish the motivation and context for learning the basic concepts that govern the mechanical behavior of materials. The purpose of this chapter is to introduce students to the "real" microstructure in engineering materials and to review the length scale of these features. We concluded **Chapter 3** with a figure relating length scale of microstructural features with different properties. Conventional teaching of mechanical behavior of materials first presents the fundamental mechanisms and then discusses applicability. Conversely, this book establishes some basis of what motivates us. Why the fundamentals are important? Which basic concepts apply to actual microstructure because of the length scale of that feature? Such foundational questions in our opinion lead to a better learning experience and promote an appreciation of the gaps in knowledge as new materials are developed. In some cases, the existing theoretical framework will be simply inadequate. In such cases students or researchers should question the existing fundamentals early on and establish a need for new models.

The generic use of the term *microstructure* should not confuse the readers. A dictionary definition (from dictionary.com) of microstructure is "the structure of a metal or alloy as observed, after etching and polishing, under a high degree of magnification." Note that the origin of this word was through the work of H. C. Sorby at Sheffield in the period of 1863 to 1885, where microstructure was examined using optical microscope. Obviously, the resolution of grains was limited by the wavelength of light. Coincidentally, the processing techniques of structural materials in that period (before 1940) were not able to produce average grains below the limit of optical microscopy. With the advent of electron microscopy, the resolution went submicron, and development of transmission electron microscopy (TEM) in the 1950s allowed observation of dislocations in materials and second-phase particles that could not be seen in optical microscope. The era of nanocrystalline materials started in the 1980s when Gleiter (1984, 1989) published pioneering research papers on this topic. Nanocrystalline materials have grains smaller than 100 nm. Of course, this name is based on the length scale. In this book, *microstructure* is used as a generic word to imply the details of structure in a material. Similarly, the word *structure* is used in both metallurgical and mechanical contexts. Each microstructural feature has a characteristic length scale, such as diameter of an atom, a grain size measured from a 2D image as a predefined (e.g., largest or average) length dimension, etc.

Engineering structural materials can be broadly classified into metallic and non-metallic materials. Another way to classify them is crystalline and non-crystalline. Both approaches are appropriate for what will be discussed in this chapter. The focus is simply to learn what are the basic features or microstructural elements. As the initial fundamentals have been mastered through introductory materials science books, the current discussion will be kept short, just to get the flow of this book going. A brief review of some keywords and key phrases is in order before actual microstructures are considered. Note that most of these will be treated in more detail in the rest of the book.

Atom: The smallest unit of material that we will discuss during deformation mechanisms; typical length scale of 2 to 3 Angstroms (recall: 1 Angstrom = 10^{-10} m). This is important for discussion of deformation mechanisms based on formation and movement of vacancies. It is also a building block for understanding dislocation-based mechanisms.

Crystal structure: The way atoms are arranged in periodic array. This is critical to discussion of dislocation-based deformation mechanisms. If a material is non-crystalline, then the dislocation theory does not apply! Crystal structure also defines planes and directions that are the basis for dislocation-based discussion. The lattice parameters for materials with simple crystal structure are typically 3 to 5 Angstrom. The value of spacing between planes of atoms is also in a similar range. Recalling the information about orientation of crystal will be important. Orientation imaging in scanning electron microscope (SEM) is a major advancement that helps in understanding crystal plasticity. The orientation imaging in SEM uses electron backscatter diffraction (EBSD) technique to map and display. Orientation image microscopy (OIM) also makes relevant microstructural quantification very rapid and takes full advantage of such information. In addition to the individual grain or phase orientation, the EBSD technique has developed to provide many other crystal orientation-based information. For example, orientation relationship between phases or grains, quantification of dislocation density based on orientation variation within the grains, and identification of special grain boundaries.

Grain and grain boundary: Grains are important units for crystal plasticity, and they are a basic space-filling microstructural feature in crystalline materials. Deformation in polycrystalline material is controlled by grains, as each grain defines a domain of atoms arranged in a particular orientation. The orientation and size of grains control the dislocation-based plasticity. The grain boundaries are regions where two grains meet and represent a planar defect where the arrangement of atoms is interrupted. The properties of grain boundaries are not obvious by just looking at the microstructure. The smallest grain size for discussion in this textbook has to have at least 8 to 10 atoms across. That means 2 to 3 nm is the finest grain size, as atomic clusters below this size will have atoms mostly associated with grain boundary regions, which we considered planar defect with lower atomic packing density. Defining the largest grain size is tricky and obviously cannot exceed the size of component, in which case it becomes a single crystal. Typically, if the grain size is bigger than ~50 to 100 μm, the material is referred to as coarse-grained material. If the grain size is below 100 nm, it is referred to as nanocrystalline material. If the grain size is between 100 and 1000 nm, the material is referred to as ultrafine-grained material. The grain boundaries are also referred to as *interfaces*.

Precipitates: Precipitates are formed from alloying elements that go into solution above the solvus temperature. A precipitate forms from the solid solution during an aging step. The precipitates strengthen the matrix when they are a few nanometers, resolvable only by electron microscopy, and lose their strengthening effectiveness when they increase in size to a micrometer. Because they precipitate out from a solid solution matrix, they often have a crystallographic relationship with the matrix. At finer scale, they are coherent and gradually become incoherent as they coarsen to a bigger size. The stability of such precipitates depends on temperature. The matrix/particle interfaces are very important for the deformation mechanism and because of the difference in chemistry can be referred to as *interphase interfaces*.

Dispersoids: Dispersoids are particles that are insoluble in the matrix. The term dispersoid is used when the alloying element is added purposefully. Dispersoids form during the solidification stage and then remain insoluble during any solid-state processing step. The finer dispersoids are of the order of 10 to 20 nm, and the coarser dispersoids can be as coarse as 100 μm. When dispersoids form from elements that are not intentionally added, that is, from impurity elements, they are referred to as *constituents* or *constituent particles*. Constituent particles also do not dissolve in the matrix and generally form in the liquid stage during solidification. They are quite coarse, 20 to 200 μm. The interface is incoherent and becomes critical for deformation and fracture. Additionally, powder metallurgical processing of materials allows addition of ceramic dispersoids in metallic alloy matrix. For example, yttria-dispersed nickel alloy is a dispersion-strengthened alloy. Unlike precipitates, which coarsen or dissolve at high temperature, these dispersoids are thermally stable. It is critical to understand this distinction as it also has implication for orientation relationship of a particle with the matrix. It is expected that in most cases, the dispersoids will be incoherent or not have an orientation relationship with the matrix. On the other hand, the precipitates, depending on the crystal structure and lattice parameter of the particle, are likely to have some form of orientation relationship.

Twins: Twin boundaries can form during annealing of low stacking fault energy alloys as well as during deformation. Twin boundaries can be coherent or incoherent. The impact of twin boundaries on mechanical properties can be very significant, and at this stage we would just consider them as a microstructural feature in low stacking fault energy alloys.

With this short summary of microstructural features, let us consider some engineering structural alloys and reflect on what can be observed with a few characterization tools.

4.1 Metallic materials

Common structural metallic materials are:

- Steels
- Aluminum alloys
- Nickel-based superalloys
- Titanium alloys

These alloys cover a wide range of microstructures and, in some cases, give many variants because of allotropic transformation. Some examples for each of these materials are discussed below.

Steels are the most widely used metallic materials and exhibit quite a range of microstructure as a function of chemistry and processing steps. Fig. 4.1 shows a few micrographs of steels with different levels of carbon and processing history. Note many different features in these micrographs. First, as carbon increases, the volume fraction of ferrite and pearlite changes. In low carbon steel, there are

50 Chapter 4 Microstructural elements in engineering structural materials

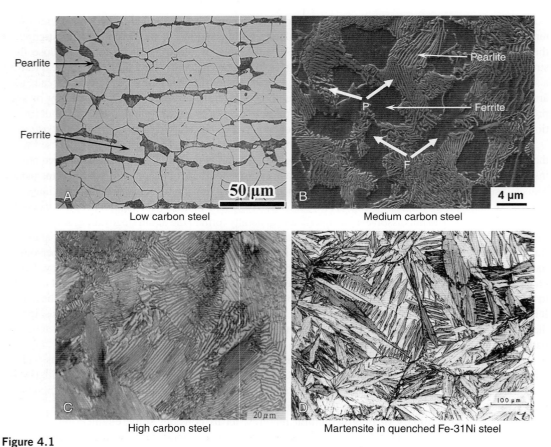

Figure 4.1
Optical micrographs of steels with different carbon level and different processing history.[1–4] Note the variations of ferrite shape and size in various steels.

large grains of ferrite with regions of pearlite. This is a point to focus on in the context of this discussion. The aim of looking at these microstructures is to figure out the important features from a mechanical behavior perspective. Remember, pearlite is not a single phase! It is a lamellar structure made of ferrite and cementite. The scale is very fine. Obvious questions are: What governs the strength of each region? How does deformation proceed in the lamellar region? How does deformation or strain

---

[1] Wang S, Nagao A, Sofronis P, Robertson IM. Hydrogen-modified dislocation structures in a cyclically deformed ferritic-pearlitic low carbon steel. *Acta Mater.* 2018;144:164-176.
[2] Hui W, Zhang Y, Zhao X, Xiao N, Hu F. High cycle fatigue behavior of V-microalloyed medium carbon steels: a comparison between bainitic and ferritic-pearlitic microstructures. *Int J Fatigue.* 2016;91(Part 1):232-241.
[3] Putatunda SK. Fracture toughness of a high carbon and high silicon steel. *Mater Sci Eng A.* 2001;297(1-2):31-43.
[4] Guimarães JRC, Rios PR. Microstructural path analysis of martensite dimensions in FeNiC and FeC alloys. *Mater Res.* 2015;18(3). https://doi.org/10.1590/1516-1439.000215.

partitions, and what would be its impact on the onset of fracture process? How does this discussion change for high carbon steel with 0.8%C, which has 100% pearlitic microstructure? The quenched microstructure is made of very fine martensitic laths. How does this impact strength and ductility? Clearly, to answer these questions for elastic and plastic deformation stages, significantly more fundamental knowledge is needed than what was covered in introductory classes. In addition, such discussion before covering the theories would hopefully be a motivation to ask more questions about applicability of specific theories to a particular microstructure. Dislocation-based plastic deformation theories would revolve around **dislocation generation, dislocation movement, and dislocation annihilation/rearrangement**. So, repeatedly, we need to ask questions pertaining to these steps and to relate the mechanisms to the microstructural scale.

Aluminum alloys are very widely used in applications where light weight (high strength to weight ratio) is important. Casting is used for low-cost production of complex components. Let us examine some typical microstructures in aluminum alloys. Fig. 4.2 shows micrographs for two different casting

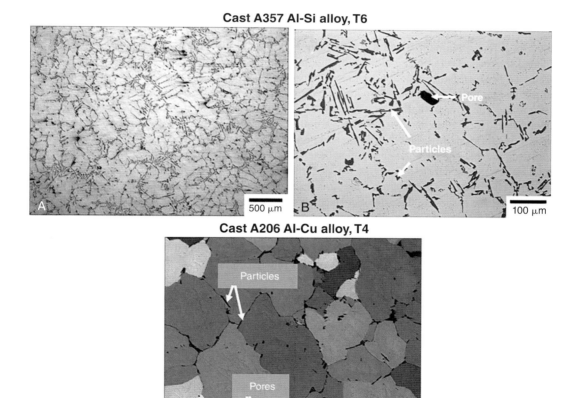

Figure 4.2

Optical micrographs of cast aluminum alloys with different heat treatment. Note the key microstructural elements in the micrographs, particularly the scales of grains, second-phase particles, and porosities.

Chapter 4 Microstructural elements in engineering structural materials

Table 4.1 Commonly used temper designations for aluminum alloys.

Temper designation	Sequence of heat treatment
F	As-fabricated condition (variable strength)
O	Annealed (lowest strength state)
W	Solution treated—applies to alloys that age significantly at room temperature
T3	Solution treated, cold worked and naturally aged to stable temper
T4	Solution treated and naturally aged to stable temper
T5	Cooled from high-temperature shaping process and artificially aged
T6	Solution treated and artificially aged
T7	Solution treated and artificially aged beyond maximum strength
T8	Solution treated, cold worked and artificially aged

alloys in different tempers. Table 4.1 gives some commonly used temper designations and has been included for ease of referencing along with the micrographs that will be discussed here. Note that both T4 and T6 treatments use solution treatment steps. We want to focus on key microstructural elements that are visible in these optical micrographs. Alloy A357 is Al-Si-Mg alloy, whereas alloy A206 is an Al-Cu alloy. The key strengthening phases in these alloys will not be resolvable at the optical microscopy level as they are too fine, in nanometer-length scale. Starting with A357, clearly, the microstructure consists of dendritic arms, large particles, and porosity. Any time we have dendritic microstructure, the microstructure becomes quite inhomogeneous. The aluminum grains in this case are free of coarse particles. The interdendritic regions are full of constituent particles and, in this case, large Si-containing particles. The higher magnification micrograph clearly reveals that the aluminum grains are around 100 μm, and that is the length scale of coarser, second-phase particles. Again, we can start asking some curiosity-driven questions based on knowledge gained in the introductory materials science and engineering class. What is the difference in modulus values of aluminum grain and second-phase particles? That would influence elastic loading of the microstructure! What is the difference between strength of the grain and interdendritic regions? How would it influence plastic deformation? Will the large particles deform with the matrix or cause early onset of microcracking in the microstructure? What is the significance of porosity during deformation? When will porosity start growing to form cracks? If we compare the microstructure of A206 alloy, the grains are even larger and there is significantly higher porosity level. The interdendritic region, or in this case grain boundaries, has large particles. Again, many questions can be raised to discuss the deformation of this alloy. Many components go through cyclic loading, and in that case fatigue properties would be important. When we discuss fracture toughness, the concept of stress concentration will become important. But how about stress concentration at these microstructural features? When does it become important? List your questions, and try to develop a sequence among those. In what order would you like to learn the fundamentals? Of course, the book is structured in one way, but if your interests are in some specific aspects, jump to those sections and come back to other topics!

Cast microstructure is full of features that limit mechanical properties. The microstructure can be homogenized and refined by thermomechanical processing of alloys. As a reminder, the generic term for the

4.1 Metallic materials

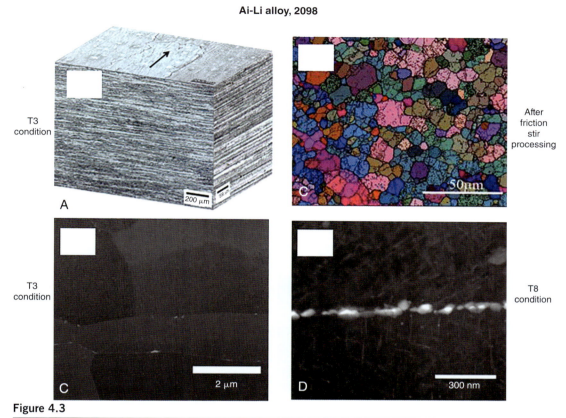

Figure 4.3

Micrographs of a wrought Al-Li alloy designed for high structural performance. (a) The 3D construction of optical micrographs shows the pancake microstructure because of rolling step. (b) Friction stir processing changes to equiaxed microstructure. This micrograph is taken using orientation imaging microscopy (OIM), and colors of grains are based on orientation. This technique allows identification of different types of grain boundaries in the micrograph. (c) Higher magnification backscattered SEM micrograph in T3 condition shows a few grain boundary particles. (d) Transmission electron microscopy (TEM) at quite high magnification shows precipitates within the grain and grain boundary particles. At this magnification, one can resolve the precipitates clearly and notice the shape, size, and distribution.

alloys after deformation processing is "wrought." The microstructure in wrought condition is quite different from what we discussed in the last paragraph. Also, the chemistry of alloys for wrought products is different, as these alloys are designed to keep in mind the thermomechanical steps. Fig. 4.3 shows micrographs of Al2098 alloys taken using different techniques at different magnifications. Also, the processing conditions are different to highlight the change in microstructural features. The optical micrograph montage (Fig. 4.3 [a]) gives the 3D nature of pancake grains produced by rolling of the alloy. At this level of resolution, details of the microstructure are not clear. If the material is processed by another technique, such as friction stir processing (details are not important at this stage), it changes the grain morphology to

54 Chapter 4 Microstructural elements in engineering structural materials

equiaxed. The use of orientation imaging microscopy (OIM) allows visualization of crystal orientation, as the grains are colored on the basis of orientation as shown in Fig. 4.3 (b). It also allows for quantification of different types of boundaries. Increasing the magnification in SEM (Fig. 4.3 [c]) allows further definition of interior of grains and grain boundaries. The final micrograph in this set (Fig. 4.3 [d]) is taken using transmission electron microscopy (TEM). At this level of resolution, the strengthening precipitates can be seen clearly. Shape, size, and distribution can be observed and to some extent quantified.

While the micrographs of aluminum alloys in Figs. 4.2 and 4.3 were from commercial alloys, Fig. 4.4 is from an experimental Al-10 wt.% Ti-2 wt.% Cu alloy. This alloy was mechanically alloyed, which was obtained by ball milling of powders. The powders were processed by hot isostatic pressing (HIP) and extruded. The first thing to observe is significant inhomogeneity in the HIPed condition, and that carries to the extruded product. The SEM micrographs are not able to resolve grains at these mag-

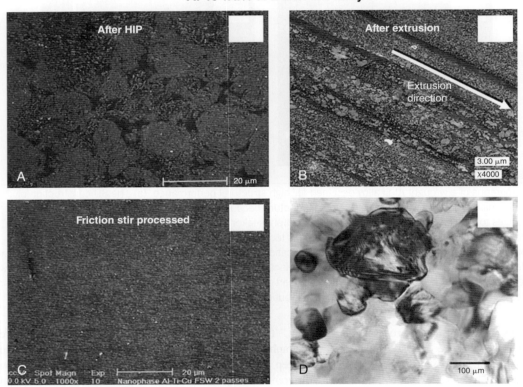

Figure 4.4

Micrographs of an experimental Al-10 wt.% Ti-2 wt.% Cu alloy. Overall microstructure after (a) hot isostatic processing (HIP) and (b) extrusion. First thing to observe is the inhomogeneity at this stage, which is eliminated at this resolution after (c) friction stir processing. The SEM micrographs are not able to resolve grains at these magnifications. (d) The TEM micrograph showing grains and particles after friction stir processing.

nifications. While overall inhomogeneity can be noted, details are missing. In the TEM micrograph, the grains are around 200 nm and particles are around 50 to 80 nm.

Note that in this set of aluminum alloys, we have looked at cast microstructure, wrought microstructure of a commercial alloy using both conventional and non-conventional processing steps, and finally an experimental nanostructured alloy. The goal is to link the microstructure with properties. These examples are offered to motivate the reader/student to start thinking about the linkage between processing and microstructure, even though that is not the primary focus.

Next, we turn to nickel-based superalloys. Nickel-based superalloys are used in very demanding structural applications. A critical use is in high-temperature turbines. No other metallic material is used at as high a homologous temperature (defined as the ratio of service temperature and melting point of alloy, both temperatures in K) as these superalloys. The superalloys have evolved during the last 50 years of use, and specific design of these alloys depends on the particular application as well as on the processing path. We just want to have a high-level glimpse of microstructure. Fig. 4.5 is a collection of micrographs and components to show the context. As mentioned earlier, nickel-based superalloys in turbine engines serve in very demanding conditions. The engine rotates at very high temperature and can get hit by foreign objects. So, the components like blades require a combination of stiffness, creep strength, fatigue strength, and fracture toughness. What type of microstructure can provide the best properties? Obviously it would be difficult to have the best of all these properties in one single alloy or microstructure! Therefore, a set of alloys and microstructures are developed to insert in the right stage of the engine. This figure depicts the change in overall microstructure of the blade from equiaxed to single crystal. What is the importance of eliminating all the boundaries for the highest level of creep performance? We need to understand the linkage between the failure processes at high temperature and grain boundaries.

Titanium alloys also exhibit complex microstructural variations because of allotropic transformation, which is cooling-rate dependent and stabilization of high-temperature β-phase through alloying addition. Fig. 4.6 illustrates key microstructural parameters observed in a wide range of titanium alloys (Banerjee and Williams, 2013). To note length scale of various features, we can focus on the distinguishing characteristics of a microstructure obtained after processing above β transus and compare that resulting microstructure for processing in α + β region. The alpha phase exists in several microstructural length scales, ranging from very fine alpha laths in beta-enriched alloys to colonies of similarly oriented laths and to the equiaxed alpha phase morphology. The β grain size also varies significantly, and the remnant β lath can be very fine. How to quantify the microstructural features to predict mechanical behavior remains a major challenge for titanium alloys. As we go through the fundamentals, again we must focus on how these apply to such microstructures.

4.2 Nonmetallic materials

Four broad categories of bulk nonmetallic materials are:

a) Amorphous materials
b) Polymers
c) Polymer-based composites (generally referred to as Composites)
d) Ceramics

Although we have not talked about amorphous materials, a significant number of materials are in this category. The range is from metallic glasses to traditional silicate-based glasses. From our perspective,

56   Chapter 4 Microstructural elements in engineering structural materials

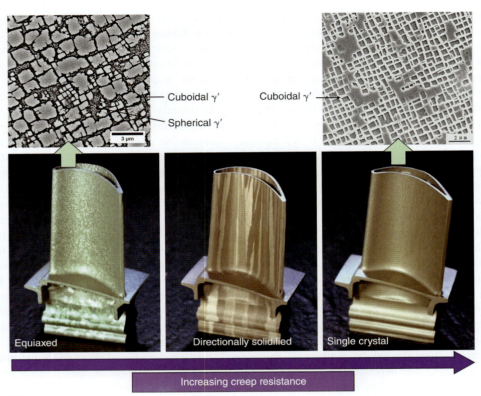

Figure 4.5
A collection of micrographs and components of nickel-based superalloys. Note the use of turbine blades with changing microstructure. All the boundaries are eliminated in the single crystal blade to enhance the use temperature. Different types of precipitates are designed to obtain best combination of properties. Details can be understood from these references.[5-7]

[5]Howmet Aerospace Technology Day 2022. https://www.howmet.com/wp-content/uploads/sites/3/2023/05/Howmet-Aerospace-Technology-Day-2022.pdf
[6]Shang Z, Wei X, Song D, et al. Microstructure and mechanical properties of a new nickel-based single crystal superalloy. *J Mater Res Technol*. 2020;9(5):11641-11649.
[7]Rakoczy Ł, Grudzień-Rakoczy M, Hanning F, et al. Investigation of the γ' precipitates dissolution in a Ni-based superalloy during stress-free short-term annealing at high homologous temperatures. *Metall Mater Trans A*. 2021;52:4767-4784.

there is not a significant microstructural feature that we want to consider for amorphous or glassy materials.

Polymers are made of molecular chains and as such do not have microstructural features that at this stage we will discuss. Polymeric molecular chains do serve a major role as matrix for polymer-based composites. Polymer-based composites have come a long way since the 1940s. Major aircraft structures are now built from polymer matrix composites. Fig. 4.7 is a picture of an Airbus A380, which

β heat treated structures

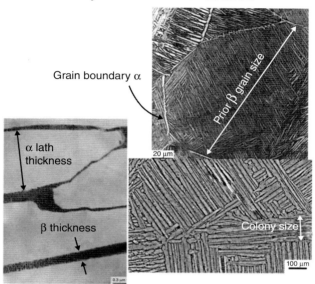

α + β heat treated structures

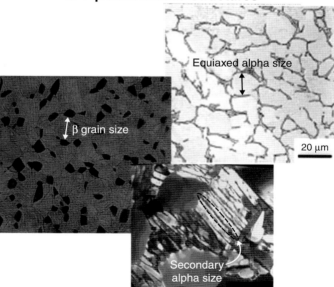

Figure 4.6

Example of key microstructural parameters in various classes of microstructures observed in a wide range of titanium alloys.

58 Chapter 4 Microstructural elements in engineering structural materials

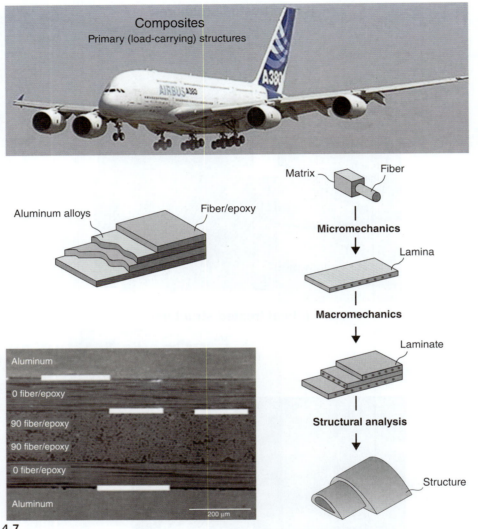

Figure 4.7

An example of composite structure made up of fibers and polymer matrix used in high-performance aircraft structures.

uses composites. The montage in Fig. 4.7 shows the composite-made polymer matrix reinforced with fibers. The major load-bearing microstructural feature in these composites is the fiber itself and how the fibers are arranged to provide required orientation for desired stiffness and/or strength. What is the mechanism of property enhancement? How does it differ from crystalline materials?

The last group that we will consider is ceramics. To start with, the introductory materials science and engineering course pointed out that ceramics have different bonding than metals, and the crystal structure is usually more complicated. A major difference in the mechanical response of ceramics is that there is not much plastic deformation. So, ceramics fail during elastic deformation, and fracture criterion

4.2 Nonmetallic materials

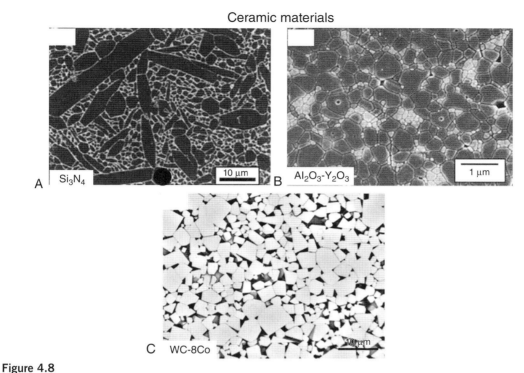

Figure 4.8

Examples of ceramic microstructures. (a) SEM micrograph of Si_3N_4 with secondary phase at the grain boundaries[8] (Bocanegra-Bernal et al., 2009). Note the morphology of grains. (b) SEM micrograph of toughened alumina[9] (Casellas et al., 2003). The alumina phase is the darker one. (c) A typical microstructure of very popular WC-Co cutting tool cermet[10] (He et al., 2018). Note the very faceted nature of WC phase.

falls under the category of brittle fracture. Even with this limitation, ceramics can be toughened to produce engineering materials. Good examples of these are cutting tools and Si_3N_4 ball bearing. Fig. 4.8 shows examples of three types of ceramics. Note that the Si_3N_4 has secondary phase at the grain boundaries. Also, note the morphology of certain grains. Certain additives during sintering of Si_3N_4 lead to formation of glassy phases along grain boundaries and promote long needle- or rod-shaped grains. This is very effective at toughening, as the long grains do "crack bridging." What is the framework for this type of toughening mechanism? What is the limit of such an approach? Similarly, notice the distribution of zirconia in alumina. What is the mechanism of toughening in this case, as the microstructural feature is quite different from Si_3N_4? The last figure in this group is that of WC-Co cermet, which is a very popular cutting tool material. Here the toughening is achieved by adding a small volume fraction of

---

[8]Bocanegra-Bernal MH, Matovic B. Dense and near-net-shape fabrication of Si3N4 ceramics. *Mater Sci Eng A*. 2009; 500(1-2):130-149.
[9]Casellas D, Nagl MM, Llanes L, Anglada M. Fracture toughness of alumina and ZTA ceramics: microstructural coarsening effects. *J Mater Process Technol*. 2003;143-144(1):148-152.
[10]He M, Wang J, He R, Yang H, Ruan J. Effect of cobalt content on the microstructure and mechanical properties of coarse grained WC-Co cemented carbides fabricated from chemically coated composite powder. *J Alloys Compd*. 2018;766:556-563.

metal. The shape and size of the ceramic phase and the volume fraction of metallic binder are the major microstructural features that govern the balance of hardness and toughness of these cutting tools.

As we have gone through many examples, several questions were raised in the text. Finding answers to those questions is the primary goal of this course. Also, understanding of limitations, what we know, and what we do not know! How to apply a theoretical framework to an engineering material of your choice? Most textbook examples are built around very simple microstructural elements. As we have seen, engineering materials possess complex microstructural features! With the initial set of four introductory chapters, hopefully this broader perspective will whet your appetite for learning, motivate you to plod through the theoretical rigor, and pique your curiosity to apply your knowledge to enable you to predict the mechanical properties of engineering materials.

4.3 Exercises

1. Connect the list of five materials that you are interested in (your answer in Chapter 1) with microstructural features.

2. Do an internet search of a material or alloy name along with the term microstructure. For example, steel microstructure. Take a typical micrograph of that material and label various microstructural features that you see. List the scale of these microstructural features.

3. From the microstructural features that you analyzed in question 2, try to connect to mechanical behavior. For the first step towards this, list a set of curiosity questions, like, how will feature "X" impact the mechanical property that you are interested in?

4. Based on the list of questions you created so far, make a list of five things that you would like to learn during this course. At the end of the semester, revisit this answer.

5. Compare the microstructural elements of two alloys or two materials. Why are they different? What are the potential reasons for these variations in the microstructures?

6. Do an internet search with keywords "microstructure" and "Ti-6Al-4V" alloy. Summarize the microstructural feature change for three processes, casting, forging, and additive manufacturing.

7. Do an internet search with keywords "microstructure" and "IN718" alloy. Summarize the microstructural feature change for three processes, casting, forging, and additive manufacturing.

Further readings

This list gives some resources for microstructures in engineering materials. Students are encouraged to browse through some of these references to increase familiarity with the microstructure. Then, hopefully some obvious keywords will pop up and initiate the thinking process regarding mechanical response.

Banerjee D, Williams JC. Perspectives on titanium science and technology. *Acta Mater*. 2013;61:844-879.

Bocanegra-Bernal MH, Matovic B. Dense and near-net-shape fabrication of Si3N4 ceramics. *Mater Sci Eng A*. 2009;500(1-2):130-149.

Campbell FC. Superalloys. In: *Manufacturing Technology for Aerospace Structural Materials*. Elsevier; 2006:211-272.

Caron RN, Staley JT. Effects of composition, processing and structure on properties of nonferrous alloys, in nonferrous alloys. In: Dieter GE, ed. *Materials Selection and Design*. Vol. 20. ASM Handbook, ASM International; 1997:383-415.

Casellas D, Nagl MM, Llanes L, Anglada M. Fracture toughness of alumina and ZTA ceramics: microstructural coarsening effects. *J Mater Process Technol*. 2003;143-144(1):148-152.

Gleiter H. Nanocrystalline materials. *Prog Materi Sci*. 1989;33:223-315.

Gleiter H, Marquardt P. Nanokristalline Strukturen—ein Weg zu neuen Materialien? *Zeitschrift für Materialkunde*. 1984;75:263-267.

He M, Wang J, He R, Yang H, Ruan J. Effect of cobalt content on the microstructure and mechanical properties of coarse grained WC-Co cemented carbides fabricated from chemically coated composite powder. *J Alloys Compd*. 2018;766:556-563.

Pollock TM, Tin S. Nickel-based superalloys for advanced turbine engines: chemistry, microstructure and properties. *J Propuls Power*. 2006;22(2):361-374.

CHAPTER 5

Simple mechanical tests

Chapter outline

5.1 Testing methods for mechanical properties 64
5.2 Hardness testing 64
    5.2.1 Macrohardness test techniques 65
    5.2.2 Microhardness test techniques 66
    5.2.3 Nanoindentation technique 68
5.3 Tensile testing 71
    5.3.1 Hollomon's equation 74
    5.3.2 Effect of temperature 78
    5.3.3 Mechanical anisotropy 78
5.4 Fatigue testing 79
5.5 Fracture toughness 81
5.6 Creep testing 84
5.7 Examples of influence of mechanics and dislocation-based plasticity theory 86
5.8 Key chapter takeaways 87
5.9 Exercise 88

Learning objectives

- An understanding of different testing methods for the properties related to five design approaches.
- An appreciation of mechanical and microstructural flow complexities at various length scales during testing.

As introduced in the previous chapters, we are interested in many different aspects of mechanical behavior that cover various deformation mechanisms and failure mechanisms. These behaviors or properties are probed using different types of loading to measure the specimen or material response. This chapter focuses on some common test methods and briefly describes these. The introduction of these techniques is done to develop an appreciation and understanding of how to gather initial information. It is also important to understand the basic steps involved in ASTM standards and why these are critical for obtaining proper values for various mechanical properties. On the one hand, these property values build toward design allowables, and on the other, proper understanding allows us to extend the testing methods to different length scales as needed for complex components. As you develop understanding of the overall mechanical behavior, an appreciation of the difference between testing for design and testing for scientific curiosity will become clear. For many fundamental studies, it is perfectly fine to use non-standard specimen geometry, but this should be done with proper understanding.

5.1 Testing methods for mechanical properties

Since mechanical properties depend on intrinsic bonding, inherent microstructure, and defects in materials, test methods are needed to measure mechanical properties properly. The elastic properties of a single phase, such as Young's modulus, come from Hooke's Law, which is a material model. Therefore, they are well defined and should not depend on specimen size, shape, etc. Any uncertainty in their measurement can come from imprecision of measuring devices, such as strain gauges, misalignment in the loading train of the testing machine, and so forth. On the other hand, properties involving plastic deformation of a material are not absolute in nature; properties of engineering materials depend on the conditions under which these tests are performed and geometry of the specimen. Common testing methods include indentation hardness, tensile, impact, creep, and fatigue testing. Changes in temperature, strain rate, the presence of stress concentration sites and of embrittling agents, and several microstructural factors, all can affect these properties.

Therefore there is need for standardization of test methods and the steps involved in interpretation of results. American Society for Testing and Materials (ASTM) produces standard testing methods for a variety of properties. It is important to be familiar with ASTM standards and then evaluate the context of a particular set of properties for a particular application or product. The use of the term "simple" in the title of this chapter does not cover the "complexities" that exist if details are considered. If one carefully considers the elastic and plastic deformation involved in that test, the complexities will become apparent. Of course, you will need to understand the basics before you can appreciate the details. So, when you go through this chapter, formulate basic deformation-related questions that you need to understand for a particular property. As you read the subsequent chapters, try to correlate the testing methods to the fundamentals of deformation mechanisms.

5.2 Hardness testing

Hardness of a material is usually defined as the resistance to plastic deformation. In Materials Science and Engineering, among all hardness testing methods, indentation hardness testing is the most widely used. In indentation testing, a hard sphere-, cone-, or pyramid-shaped indenter is forced to press against the surface of the specimen by application of a fixed load (in most cases); subsequently hardness values are obtained either from depth or lateral dimensions of the indentations thus created. The hardness values could be empirically related to yield strength or tensile strength. Hardness testing is a quick, simple quality control tool. In some cases, the hardness testing has been converted to a portable instrument so that we can use on shop floors and at construction sites with relative ease. Also, hardness testing is not typically considered a destructive test technique like the ones that will be discussed in subsequent sections. Rather the hardness testing is considered semidestructive or nondestructive.

Earlier, two major types of indentation hardness techniques were developed: *macrohardness* and *microhardness*. Of late, a method known as *nanoindentation* technique has become popular. Here each of these techniques and their variants are discussed in brief. Detailed hardness conversion tables can be found in the ASTM E140 standard.

5.2.1 Macrohardness test techniques

In the group of macrohardness test techniques, Brinell, Rockwell, and macro-Vickers are the most prominent. Generally, the larger loads used in these techniques give rise to larger size indentations on the sample surface compared to microindentation and nanoindentation techniques.

J.A. Brinell of Sweden in 1900 developed an indentation-based hardness technique that became so popular that it became known as the Brinell test. Generally, this test technique uses a 10 mm diameter ball made of either ordinary high carbon steel or specially heat-treated high carbon steel (Hultgren ball) with the application of 3000 kgf (kilogram-force is the standard unit used for hardness testing, although it is not a SI unit). Under certain circumstances, 500 kgf load is used for soft materials to avoid making too deep an indentation, whereas tungsten carbide indenter is used for hard materials to avoid indenter distortion. The Brinell hardness number (BHN) is given by the following relationship,

$$\text{BHN} = \frac{2P}{\pi D \left(D - \sqrt{D^2 - d^2}\right)} \qquad (5.1)$$

where P is the applied load (in kgf), D is the ball diameter (in mm), and d is the indentation diameter (in mm). The indentation diameter is measured after unloading from a number of measurements using a measuring microscope, and then Eq. (5.1) is used to determine BHN. BHN has a unit of kgf/mm². Useful relationships between BHN and ultimate tensile strength (UTS) include the one shown below for heat-treated plain carbon steels and some other varieties of steels,

$$\text{UTS (in MPa)} = 3.4 \times \text{BHN}. \qquad (5.2)$$

In 1908, Professor Ludwig of Vienna, Austria, first described a method by which hardness of a material can be measured by a *differential depth measurement* technique. This mode of hardness test technique consisted of measuring the increment of depth of a cone-shaped diamond indenter (known as Brale) or spherical steel indenters of various diameters forced into the material by a minor load and a major load. This method of hardness measurement is known as the *Rockwell hardness test*. The test first uses a minor load of 10 kgf to set the indenter onto the material surface, and then a major load (as determined by the particular scale chosen) is applied and the penetration depth is shown instantly on the scale with 100 divisions (usually each division corresponds to 0.00008 inches or 0.00203 mm). Although the required surface preparation is minimal depending on the hardness of the material, several Rockwell scales represent a particular combination of the indenters and major loads used; some of them are included in Table 5.1. Note that a *Superficial Rockwell* test mode uses lower loads compared to the standard Rockwell tests and is used for thinner materials or for probing surface-hardened materials. The martensite formed in eutectoid (0.77 wt.% C) plain carbon steel can have a very high hardness (close to 65 R_c).

Table 5.1 Some standard Rockwell scales.

Rockwell hardness scale	Indenter	Major load (kgf)
A	120° diamond cone (Brale)	60
B	1/16 in. (~1.6 mm) diameter steel ball	100
C	Brale	150
D	Brale	100

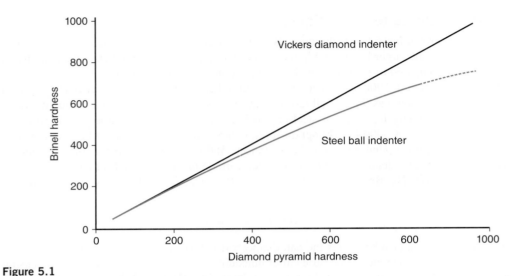

Figure 5.1

Comparison between Vickers hardness and Brinell hardness.

Vickers hardness is another macrohardness test technique that uses a diamond pyramid indenter known as Vickers indenter. This instrument load can range from 1 kgf to 120 kgf. Generally, a load of 30 or 50 kgf is most appropriate. The plot in Fig. 5.1 shows that BHN and Vickers hardness number (VHN) compare well for less hard materials. However, as material hardness increases, the BHN values deviate from the VHN ones because the steel ball indenter begins to distort and then gives erroneous values. More details about this testing are given in the next section on microhardness test techniques, since both macro- and micro-Vickers principles are quite similar.

5.2.2 Microhardness test techniques

Microhardness technique is generally used to measure indentation hardness of very small objects, thin-sheet materials, surface-hardened materials, electroplated materials, structural phases in multiphase alloys, and so forth. Generally, we can use two standard indenters for microhardness tests. Vickers indenter, a square-based diamond pyramid with 136° apex angle, as shown in Fig. 5.2 [a]. Knoop indenter is also a pyramid-shaped indenter albeit with different apex angles (an included transverse angle of 130° and a longitudinal angle of 172°30′ such that it creates an indentation with a long-to-short diagonal ratio of approximately 7:1) (Fig. 5.2 [b]).

The Vickers indenter is pressed into a smooth, polished surface of the specimen. The load, which is up to 1000 gf (1 kgf), is applied for a predetermined time and then removed. The average value of the length of the two diagonals of the indentation is measured, and the VHN is calculated as the ratio of the applied load to the surface area using the relationship,

$$\text{VHN} = 1.854 \frac{P}{D^2} \tag{5.3}$$

where P is the load (in kgf) and D is the arithmetic mean of the two diagonal lengths (in mm). So, VHN has actually a unit of pressure or stress, kgf/mm². Nowadays, VHN is popularly expressed in

5.2 Hardness testing

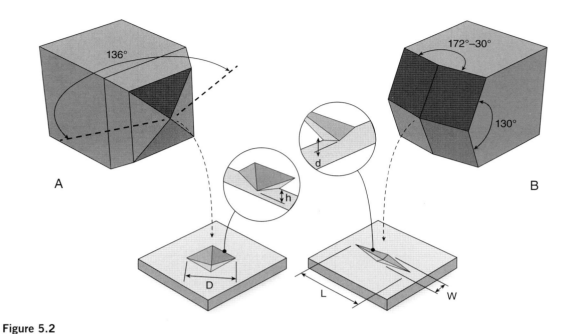

Figure 5.2

Schematic views of the microhardness indenter tips and associated indentations: (a) Vickers indenter and (b) Knoop indenter.

SI units of either GPa or MPa by multiplying with 9.81 m/s². After a specified load and time, precision objectives are used to view the indentation, and using scales attached on the eye-piece, the size of the two diagonal lengths of the square-shaped impression can be measured. Then this square length is used to measure VHN from Eq. (5.3). Standard charts are also available. Generally, indentation should be made away from the edge of the sample as much as possible. Thickness of the specimen should be at least 1.5 times the diagonal length. Indentations should be spaced approximately 5–10 indents apart so that their individual deformation zones are not affected or do not overlap. Generally, one takes several indents from a region to measure hardness and then reports the mean with its associated measurement error.

Principles of Microhardness Testing with Knoop Indenter: In all indentation hardness tests, when the applied load is removed, some elastic recovery always occurs. The amount of recovery and the distorted shape depend on the size and precise shape of the indenter. Due to the unique shape of the Knoop indenter, elastic recovery of the projected impression occurs in a transverse direction, i.e., shorter diagonal length rather than long diagonal. Therefore the measured longer diagonal length will give a hardness value close to what is given by the unrecovered impression. The Knoop hardness number (KHN) is given by the load divided by the unrecovered projected area of the impression. Note that the area referred here is the projected area and not the surface area of the indentation as in the Vickers and Brinell hardness techniques. Hence, KHN is given by: KHN = P/A_p = P/CL^2, where P is the applied load in kgf, A_p is the unrecovered projected area of indentation (mm²), L is the longer diagonal length (in mm), and C is the Knoop indenter constant that relates the longer

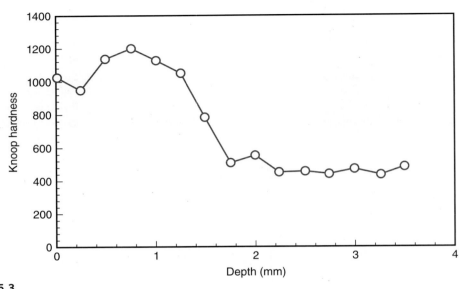

Figure 5.3

Knoop hardness profile in 430 SS and EX-15 carburized steel specimens.[1]

diagonal length to the unrecovered projected area (generally 0.07028). Standard hardness charts are available where hardness values are provided against load and long diagonal values. Note that KHN is not independent of the load used. So, while reporting hardness numbers, the load used should also be reported (especially when the load is less than 300 gf). Knoop hardness technique is much more sensitive to specimen surface preparation than Vickers hardness in the low load range. Main advantages of Knoop hardness technique include its ability to measure near-surface hardness and hardness gradients. Fig. 5.3 shows a generic Knoop microhardness profile from the surface toward the core in a surface-hardened alloy. The data indicates that the case is much harder than the core. For instance, automotive gears are case-hardened to impart wear and tear resistance to the gear teeth areas while keeping the core tough.

5.2.3 Nanoindentation technique

During the last few decades (since 1980s), very significant developments have taken place in exploring microplasticity and localized deformation mechanisms using instrumented indentation techniques. The nanoindentation technique was developed primarily to measure modulus and hardness on a localized scale. Microhardness cannot measure hardness of materials where the indentation is roughly less than 1 μm because the optical microscope attached to the system to measure the indentation size loses resolution in that range. Nanoindentation was first developed to measure hardness of thin conduction lines or thin films in microelectronic applications. The load range used in nanoindentation generally

---

[1] Renner P, Raut A, Liang H. High-Performance Ni-SiC Coatings Fabricated by Flash Heating. Lubricants. 2022; 10(3):42.

stays below nN level, thus leading to an indentation size that is usually less than a micron. In modern designs of nanoindenters, an atomic force microscope (AFM) can be attached to the nanoindenter to image the indentation directly. Modern nanoindenters have nN load resolution and sub-Angstrom depth resolution. Because of the high resolution of the load-displacement data, fundamental events, such as activation of dislocation sources, initiation of deformation instability, phase transformations, and twinning, can be analyzed by performing a carefully designed nanoindentation test. Since the Vickers or Knoop indenter is a four-faced pyramid, the four planes do not generally meet at a point and thus do not bear the sharpness necessary before nanoindentation. Unlike Vickers or Knoop indenters, the most widely used nanoindenter diamond tip is a three-faced pyramid with an included apex angle of 142°. This nanoindenter tip as shown in Fig. 5.4 (a) is known to have the required sharpness and Berkovich geometry, which is named after Soviet scientist E.S. Berkovich, who in 1950 first described the indenter shape. A top-down atomic force microscopy (AFM) image of the corresponding indenter tip depicting the threefold symmetry is shown in Fig. 5.4 (b).

Generally, each nanoindentation test is an effort to find the hardness and elastic modulus of a loading and unloading cycle. Fig. 5.5 shows a typical load-displacement (P-h) curve. The experimental unloading curve has a clear curved feature and can be approximated well by the following power-law type equation,

$$P = \alpha (h - h_f)^m \tag{5.4}$$

where h_f is the final contact depth upon unloading, and α and m are power-law fitting constants. Table 5.2 shows different values of α and m for different materials using a Berkovich nanoindenter tip.

The nanohardness (H_{nano}) typically measured is given by,

$$H_{nano} = P_{max}/A_p \tag{5.5}$$

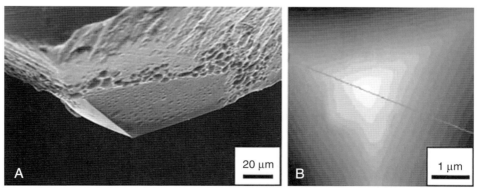

Figure 5.4

Views of the Berkovich diamond geometry commonly used in nanoindentation testing. (a) A profile view, as observed in a scanning electron microscope, showing the pyramidal diamond tip. (b) Top-down atomic force microscopy image showing the threefold pyramidal symmetry of the Berkovich indenter.[2]

---

[2]Schuh CA. Nanoindentation studies of materials. *Materials Today*. 2006;9(5):32-40.

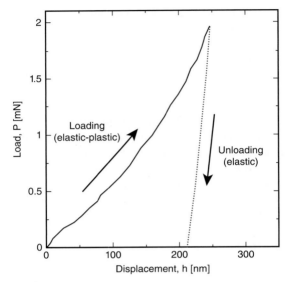

Figure 5.5
An example of load-displacement curve obtained in a nanoindentation test.[3]

Table 5.2 Different values of α and m for various materials using a Berkovich nanoindenter tip.

Material	α (mN/nm^m)	m	Correlation coefficient, R
Aluminum	0.265	1.38	0.999938
Tungsten	0.141	1.51	0.999986
Silica	0.0215	1.43	0.999985
Sapphire	0.0435	1.47	0.999998

where P_{max} is the maximum load and A_p is the projected contact area at P_{max}. The contact area at the maximum load is calculated from the contact depth, h_c, at the full load following the relationship,

$$A_p = 24.56\, h_c^2 \tag{5.6}$$

for an ideal Berkovich tip. If the tip is not perfect, additional terms need to be added to the right-hand side of Eq. (5.6).

Generally, the Oliver-Pharr method, which was developed in 1992, is used to analyze the nanoindentation data based on the Hertzian theory of elastic contact. Hertz's theory states that the specimen's unloading is assumed to be purely elastic between two elastic spheres. The theory works for conical and pyramidal indenters, with the reduced elastic modulus determined by,

$$E_r = \left(\frac{\sqrt{\pi}}{2}\right)\left(\frac{S}{\sqrt{A_p}}\right) \tag{5.7}$$

---

[3]Schuh CA. Nanoindentation studies of materials. *Materials Today*. 2006;9(5):32-40.

where S is stiffness measured by the slope $\frac{dP}{dh}$ of the unloading curve (as shown in Fig. 5.5) and A_p is the projected area. The reduced modulus is basically an expression of effective modulus which is given by,

$$E_r = \frac{1}{\left[\frac{(1-v_1^2)}{E_1} + \frac{(1-v_2^2)}{E_2}\right]} \tag{5.8}$$

where E_1 and E_2 are elastic moduli, and v_1 and v_2 are Poisson's ratios of sample and indenter, respectively. For diamond tip, $E_2 = 1140$ GPa and $v_2 = 0.07$. So, once E_r value is established, Eq. (5.8) can be used to evaluate E_1.

Properties such as creep behavior of materials have also been investigated by nanoindentation technique. Electronic industries have shown interest in incorporating nanoindentation technique to test micro- and nanodevices in their processes as a quality control tool. Nanoindentation is also being used in biomedical research.

5.3 Tensile testing

Tensile testing provides an assessment of the *short-term* mechanical behavior (strength and ductility) of a material under quasistatic uniaxial tensile stress. If the material is ductile, we can evaluate a number of properties (yield strength, UTS, percentage elongation to failure, percentage reduction in area, tensile toughness, and so forth) from this test. This test provides important design data for static load-bearing components. The standard guidelines for tensile testing can be found in the ASTM Standard E-8 (Standard Test Methods for Tension Testing of Metallic Materials). A full-size standard tensile specimen design is shown in Fig. 5.6 (a). The tensile specimen shape could be cylindrical or flat with rectangular cross-section. Subsize specimens are used when there is an interest in studying neutron-irradiated materials or the processing system is not capable of producing large enough volume of materials or there is an interest in obtaining localized properties to correlate with local microstructure. A good example of the need for mini-tensile specimens is to characterize various zones of a welded material. The specimen is loaded onto the tensile test frame as shown in Fig. 5.6 (b). The specimen is then stretched in the tensile direction at a constant extension rate (*aka* crosshead speed). The elongation of the specimen is measured by strain gage or extensometer and the load by load cell. The load and displacement data are recorded on a computer, and then further analyses are carried out to calculate the mechanical properties.

To a good approximation, to compare elongation measurements in different-sized specimens, the specimens must be geometrically similar,

$$\frac{L_G}{\sqrt{A_G}} = \text{constant} \tag{5.9}$$

where A_G and L_G are the cross-sectional areas and lengths of the gauge section of two different tensile specimens. This relation was first suggested by Barba in 1880, and it is referred to as Barba's law.

Chapter 5 Simple mechanical tests

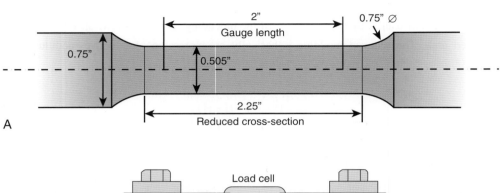

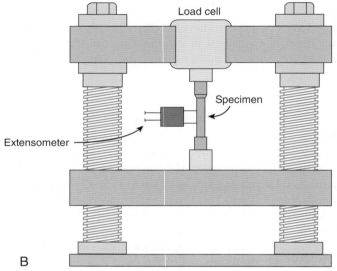

Figure 5.6
(a) The design of a standard tensile specimen, and (b) a schematic of tensile testing equipment.

A typical engineering stress-strain curve is shown in Fig. 5.7 (a). Engineering stress-strain curves are constructed from load-elongation data obtained from a tension test. Engineering stress and strain are based on the original dimensions of the gauge section. Engineering stress (S) is given by,

$$S = \frac{P}{A_o} \tag{5.10}$$

where P is the load and A_o is the original cross-sectional area of the tensile specimen gauge. Engineering strain (e) is defined as the change in length (ΔL) over its original gauge length (L_o),

$$e = \frac{\Delta L}{L_o} = \frac{L_f - L_o}{L_o}. \tag{5.11}$$

Yield strength is the stress required to produce a small, specified amount of plastic strain, generally 0.2%. This is also known as the offset yield strength or proof stress. However, if the stress-strain curve

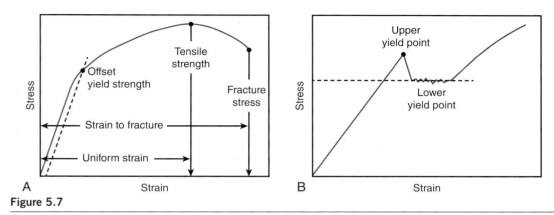

Figure 5.7

Two types of engineering stress-strain curves: (a) continuous and (b) discontinuous (presence of two yield points – typical in low carbon steels).

shows discontinuous yielding (well-known yield point phenomenon in low carbon steels), as shown in Fig. 5.7 (b), two yield stresses can be defined – the peak stress from which stress drop occurs is called upper yield point, and after drop, the stabilized stress is called the lower yield point. Tensile strength or UTS, S_u, is defined as maximum load (P_{max}) divided by the original cross-sectional area of the specimen,

$$s_u = \frac{P_{max}}{A_o}. \tag{5.12}$$

On the other hand, ductility is a measure of the degree of plastic deformation that has been sustained until fracture. Even though ductility is an important mechanical property, it is still quite qualitative and subjective. Conventional measures of ductility are expressed in two ways: engineering strain at fracture (percentage elongation to failure) and percentage reduction in area.

Engineering strain at fracture (e_f) is given by,

$$e_f = (L_f - L_o)/L_o \tag{5.13}$$

where L_f is the final gauge length at fracture.

Reduction in area is calculated by measuring the change in the cross-sectional area at fracture over the original cross-sectional area, and it is given by,

$$RA = (A_o - A_f)/A_o \tag{5.14}$$

where A_f is the final cross-sectional area. The reduction in area is considered more sensitive to the actual response of the material and is considered true property. Both ductility properties are often expressed in "percentage."

Ductility is a microstructure-sensitive property that is of interest in three ways. Ductility can be used to indicate the extent to which a metal can be deformed without fracture in metalworking operations. A highly ductile material implies a "forgiving" material and is likely to deform locally without fracture. A designer often needs to know the relevant ductility of a material. For most load-bearing structural applications, a minimum of 5% elongation is generally desired. Sometimes, it may also serve as an indicator of changes in impurity level or processing conditions. Ductility measurements may be specified to assess material "quality" even though no direct relationship exists between ductility measurement and design allowables for application or service.

While engineering stress-strain curve could provide adequate data for engineering design, large plastic deformation situations (such as metal deformation processing) require the concept of true stress and true strain. Engineering stress-strain curves do not provide a true indication of the deformation characteristics of a material because the curves are based entirely on the original dimensions of the specimen, and these dimensions change continuously during the test. We can construct a true stress-strain curve, *aka* flow curve, from a given engineering stress-strain curve. The following relations can be used to convert engineering stress to true stress and engineering strain to true strain. True or natural stress (σ) is given by,

$$\sigma = \frac{P}{A} = \frac{P}{A_o}(e+1) = S(e+1). \tag{5.15}$$

Note that the above relation becomes possible on the assumption that plastic deformation does not involve any volume change. True or natural strain (ε) based on instantaneous gauge length (L_i) is given by,

$$\varepsilon = \int_{L_o}^{L_i} \frac{dL}{L} = \ln(L_i/L_o) = \ln(e+1). \tag{5.16}$$

Fig. 5.8 shows a comparison between true stress-strain and engineering stress-strain curves. Note that for the initial period of tensile deformation, both curves follow the same path. However, they deviate from each other as the deformation becomes larger. Total plastic deformation is composed of two components: uniform plastic deformation (up to the point of maximum load corresponding to M on the engineering stress-strain curve and M' on the true stress-strain curve in Fig. 5.8) and nonuniform plastic deformation. Nonuniform plastic deformation will be discussed in greater detail in a later section.

5.3.1 Hollomon's equation

Hooke's law ($\sigma = E\varepsilon$) is not applicable beyond the elastic deformation regime. The flow curve of many metals in the region of uniform plastic deformation (the part of strain hardening regime) can be expressed by the simple power law type relationship,

$$\sigma = K\varepsilon^n \tag{5.17}$$

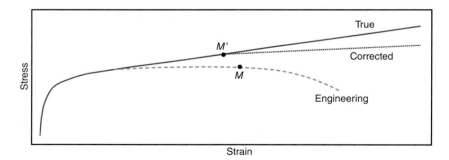

Figure 5.8

Engineering stress-strain curve vs. true stress-strain curve.

5.3 Tensile testing

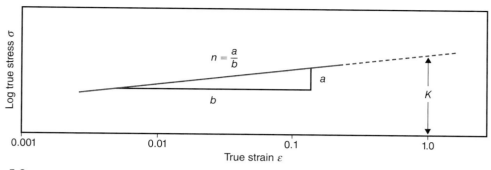

Figure 5.9
Log-log plot of true stress vs. true strain – the slope yields strain hardening exponent (*n*).

where *n* is the *strain hardening exponent* ($n = 0$ for a perfectly plastic solid and $n = 1$ for a perfectly elastic solid, usually varying between 0.1 and 0.5), and *K* is the *strength coefficient* that is basically the true stress at a true strain of unity. While *K* does not have a physical meaning, *n* does represent the extent of uniform plastic deformation and a measure of material's work-hardening capability. Strain hardening exponent is also sensitive to microstructure, temperature, and other environmental factors (such as irradiation). An important point to note is that true stress/true strain data in the region between yield stress and point M' as noted in Fig. 5.8 need to be gathered and plotted in double logarithmic plot (Fig. 5.9). Hollomon's equation has shown certain issues as the validity of the *n* evaluated is drawn into question. That is why researchers have tried to calculate the strain hardening exponent using a variety of other forms of equations (e.g., Ludwik, Ramberg-Osgood, Voce). Table 5.3 lists strain hardening exponent and strength coefficients for two metallic materials (plain carbon steel and copper). For example, ASTM E646 standard describes the evaluation process of tensile strain-hardening exponent of metallic sheet materials that exhibit a continuous yielding in the stress-strain curve in the plastic region. Metallic samples tested by this method should have thicknesses of at least 0.13 mm but not greater than 6.4 mm.

If we follow the shape of the engineering stress-strain curve as shown in Fig. 5.8, we see a continuous decrease in engineering stress beyond the ultimate tensile stress, which gives the impression that the material work-softens! But in reality that is not the case. Recall that the original cross-sectional area was used to calculate engineering stress. After reaching the maximum load (or UTS) of a material, a geometrical

Table 5.3 Strain hardening exponent measured at room temperature for a low-carbon steel and copper.

Metal	Condition	*n*	*K* (MPa)
0.05% C-steel	Annealed	0.26	530
Copper	Annealed	0.54	320

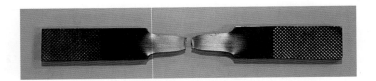

Figure 5.10
A view of a failed tensile specimen following necking. Note that while external change of cross-section can be measured in a straightforward manner using optical imaging, internal cavities and microcracks that form after necking cannot be easily probed.

instability known as necking (localized deformation) starts in the specimen. A part of a tensile specimen undergoing necking is shown in Fig. 5.10. In fact, the specimen continues to strain-harden throughout deformation until fracture, as shown by the true stress-strain curve in Fig. 5.8. However, constructing this part of the true stress-strain has always been difficult since exact calculation of true stress requires exact geometry of the neck developed. Once the necking starts, almost all deformation becomes localized in the necked region and the local stress-state is no more uniaxial. Exact true strain calculation should be based on the cross-sectional area as given by $\ln(A_o/A_{neck})$, where A_{neck} is the cross-sectional area at the neck. Correction methods such as the Bridgeman method can be implemented to correct true stress-strain curves. But because of the difficulty in characterizing necked region, for most purposes, researchers tend to ignore the postnecking part of the flow curve. The postnecking region can also contain internal voids or cavities, which is not easy to quantify for correcting the stress-strain curve. It is important to think of these two contributions as external loss of cross-section area and internal loss of cross-section area. From a metal working standpoint, the deformation after necking is no longer uniform and should be avoided. For example, in sheet metal forming, if the deformation exceeds uniform elongation, it will create undesired thickness variation in the component.

Necking generally starts at the maximum load during tensile deformation of a ductile material. A material with no strain hardening capability becomes plastically unstable, with necking developed as soon as yielding begins. On the contrary, a ductile engineering material undergoes strain hardening, which increases the load-carrying capacity of the specimen with increasing strain. Since this happens at the maximum load (a constant for a given tensile test), this condition of such plastic instability leading to localized deformation can be described mathematically by the following relationships,

$$dP = 0,$$
$$\text{or } d(\sigma A) = 0,$$
$$\text{or } \sigma\, dA + A\, d\sigma = 0;$$
(5.18)

$$\text{or,}\quad \frac{d\sigma}{\sigma} = -\frac{dA}{A}.$$
(5.19)

Eq. (5.19) confirms that the onset of necking involves fractional increase in stress that is exactly balanced by the fractional decrease in the cross-sectional area of the tensile specimen. An important result obtained from similar analysis is that the strain hardening exponent value is either equal (cylindrical specimens with circular cross-section) or proportional (flat specimens with rectangular cross-section) to the true uniform strain, $n = \varepsilon_u$ (refer to Fig. 5.7 and Eq. [5.18]). This is why when the strain hardening exponent decreases, it adversely affects ductility.

5.3 Tensile testing

Stress-strain curves are affected by the change in temperature and strain rate. The rate at which strain is applied to a specimen, called strain rate $\left(\dot{\varepsilon} = \frac{d\varepsilon}{dt}\right)$, can have an important influence on flow stress. Strain rate is expressed in the unit of s^{-1}. The term flow stress is used to describe true stress at any given strain and a generic symbol, σ, is used. In this book, we also use this symbol interchangeably for true stress. In a tensile test, a constant crosshead speed is generally used. So, to calculate the imposed strain rate in a test, the following expression in Eq. (5.20) is used to relate strain rate to the crosshead speed. An engineering strain rate ($\dot{e}$) based on the original gauge length (L_o) as shown in Eq. (5.20a) can be constant during a test conducted at a constant crosshead velocity of v. However, the true strain rate ($\dot{\varepsilon}$) changes because the gauge length continuously increases during the tensile test; the corresponding expression is shown in Eq. (5.20b). Depending on the range of strain rates used in mechanical testing, there are various tests as summarized in Table 5.4.

$$\dot{\varepsilon} = \frac{v}{L_o} \tag{5.20a}$$

$$\dot{\varepsilon} = \frac{v}{L} \tag{5.20b}$$

Flow stress increases with increasing strain rate. The effect of strain rate becomes more important at elevated temperatures. The following equation shows the relation between flow stress and true strain rate at a constant strain and temperature:

$$\sigma = C\dot{\varepsilon}^m \tag{5.21}$$

where C is a constant and m is strain rate sensitivity (SRS). The exponent m can be calculated from the slope of the double-logarithmic plot of true stress vs. true strain rate. The SRS is quite low (<0.1) at room temperature or low homologous temperature (T/T_m, where T_m is the absolute melting temperature; note that the room temperature will be elevated temperature for an element with low melting temperature as described in the next subsection), but it increases as the temperature becomes higher, with a maximum value of 1 when the deformation is known as Newtonian viscous flow. Fig. 5.11 shows flow stress (at 0.2% strain) vs. strain rate on a double logarithmic plot for an annealed 6063 Al-Mg-Si alloy. In superplastic materials, the SRS is higher (0.4–0.6). But these materials require finer grain diameter (<10 μm) and temperatures at or above half of the melting temperature (in K). Superplastic materials exhibit higher than normal ductility (as a rule of thumb, equal to or more than 200% for metallic materials).

Table 5.4 A list of mechanical tests probing different strain rate ranges.

Range of strain rate (s^{-1})	Condition of type of test
10^{-8} to 10^{-5}	Creep tests
10^{-5} to 10^{-1}	Tension test with hydraulic or screw-driven machines
10^{-1} to 10^{2}	Dynamic tension or compression test
10^{2} to 10^{4}	High-speed testing using impact bars
10^{4} to 10^{8}	Hypervelocity impact using gas guns or explosively driven projectiles

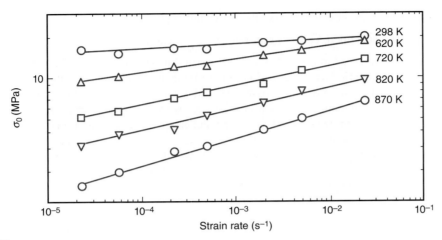

Figure 5.11
Flow stress vs. strain rate for an annealed 6063 Al alloy at different temperatures. Note that strain rate sensitivity increases with increasing temperature. The slope of the fitted lines at each temperature increases as the temperature increases.[4]

5.3.2 Effect of temperature

In general, strength decreases and ductility increases with increasing temperature (Fig. 5.12). However, metallurgical changes such as precipitation, strain aging, recrystallization, particle, and grain coarsening may occur in certain temperature ranges and can alter the general trend. The best way to compare mechanical properties of different materials at various temperatures is in terms of the ratio of the test temperature to melting temperature, expressed in K. This ratio is referred to as the *homologous temperature* (T_h). For example, for lead (melting point 327°C), room temperature is equivalent to a T_h of 0.5. On the other hand, tungsten, which has a very high melting point (3422°C), the homologous temperature corresponding to room temperature would be quite small (0.08). With the higher homologous temperature, plastic deformation becomes thermally activated. At 0.4–0.5, the diffusion process becomes predominant, and diffusion-assisted plastic deformation processes become important.

5.3.3 Mechanical anisotropy

Variation of mechanical properties in different directions may be beneficial or detrimental depending on the situation. *R*-parameter of sheets can be evaluated by conducting tensile testing along the rolling direction. *R*-parameter is known as transverse contractile ratio,

$$R = \frac{\ln(w_o/w_f)}{\ln(t_o/t_f)} \tag{5.22}$$

---

[4]Modified from Mignogna R., D'antonio C, Maciag R, Mukherjee K. The mechanical behavior of 6063 aluminum. *Met Trans.*, 1 (1970) 1771-1772.

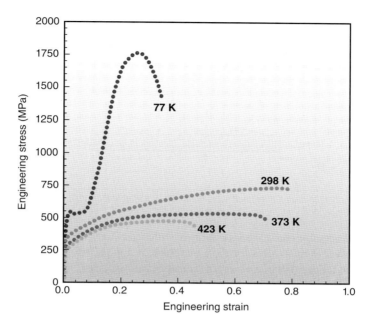

Figure 5.12

Changes in engineering stress-strain curves of stainless steel 304 with temperature.[5,6]

where w_o and w_f are the original and final widths of the tensile gauge section, respectively, and t_o and t_f are the original and final thicknesses of the gauge section, respectively. Schematic of the width and thickness strains for *R*-parameter calculation is shown in Fig. 5.13. A parameter termed *P*-parameter is also known as R_{90}. This parameter basically represents the transverse contractile ratio when it is measured in the direction perpendicular to the rolling direction but in the sheet plane. Higher values of *R* and *P* are good for sheet formability. This means that thickness strain can be low without the risk of failure because of excessive thinning during the sheet-forming process. For example, in deep drawing, a certain type of anisotropy provides resistance to thinning. Textured materials especially appear to have preferred orientation that can affect *R* and *P* parameters. For more information on these parameters, refer to ASTM standard E517.

5.4 Fatigue testing

Fatigue testing is performed to understand material mechanical behavior under cyclic or fluctuating stress, i.e., varying over time. That is why fatigue testing is described as *dynamic* testing. This kind of

---

[5]Cullen GW, Korkolis YP. Ductility of 304 stainless steel under pulsed uniaxial loading. *Int J Solids Struct*. 2013;50(10):1621-1633.
[6]Qin S, Yang M, Yuan F, Wu X. Simultaneous improvement of yield strength and ductility at cryogenic temperature by gradient structure in 304 Stainless Steel. *Nanomaterials (Basel)*. 2021;11(7):1856.

80 Chapter 5 Simple mechanical tests

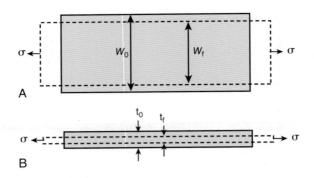

Figure 5.13

Width and thickness strains in uniaxial tension test on a sheet metal specimen.

stress situation could actually lead to fatigue failure, which accounts for a substantial fraction of component failures. At lower temperatures for most materials, if applied stress is kept below a static stress equivalent to UTS, no failure will occur, and the material can withstand load for an extended period of time without failure. But if that stress is reversed somewhat and brought to the static stress level and this cycling is continued, eventually the material will fracture due to fatigue with cycle life dependent on the material, mean stress, and stress range. Mean stress (σ_m) is the algebraic average of the maximum and minimum stress in the cycle. The stress range ($\Delta\sigma$) is basically the difference between the maximum stress and the minimum stress used. With higher stress range and mean stress, fatigue life is shortened. We will discuss these aspects in detail in **Chapter 10**.

Fatigue tests are done in two ways. Low-cycle fatigue testing is carried out from tens of cycles to a few thousand cycles ($<10^4$ cycles). Generally, for low cycle fatigue, large strain ranges are involved. Thus the specimen can accumulate macroscopic plastic strain. In that case, fatigue life correlates well with the plastic strain range and not with the stress-related parameters. On the other hand, high cycle fatigue occurs over millions of cycles in the elastic strain regime, albeit microscopic deformation may be plastic. Thus high cycle fatigue lives are correlated to the elastic strain range and stress range. Stress concentrations at local level or inhomogeneities at the microstructural level promote nucleation of cracks. Also note that fatigue cracks tend to start at the surface unless there is a critical subsurface defect in the component or specimen. The crack thus created propagates slowly under the applied stress cycle in the direction more or less perpendicular to the stress axis. After reaching a critical size, it grows rapidly and results in final fracture.

Fatigue failure usually results in characteristic fracture surface. Low magnification image shows the crack growth region and appears as bright and burnished. This is where the two mating crack surfaces grind for a longer period of time. This area of the fracture surface appears as a collection of ridges, with each representing the distance traveled by the crack per cycle. These ridges are called fatigue striations. As an example, Fig. 5.14 (a) shows the striations in a 1018 steel tested in fatigue. The corresponding microstructure comprising cell-like substructures beneath the striations are shown in Fig. 5.14 (b). Depending on the environment of testing, the crack surface appearance may change and striations may become not so visible. So, care must be taken while analyzing a fracture and concluding the cause of such failure.

Material response under fatigue is quite different from what is obtained under monotonic loading. Generally, under monotonic loading, strain hardening is present in all metallic materials. However,

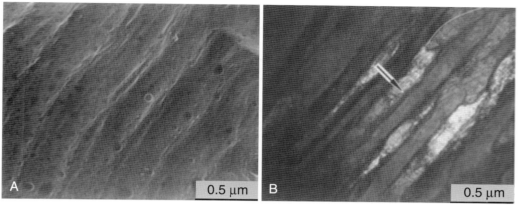

Figure 5.14

(a) The appearance of striations formed in 1018 grade low carbon steel at crack growth rate of 10^{-3} mm/cycle, and (b) the underlying dislocation arrays beneath the fatigue striations.[7]

under fatigue either "work softening" (if the stress required to maintain a specific strain range decreases) or "work hardening" (the opposite) takes place. Generally, hard materials work soften and soft materials work harden under cyclic deformation. An important point to remember is the different variants of fatigue testing with several different terminologies, to address more complicated service requirements such as creep-fatigue, thermomechanical fatigue, and corrosion fatigue. Various ASTM fatigue standards pertain to fatigue test methodologies, apparatus, and fracture mechanics-based fatigue crack growth testing.

5.5 Fracture toughness

Present-day engineering design often assumes that some form of flaw already exists in engineering materials. So, when this type of material is contemplated for use in load-bearing service, the design approach should be based on that. Thus a measure of fracture toughness as an inherent mechanical property of materials is important, as is yield stress. While tensile toughness and impact toughness can provide some measure of toughness, fracture toughness is the only reliable estimate of toughness for mechanical design. Mode-I fracture toughness (K_{Ic}) meets that definition with certain limitations. As the analysis involved still depends on linear elastic fracture mechanics theories, the testing procedures followed are applicable to materials with limited ductility, such as high-strength steels, some titanium and aluminum alloys, and, of course, other brittle materials like ceramics.

The elastic stress field around the crack tip can be described by a single parameter known as "stress intensity factor (K)" and depends on factors such as the geometry of the crack-containing specimen, crack size and its location, and the value and distribution of applied loads. A reasonable assumption is that an unstable rapid failure would occur if a critical value of K is reached. The three general modes

---

[7]Cai H, McEvily AJ. On striations and fatigue crack growth in 1018 steel. *Mate Sci Eng A.* 2001;314:86-89.

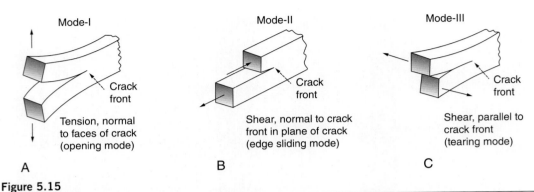

Figure 5.15

Three standard modes of loading in fracture toughness testing: (a) mode-I (opening mode), (b) mode-II (sliding mode), and (c) mode-III (tearing mode).

of testing are shown in Fig. 5.15. In the opening mode (mode-I), the displacement is perpendicular to the crack faces. In mode-II (sliding mode), the displacement is made parallel to the crack faces but perpendicular to the leading edge. In mode-III (tearing mode), the displacement is parallel to the crack faces and the leading edge.

In reality, opening mode (i.e., mode-I) is the most important. That is why conventional tests for fracture toughness are done under mode-I (i.e., opening mode of loading), and the critical value of K is called K_{Ic}, the plain strain fracture toughness. For a given type of loading and geometry, the relation is

$$K_{Ic} = Y\sigma\sqrt{\pi a_c} \qquad (5.23)$$

where Y is a parameter dependent on specimen and crack geometry, a_c is critical crack length, and σ is applied stress. If K_{Ic} and yield strength are known, the maximum crack length tolerable can be computed. In other words, maximum allowable stress can be computed for a given crack size provided the K_{Ic} value of the material is known. K_{Ic} generally decreases with decreasing temperature and increasing strain rate, and vice versa. It is also strongly dependent on metallurgical variabls such as heat treatment, crystallographic texture, impurities, inclusions, and grain size.

A notch in a thick plate is far more damaging than in a thin plate because the notch leads to a plane strain state of stress with a high degree of triaxiality. Fracture toughness measured under plane strain conditions is obtained under maximum constraint and becomes independent of specimen size. Thus the plane-strain fracture toughness is designated as K_{Ic} and becomes a true material property. A mixed mode (ductile-brittle) fracture occurs with thin specimens. Once the specimen has the critical thickness, the fracture surface becomes flat and fracture toughness reaches a constant minimum value with increasing specimen thickness (Fig. 5.16). The minimum thickness (B for breadth) to achieve plane strain condition is given by:

$$B \geq 2.5(K_{Ic}/\sigma_{YS})^2 \qquad (5.24)$$

where σ_{YS} is the 0.2% yield stress. Generally, for ductile materials, the stress-intensity factor-based approach falls out of favor since they are good for brittle or brittle-like materials. In those cases, J-integral approach or crack opening displacement calculation needs to be used.

5.5 Fracture toughness

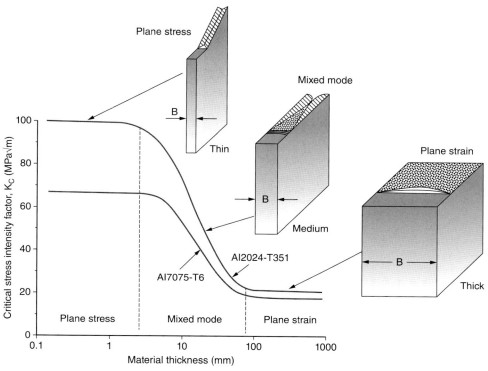

Figure 5.16

The critical stress intensity factor as a function of plate thickness. The thickness must be sufficiently large for achieving plane strain condition.

Three different types of specimen designs available for determining fracture toughness are shown in Fig. 5.17. Compact tension (CT) specimen is one of the most used designs. There are other ways of evaluating fracture toughness. For ceramics, creating a sharp notch in a ceramic material is a problem, and even if one can be created, it will likely grow catastrophically to failure upon loading.

A reasonable estimate of fracture toughness in ceramics can be measured through indentation testing. Upon indentation, cracks are created at the corners of the indentation. One example of an indentation is seen in Fig. 5.18. While there are different mathematical ways of expressing indentation fracture toughness, one widely used one is shown here. After the crack lengths are measured, the following relation can be used to calculate the K_{Ic} value.

$$K_{Ic} = \alpha \left(\frac{E}{H}\right)^{\frac{1}{2}} \left(\frac{P}{d^{1.5}}\right) \tag{5.25}$$

where E is the elastic modulus, H is hardness, P is the indentation load, and $2d$ is the total indentation crack length. While this technique can provide an initial assessment of fracture toughness, it is not without limitations.

Chapter 5 Simple mechanical tests

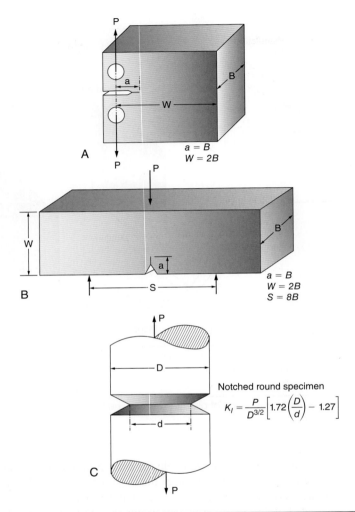

Figure 5.17
Three different types of plain strain fracture toughness specimen designs: (a) compact tension (CT), (b) single edge notched bend (SENB) specimen, and (c) notched round specimen.

5.6 Creep testing

At temperatures above 0.4–$0.5 T_m$ (T_m is the melting point of the material), plastic deformation occurs as a function of time at constant load or stress. The phenomenon is known as creep, which is nothing but time-dependent plastic deformation. Generally, a creep test is carried out under uniaxial tensile stress. However, different variants of creep tests include compressive creep, double shear creep, impression creep, and indentation creep. A typical creep curve shows creep strain as a function of time at a constant load (or a constant stress – depending on the availability of such instrument setup to keep stress

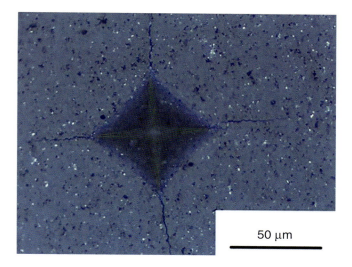

Figure 5.18

An example of crack formation at an indentation created by Vickers instrumented indentation test with a peak load of 50 N in a Si_3N_4 specimen.[8]

constant during creep deformation). Fig. 5.19 (a) shows creep equipment with a furnace and strain measuring device, a linear variable differential transformer (LVDT). A typical creep curve (Fig. 5.19 [b]) has three distinct stages: (i) primary stage (transient creep) – strain hardening during plastic deformation is more than recovery (softening) exhibiting decreasing strain-rate with time; (ii) secondary or steady-state stage (minimum creep rate) – the rate of strain hardening and recovery balance each other; and (iii) tertiary stage – characterized by an accelerating creep rate where failure mechanisms predominate. The third stage of tertiary creep is often considered the fracture stage rather than deformation. An actual example of creep rate vs. time plot for a Grade 91 steel (9Cr-1MoVNb) specimen is shown in Fig. 5.19 (c), where the steady-state creep rate occurs as a minimum rate that is typical for creep under constant load. Figs 5.20 (a and b) show the effect of temperature and applied stress on creep curves. Increasing temperature and applied stress have a similar effect, increasing the creep rates and vice versa.

Creep testing can be continued to last until rupture. This fracture can occur via either transgranular or intergranular mode. Transgranular creep fracture involves void nucleation and growth within the grains, and they interlink to form cracks. Their fracture surface appears quite like a low-temperature dimpled surface. This appears to occur at somewhat higher stress levels. For an intergranular fracture, the voids (also known as cavities) are formed on the grain boundaries, and they interlink through the grain boundaries and finally cause intergranular creep fracture through grain boundary cavitation. In both cases, the nucleation of voids is promoted at the particle-matrix interfaces. In this case, the fracture surface reveals distinct grain boundary facets. More details on creep deformation and failure mechanisms will be discussed in **Chapter 11**.

---

[8]Sun L, Ma D, Wang L, Shi X, Wang J, Chen W. Determining indentation fracture toughness of ceramics by finite element method using virtual crack closure technique. *Eng Fract Mech*. 2018;197:151-159.

Chapter 5 Simple mechanical tests

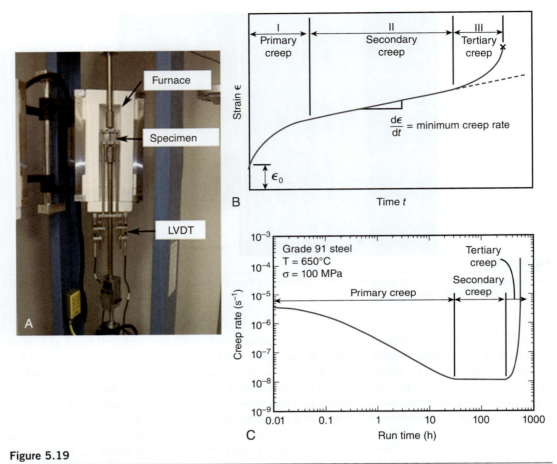

Figure 5.19
(a) A view of a creep tester with specimen loading, (b) a typical creep curve showing different stages of creep, and (c) creep rate vs. strain plot for a modified 9Cr-1Mo steel (Grade 91 steel) tested at a temperature of 650°C and 100 MPa.[9]

5.7 Examples of influence of mechanics and dislocation-based plasticity theory

Throughout the book the impact of the mechanics and microstructural approach is highlighted. First the tensile test is considered. In a mechanics-based approach, an invariant line, which does not change length during specimen deformation, is at 52°. Based on this, shear localization should take place along this invariant line. On the other hand, a dislocation plasticity-based approach puts the shear band formation at 45°.

---

[9]Shrestha T, Basirat M, Charit I, Potrirniche GP, Rink KK, Sahayam U. Creep deformation mechanisms in modified 9Cr-1Mo steel. *J Nucl Mater*. 2012;423(1-3):110-119.

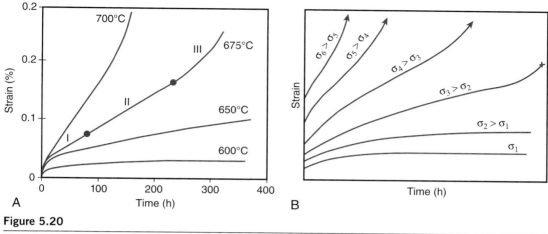

Figure 5.20
The effect of (a) temperature and (b) stress on creep curves.

Another quick example to ponder is stress concentration. We begin again with a tensile test. When a smooth cylindrical tensile specimen is being pulled, the gage region does not have a stress concentration. (The concept of stress concentration is introduced formally in **Chapter 9**.) The basic theory for plastically deforming materials limits stress concentration to a maximum value of 3. When the material reaches its ultimate tensile stress, geometrical instability like necking or an internal failure process like void formation starts. The mechanics-based analysis of tensile test predicts the onset of failure mechanism when plastic strain equals the value of the strain hardening exponent. As you think of what happens inside the material, consider this. At the point where internal void nucleates, what is the stress concentration? Just before the point of void nucleation, the stress distribution internally is such that there is no large stress concentration. However, as soon as the void nucleates, the stress distribution would abruptly change to reflect the stress concentration. In a similar vein, most engineering materials contain large constituent particles that are elastically harder than the matrix. Such particles will result in stress concentration around the particle. That would promote the onset of local plasticity at a much lower applied stress than the nominal yield strength of material. The question is, what is the impact of this plasticity during fatigue and creep testing? We will address these questions in the subsequent chapters.

5.8 Key chapter takeaways

This chapter gave a simple account of various mechanical test methods such as hardness, tensile testing, fatigue testing, fracture toughness testing, and creep testing. There are different purposes for each test, providing us with important information about the mechanical behavior of materials. In real-world applications, understanding one or a combination of these properties is important. Sometimes, there may be a tradeoff. Microstructural design is crucial in tailoring these properties. Note that many other test methods including impact testing and their variants are not discussed or highlighted here.

5.9 Exercises

1. In a Brinell hardness test, a load of 3000 kgf is applied on a sample. If the diameter of the Brinell indenter is 10 mm and the Brinell hardness number obtained is 200, what is the corresponding indentation diameter?

2. If the hardness of a material is 145 BHN, what would be the ultimate tensile strength?

3. You are doing Vickers microhardness testing. Instead of standard load of 0.5 kgf, you used 0.05 kgf. Would you get the same Vickers hardness value? If not, why not?

4. Vickers indentation testing is carried out on a sample under a load of 500 gf for 15 s. If the average diagonal length of the indentation created is 10 µm, what is the Vickers hardness of the sample?

5. Generally, macrohardness test techniques are not advisable for thin specimens. Why?

6. A test bar of a metal with 12.85 mm diameter and 50 mm gage length is loaded elastically with 156 kN tensile load and is stretched by 0.356 mm. Its diameter becomes 12.80 mm under load. Assuming that the metal is still under elastic region, what is the elastic modulus of the metal?

7. A tensile test was performed on a metal specimen with 25.4 mm gauge length. The specimen revealed 1.3 mm increase in length. (a) Calculate the engineering and true strains. (b) What would have been these strains if the original gauge length of the specimen had been 50.8 mm?

8. Show that engineering strain (e) and true strain (ε) are related by the following relation:

$$\varepsilon = \ln(1 + e).$$

9. A brass specimen of 5 mm diameter and 25 mm length tensile tested revealed the final length to be 28 mm with the final diameter of 4.1 mm. Calculate the percentage elongation to failure and percentage reduction in area.

10. Let's assume that the whole true stress (σ) – true strain (ε) curve can be described by Hollomon's equation: $\sigma = K\varepsilon^n$ where K is the strength coefficient and n is the strain hardening exponent. Derive an expression for tensile toughness (tensile toughness can be defined as the area below the stress-strain curve). You can use appropriate upper limit (ε_f) and lower limit (0) of true strain. Neglect the elastic deformation region.

11. Find the ultimate tensile strength of a material with a diameter of 3.2 mm, strain hardening exponent $n = 0.155$, and constant $k = 165$ ksi.

12. A tensile test performed on a particular alloy at 620°C to a strain of 0.1 at a strain rate of 3.3×10^{-5} s^{-1} gave a flow stress value of 150 MPa, and a sudden change of the strain rate at that strain (0.1) to 2.1×10^{-4} s^{-1} resulted in enhancing the flow stress to 250 MPa. Calculate the strain-rate-sensitivity (m) of the material.

13. Lead (Pb) can creep substantially at room temperature. Why?

14. Show a schematic of a typical engineering stress-strain curve for a ductile material and a stress-strain for brittle material. Is brittleness or ductility an absolute property?

5.9 Exercises

15. *Thought Experiment*: Fatigue cracking starts at the surface and propagates under tensile stress. Present some ideas through which fatigue life can be increased.

16. (a) During a tensile test, the strain rate suddenly doubled. It caused the load to rise by 1.5%. Assume that the strain rate ($\dot{\varepsilon}$) dependence of flow stress (σ) is described by $\sigma = C\dot{\varepsilon}^m$, where C is a constant and m is the strain rate sensitivity. What is the value of m? State any assumptions you made in answering the question.

    (b) Generally, increasing strain rate leads to an increase in flow stress. Explain why based on dislocation theory. (Hint: Orowan's equation and Taylor's equation.)

    (c) Show schematically the effect of strain rates on true stress-strain curves (including the elastic part).

17. If the crosshead speed of the tension testing machine is 5 mm s^{-1} and the gage length of the specimen is 12.5 mm, what would be the initial true strain rate achieved during the tensile test?

18. (a) Describe Hollomon's equation. (b) A round tensile specimen has been tested in uniaxial tension. It was found that the true uniform strain was 0.25. What is the strain-hardening exponent? Justify your answer.

19. (a) Name three macrohardness tests and two microhardness tests. (b) Suppose you are doing Vickers microhardness test at a load of 500 gf that generated a square-shaped indentation with an average diagonal length of 25 μm. What is the Vickers hardness number (VHN)?

20. (a) What are the two types of fracture? (b) Compare these two fracture types.

21. (a) Name two types of impact tests. (b) How does impact test give conservative estimate of impact property? (c) Define ductile-brittle transition temperature.

22. (a) Describe the characteristics of a fatigue fracture surface. (b) How are the S-N curves constructed from fatigue tests? (c) How can fatigue failures be prevented?

23. (a) Define creep. (b) Illustrate a standard creep curve showing different stages of creep. (c) Give Norton's law and show how stress exponents can be obtained using this equation.

Further readings

Cai H, McEvily AJ. On striations and fatigue crack growth in 1018 steel. *Mate Sci Eng A*. 2001;314:86-89.
Cullen GW, Korkolis YP. Ductility of 304 stainless steel under pulsed uniaxial loading. *Int J Solids Struct*. 2013;50(10):1621-1633.
Dieter GE. *Mechanical Metallurgy*. 3rd ed. McGraw-Hill Publishers; 1986.
Mignogna R., D'antonio C, Maciag R, Mukherjee K. The mechanical behavior of 6063 aluminum. *Met Trans*. 1970;1:1771-1772.
Qin S, Yang M, Yuan F, Wu X. Simultaneous improvement of yield strength and ductility at cryogenic temperature by gradient structure in 304 Stainless Steel. *Nanomaterials (Basel)*. 2021;11(7):1856.
Shrestha T, Basirat M, Charit I, Potrirniche GP, Rink KK, Sahayam U. Creep deformation mechanisms in modified 9Cr-1Mo steel. *J Nucl Mater*. 2012;423(1-3):110-119.
Sun L, Ma D, Wang L, Shi X, Wang J, Chen W. Determining indentation fracture toughness of ceramics by finite element method using virtual crack closure technique. *Eng Fract Mech*. 2018;197:151-159.
Thomas H. *Courtney, Mechanical Behavior of Materials*. 2nd ed. Waveland Press; 2005.

CHAPTER 6

Elastic response of materials: stiffness limiting design

Chapter outline

6.1 Need for discussion on elastic response of materials 92
6.2 Development of elasticity theory 96
    6.2.1 Definition of stress 97
    6.2.2 Definition of strain 100
    6.2.3 Transformation of axes 102
    6.2.4 Simple Hooke's law (various moduli—Young's modulus, shear modulus and bulk modulus; and Poisson's ratio) 105
    6.2.5 Resilience 108
    6.2.6 Anisotropy in elastic property 110
    6.2.7 Elastic behavior of a polycrystal as a single crystal average 112
6.3 Design of high stiffness composite materials 114
    6.3.1 Particulate-reinforced composites 116
        6.3.1.1 Constant strain approach *118*
        6.3.1.2 Constant stress approach *119*
    6.3.2 Fiber-reinforced composites 119
6.4 Key chapter takeaways 122
6.5 Exercises 122

Learning objectives

- Basic concept of elasticity.
- An understanding of load transfer-based strengthening mechanism.
- Design of high stiffness composite materials.

This is the first of five chapters focused on various design approaches and the underlying fundamentals. The first of these is stiffness limiting design. Conceptually this is also the simplest as it only depends on the elastic properties of materials. The elastic properties are directly derived from atomic bonds and intrinsic to the atomic arrangement in the material. The elastic properties at the primary level are microstructurally insensitive. For example, if a pure metal is thermomechanically processed to refine the grain size, its strength-ductility changes, but not the modulus. Similarly, if we age a solution-treated aluminum alloy to increase strength through precipitation of a second phase, the modulus change is typically minor. In the later parts of this chapter, we will discuss composite strengthening, which is based on bringing phases together to alter the modulus values. The stiffness limiting design implies that the structure will remain in the elastic regime, and therefore the transition from elastic behavior to plastic deformation sets the upper limit of design load for such structures. While higher modulus materials are desirable for applications where we want low deflection of structure, lower modulus materials are attractive for applications where large elastic strains are needed.

6.1 Need for discussion on elastic response of materials

Theory of elasticity involves a set of principles and equations describing interplay between load and deformation or stress and strain when the extent of deformation is quite small. But at the core, this fundamental behavior could be as simple as Hooke's law, which states that the stress is directly proportionality to strain or a linear relationship holds. Because of this type of linearity in relation, crystalline materials exhibit *linearly elastic* behavior during the initial part of deformation. Now recall from **Chapter 5**, where we first discussed tensile testing and stress/strain definitions. Hooke's law is the relation that is often invoked to describe this region as given in the following relation:

$$\sigma = E\varepsilon \tag{6.1}$$

where σ is the true stress and ε is the true strain in tension or compression. Hooke's law is one of the most fundamental constitutive equations that we quote to start discussion on mechanical properties. The proportionality constant (E) is known as modulus of elasticity or elastic modulus or Young's modulus. For small strains such as found in the elastic regime, the stress and strain could be simply engineering stress/strain, as the linear regions of both engineering and true stress-strain curves basically superimpose on each other.

Linear elastic behavior is very significant in engineering because by designing machines and structures to serve in this region, we can prevent them from accumulating permanent deformation via yielding. Of course, when civil engineers build bridges or buildings, they rely heavily on the linear elastic properties of steels. Similarly, when a mechanical engineer designs a machine, many of the design considerations would consider the linear elastic behavior of materials.

Continuum theories including the theory of elasticity are generally based on the assumption that materials are homogeneous, isotropic, and continuous. A *homogeneous* object is one that is chemically of the same constitution from one point to another. A *continuous* object refers to a body devoid of any void or empty spaces. An *isotropic* material has the same properties irrespective of test directions. But bodies can be anisotropic, too. Later in the chapter, elastic properties will be revealed as being quite anisotropic, depending on crystal symmetry and structure.

Elastic properties can be quite varied (Fig. 6.1; Table 1.1). The elastic moduli of metals and ceramics range between 10 and 600 GPa. As the discussion of elasticity equations progresses, a basic idea of the origin of elastic properties in materials will become apparent. Actually, elastic modulus is considered a structure-insensitive property. That is, it does not depend on microstructure. That also means that cold working, heat treatment, and small additions of alloying elements are not able to change elastic moduli. It actually correlates well with the melting point and gives credence to the fact that the origin of elastic properties comes from the stretching of atomic bonds. The material can recover its shape in the elastic regime without any permanent damage because the atomic bonds in materials get stretched or compressed upon loading and relax upon unloading, thus going back to the original

6.1 Need for discussion on elastic response of materials

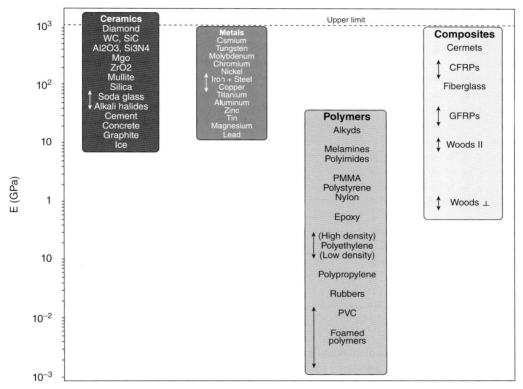

Figure 6.1
A bar chart showing the extent of elastic moduli range in all four broad categories of materials; namely, ceramics, metals, polymers and composites.[1]

condition: Thus the necessity of getting acquainted with a concept that is often used to explain various physical properties in materials.

The physics of linear elasticity is well-established. High school studies featured springs and spring constant. The atomic bonds can also be related to spring constant at the atomic level. More precisely, elastic modulus (E) actually can be related to spring constant (k_S) and interatomic spacing (r_o) and is shown by the following equation:

$$E = \frac{k_S}{r_0} \tag{6.2}$$

The interatomic spacing of materials does not vary that much—usually by a factor of two. Therefore elastic modulus depends more on the value of k_S which varies much more among materials. For

---

[1] Ashby MF, Jones DRH. *Engineering Materials 1 – An Introduction to Their Properties and Applications*. Pergamon Press; 1980.

Chapter 6 Elastic response of materials: stiffness limiting design

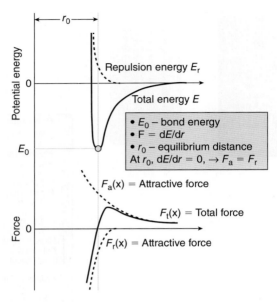

Figure 6.2

A typical potential well indicating bonding energies and forces acting between two interacting atoms.[2]

example, the spring constants of covalently bonded solid materials range between 20 and 200 N/m. On the other hand, spring constants with metallic bonding and ionic bonding mostly vary between 15 and 100 N/m. Even though the covalently bonded giant molecular chains show greater spring constants, the elastic moduli of polymers tend to be much less because of the quite small spring constants (0.5–2 N/m) created by weak Van der Waals forces acting between the polymer chains.

The moduli values depend on atomic bond strength. This bond strength is, in turn, actually a measure of the potential energy between atoms. A typical plot of potential energy variation as a function of the interatomic distance is shown in Fig. 6.2. The two main types of forces that act between two atoms are an attractive force (F_A) and a repulsive force (F_R). The magnitude of both forces varies as a function of the interatomic distance. When these two types of forces are balanced, the net force becomes zero, and a state of mechanical equilibrium is achieved. The corresponding interatomic separation distance is termed the equilibrium separation distance, r_o. Of course, in real materials, there will be more interaction forces imparted by the other neighboring atoms. Figs. 6.3 (a) and (b) show the same plots of force and potential energy versus interatomic distance, respectively. In fact, the slope of the force-interatomic distance curve $\left(\dfrac{dF}{dr}\right)$ as shown in Fig. 6.3 (a) at r_o gives a measure of elastic modulus. Note that the slope is steeper for stiffer materials (i.e., material with high elastic modulus), whereas the slope is lower for flexible materials (i.e., material with low elastic modulus). The same point can be made with the help of the potential well concept (potential energy vs. interatomic distance). As shown in Fig. 6.3 (b), E_a

---

[2]Padmabathi DA. Potential energy curves & materials. *Mater Sci Appl.* 2011;2:97-104.

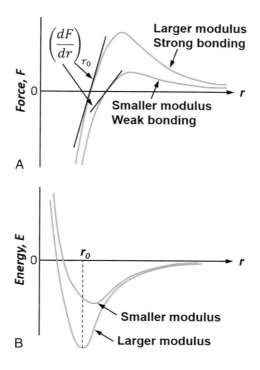

Figure 6.3

(a) Force vs. interatomic separation distance plot showing that slope determined the extent of elastic modulus. (b) Potential energy vs. interatomic separation distance for two materials with higher elastic modulus and lower elastic modulus.[3]

decreases as the energy minima decreases (i.e., lowering of bond strength), and vice versa. On the other hand, E_a becomes highest for ceramics and high for most metals, and generally lower for polymers deriving from the difference in atomic bond types. However, polymers with highly orientated molecular chains tend to have greater modulus. So, basically, the curvature of potential energy vs. distance curve provides a good explanation of the origin of elastic modulus on a basic level.

Interestingly, many physical properties, including mechanical properties, can be explained using the potential well concept. For example, the melting point of a material is governed by the depth of the potential well. This can be thought of as follows: Melting point varies with atomic bond strength; the higher the bond strength, the higher the melting point. This is also generally found in terms of elastic modulus. As a simple example, tungsten's melting point is 3422°C, and its elastic modulus is 411 GPa, whereas aluminum has a melting point of 660°C, with elastic modulus of only 69 GPa!

Almost always, the elastic modulus of materials decreases with increase in temperature. This phenomenon can be explained physically and is observed experimentally. Thermal energy applied to materials in the form of higher temperature adds to atomic vibrations. However, the energy is still small compared to

---

[3]Padmabathi DA. Potential energy curves & materials. *Mater Sci Appl.* 2011;2:97-104.

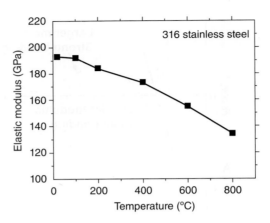

Figure 6.4

The variation of elastic modulus of 316 stainless steel with temperature.[4]

the crystal potential energy at absolute zero. The equilibrium interatomic separation distance does change with temperature, and often linearly. In most cases, this distance increases linearly with temperature, as manifested by the thermal expansion of crystalline materials, which leads to the decrease in elastic modulus. Eq. (6.3) represents a relation of the change in elastic modulus with temperature:

$$\frac{E_T}{E_O} = \left[1 - \alpha \frac{T}{T_m}\right] \qquad (6.3)$$

where E_T is the elastic modulus at temperature T, T_m is the melting temperature, E_o is the modulus at absolute zero, and α is a constant. Actual elastic modulus values of 316 stainless steel as a function of temperature show clearly that with increase in temperature, elastic modulus drops to 134 GPa at 800°C as compared to 193 GPa at room temperature (Fig. 6.4). Regardless of steel compositions, room temperature elastic modulus of steels tends to lie in the range of 193 to 206 GPa.

Concept Reinforcement—Temperature dependence of Young's modulus: There is a simple way to remember the fact that elastic modulus value decreases with temperature! A liquid phase cannot sustain tensile stress. So, the Young's modulus for the liquid phase would be zero (0). At 0 K, all elements are solid and the bond strength is highest. So, the Young's modulus value for an element will be highest at 0 K. Clearly this means that the Young's modulus value would decrease with temperature.

6.2 Development of elasticity theory

A standard course of "Mechanics of Materials" or "Strength of Materials" presents quite a bit of Theory of Elasticity. The aim here is not to delve deep into deriving all the associated theories and equations available in the literature but rather give a concise description of certain essential concepts to understand the topics we cover in this book.

---

[4]Raw data taken from Davis JR. *ASM Specialty Handbook Stainless Steel*. ASM International; 1994.

6.2 Development of elasticity theory

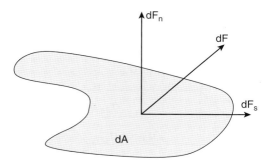

Figure 6.5
Force components dF_n and dF_s of dF act on an infinitesimal area dA.

In **Chapter 5,** we discussed in brief the definitions of stress and strain in the context of tensile testing. Now the discussion becomes more complex. Instead of a uniaxial state of stress, imagine the state of stress at a point in a reference frame of a three-axes orthogonal Cartesian coordinates system.

6.2.1 Definition of stress

What is stress? Consider the arbitrary cross-section shown in Fig. 6.5. Stress (σ) is generally defined as the intensity of the load or force at a point. So, basically each differential area (dA) experiences dF force. Force dF can be readily resolved into two components: one component—dF_n—that acts normal to the cross-sectional area and the other—dF_s—that lies on the plane itself. Thus we can now define normal stress (σ) and shear stress (τ) by the following equations:

$$\sigma = \frac{dF_n}{dA} \text{ and } \tau = \frac{dF_s}{dA} \tag{6.4}$$

Generally, σ and τ vary over the cross-sectional area.

Often the state of stress needs to be described at a point in a 3D reference system. The usual way is to use x-y-z Cartesian coordinates. However, a numerical orthogonal coordinate system (such as 1-2-3) is also used and eases understanding stress-strain relationships in certain elastic analyses, as is the case for anisotropic elastic modulus.

Generically, σ_{ij}, Cauchy stress, is used to represent the state of stress with the first subscript index (i) referring to the plane on which the stress component acts, and the second subscript index (j) refers to the direction along which the stress is applied. A complete state of stress is described in a matrix form:

$$\sigma = \begin{bmatrix} \sigma_{11} & \sigma_{12} & \sigma_{13} \\ \sigma_{21} & \sigma_{22} & \sigma_{23} \\ \sigma_{31} & \sigma_{32} & \sigma_{33} \end{bmatrix} = \begin{bmatrix} \sigma_{xx} & \sigma_{xy} & \sigma_{xz} \\ \sigma_{yx} & \sigma_{yy} & \sigma_{yz} \\ \sigma_{zx} & \sigma_{zy} & \sigma_{zz} \end{bmatrix} \tag{6.5}$$

In x-y-z 3D Cartesian coordinates, Fig. 6.6 shows the orientations of various stress components around an infinitesimal cube. Since the cube has 6 faces, and each face has 3 stress components (1 normal stress and 2 shear stresses), there are (6 × 3) = 18 stress components. Because of the static

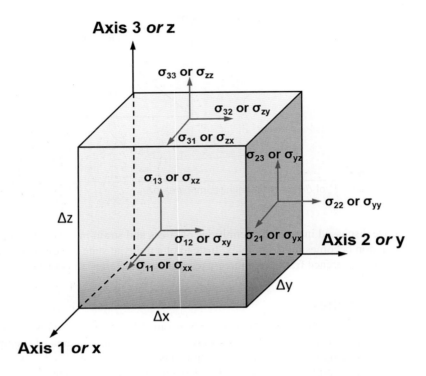

Figure 6.6

Description of a general 3D state of stress in a Cartesian (x-y-z or 1-2-3) coordinate system. Note that the stresses on the negative faces are not shown to maintain clarity.

equilibrium conditions (i.e., no net forces and no net moment), the total number of independent stress components actually becomes only 6, consisting of 3 normal and 3 shear stress components. It is customary to show the stress matrix in the form of a symmetric 3 × 3 matrix. When $i = j$, the stress component σ_{ii} is considered a normal stress component, but when $i \neq j$, the component σ_{ij} is considered a shear stress component. Summation of moments around each axis for equilibrium results in the condition $\sigma_{ij} = \sigma_{ji}$. In other words, $\sigma_{12} = \sigma_{21}$ (or, $\sigma_{xy} = \sigma_{yx}$), $\sigma_{13} = \sigma_{31}$ (or, $\sigma_{xz} = \sigma_{zx}$), and $\sigma_{23} = \sigma_{32}$ ($\sigma_{yz} = \sigma_{zy}$). Hence, the stress matrix shown in Eq. (6.5) is symmetric. Note that stress matrix forms can be expressed in many other ways. So, we should not get confused by other notations that may differ among various reference sources.

The discussion now focuses on the sign convention used for representing stresses. Normal stresses are easy. A normal stress is considered positive in tension, that is, σ_{xx} (or σ_{11}) = 150 MPa is a tensile stress, but it is considered compressive in nature when it is negative, e.g., $\sigma_{11} = -150$ MPa. With regard to shear stresses, the situation is not straightforward. However, referencing Fig. 6.7 facilitates our understanding. The shear stress σ_{ij} is considered positive if both i and j are positive or both negative. Conversely, σ_{ij} is negative if either i or j is negative. The shear stresses, as shown by arrows in Fig. 6.7

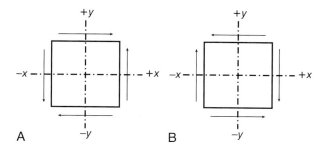

Figure 6.7

The sign convention used for shear stress: (a) positive and (b) negative.

(a), are positive, whereas the shear stresses shown by Fig. 6.7 (b) are negative. Pick any shear stress in Fig. 6.7 (a) and then follow at which side (positive or negative) of the plane and the direction (positive or negative) it is acting. If both are positive or negative, the shear stress must be positive (see Fig. 6.7 [a]). However, if the plane and the direction are situated with respect to the coordinate axes such that they are opposite in sign, the shear stress is considered negative, as it is in Fig. 6.7 (b).

The concept of tensor is used to describe a variety of physical properties. The concept is particularly useful in simplifying complex problems. Stress (σ_{ij}) is also called a tensor of rank 2. A tensor with rank n can be expressed in terms of matrix notation and would contain 3^n components. For example, stress being a rank 2 tensor would contain 3^2 or 9 components, whereas force (often denoted by F_j) is a tensor of rank 1 with 3^1 (or 3) components. Elastic modulus is a tensor of rank 4. Understandably, it is more complicated in the tensor form than force, stress, or strain. This will be treated in **Section 6.2.4**. Since the concept of tensor algebra is used only sparingly in this book, an in-depth knowledge is not needed. The interested reader is encouraged to consult the many other books in mechanics that deal with this topic in detail.

The Cauchy stress tensor (σ_{ij}) can be divided into a hydrostatic or mean stress tensor (H_{ij}), which involves pure tension and/or compression, and a deviator stress tensor (D_{ij}), which includes shear stresses in the system. Mathematically, we can write:

$$\sigma_{ij} = H_{ij} + D_{ij}, \tag{6.6a}$$

$$H_{ij} = \begin{bmatrix} \sigma_m & 0 & 0 \\ 0 & \sigma_m & 0 \\ 0 & 0 & \sigma_m \end{bmatrix} \text{ where } \sigma_m = \frac{\sigma_{11} + \sigma_{22} + \sigma_{33}}{3}. \tag{6.6b}$$

and

$$D_{ij} = \begin{bmatrix} \sigma_{11} - \sigma_m & \sigma_{12} & \sigma_{13} \\ \sigma_{21} & \sigma_{22} - \sigma_m & \sigma_{23} \\ \sigma_{31} & \sigma_{32} & \sigma_{33} - \sigma_m \end{bmatrix}. \tag{6.6c}$$

The hydrostatic component of the stress tensor does not depend on the orientation of the axes and thus is an invariant. The hydrostatic component creates only elastic volume change and no plastic deformation. The yield strength of metallic materials has been demonstrated through experiments not to depend on hydrostatic stress, although fracture strain can be strongly affected by hydrostatic stress. On the other hand, the stress deviator is generally useful in formulating theories of yielding, as they involve shear stresses.

6.2.2 Definition of strain

When a body is subjected to a state of stress, the body would undergo translations, rotations, and deformation. Dilatation (i.e., change in volume) and distortion (change in shape) constitute deformation in a body. However, strain is understood such that the effects of translation and rotation are not considered in the definition of strain as a direct measure of deformation. The effects of translation and rotation fall under the purview of a different subject, called *Dynamics*. At a basic level, normal and shear strains have already been defined with respect to uniaxial tensile tests (Chapter 5). Here we will try to understand strain from the concept of displacement. Also, further assumptions are that strain is infinitesimal, as is true in elastic deformations. Let's say that u, v, and w are the displacements in x, y, and z directions, respectively. To keep things simple, a 2D Cartesian system in x-y is assumed. Looking at Fig. 6.8, OM and ON are two lines on a body that when subjected to a stress take up new positions OM' and ON'.

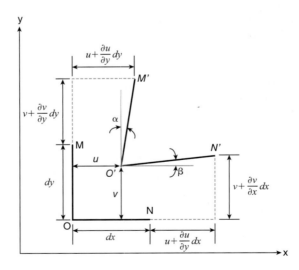

Figure 6.8

Lines OM and ON are located on a body in a strain-free state. Due to loading the configuration, MON is changed to M'O'N'.

6.2 Development of elasticity theory

Normal strain is usually denoted by the ratio of the change in length and the original length. By following that definition, the normal strain in x-direction can be given by

$$\varepsilon_{xx} = \frac{O'N' - ON}{ON} = \frac{\left[\left(dx + u + \frac{\partial u}{\partial x}dx - u\right) - dx\right]}{dx} = \frac{\partial u}{\partial x}. \tag{6.7a}$$

Similar to Eq. (6.7a), the normal strain along y-direction can be given as

$$\varepsilon_{yy} = \frac{\partial v}{\partial y}. \tag{6.7b}$$

If γ_{xy} is defined as the decrease of angle in the right angle of $\angle MON$. So, γ_{xy} can be written as the sum of α and β, given they are very small and given in radian.

$$\gamma_{xy} = \alpha + \beta = \frac{u + \frac{\partial v}{\partial y}dy - u}{dy} + \frac{v + \frac{\partial v}{\partial x}dx - v}{dx} = \frac{\partial u}{\partial y} + \frac{\partial v}{\partial x} = \gamma_{xy} \tag{6.7c}$$

So, Eqs. (6.7a–c) give the normal and shear strains in the x-y coordinates. Now further strain terms can be obtained by considering w displacement in z-direction. The remaining strain components are

$$\varepsilon_{zz} = \frac{\partial w}{\partial z}; \tag{6.7d}$$

$$\gamma_{xz} = \gamma_{zx} = \frac{\partial u}{\partial z} + \frac{\partial w}{\partial x}; \tag{6.7e}$$

$$\gamma_{zy} = \gamma_{yz} = \frac{\partial v}{\partial z} + \frac{\partial w}{\partial y}. \tag{6.7f}$$

Strain can also be written as a tensor, but the definition of shear strain needs to be changed. Thus far familiar shear strain is actually called *engineering shear strain* and is symbolized by γ. Tensorial or mathematical shear strain is, in fact, one-half of the engineering shear strain. Hence, all tensorial shear strains can be expressed by the following relationships:

$$\varepsilon_{xy} = \varepsilon_{yx} = \frac{1}{2}\gamma_{xy} = \frac{1}{2}\left(\frac{\partial u}{\partial y} + \frac{\partial v}{\partial x}\right); \tag{6.8a}$$

$$\varepsilon_{yz} = \varepsilon_{zy} = \frac{1}{2}\gamma_{yz} = \frac{1}{2}\left(\frac{\partial w}{\partial y} + \frac{\partial v}{\partial z}\right); \tag{6.8b}$$

$$\varepsilon_{zx} = \varepsilon_{xz} = \frac{1}{2}\gamma_{xz} = \frac{1}{2}\left(\frac{\partial u}{\partial z} + \frac{\partial w}{\partial x}\right). \tag{6.8c}$$

Generic form of strains can be given by the following:

$$\varepsilon_{ij} = \frac{1}{2}\left(\frac{\partial u_i}{\partial x_j} + \frac{\partial u_j}{\partial x_i}\right) \tag{6.8d}$$

The whole strain tensor is given just like the stress tensor:

$$\varepsilon_{ij} = \begin{bmatrix} \varepsilon_{xx} & \varepsilon_{xy} & \varepsilon_{xz} \\ \varepsilon_{yx} & \varepsilon_{yy} & \varepsilon_{yz} \\ \varepsilon_{zx} & \varepsilon_{zy} & \varepsilon_{zz} \end{bmatrix} = \begin{bmatrix} \varepsilon_{11} & \varepsilon_{12} & \varepsilon_{13} \\ \varepsilon_{21} & \varepsilon_{22} & \varepsilon_{23} \\ \varepsilon_{31} & \varepsilon_{32} & \varepsilon_{33} \end{bmatrix}. \qquad (6.9)$$

Only normal strains produce volume changes.

$$\varepsilon_{xx} + \varepsilon_{yy} + \varepsilon_{zz} = \varepsilon_{11} + \varepsilon_{22} + \varepsilon_{33} = \frac{\Delta V}{V_0}. \qquad (6.10)$$

Just like stress tensor, a strain tensor is composed of hydrostatic strain tensor (H_{ij}) and deviatoric strain tensor (D_{ij}).

$$\varepsilon_{ij} = H_{ij} + D_{ij} = \begin{bmatrix} \varepsilon_m & 0 & 0 \\ 0 & \varepsilon_m & 0 \\ 0 & 0 & \varepsilon_m \end{bmatrix} + \begin{bmatrix} \varepsilon_{11} - \varepsilon_m & \varepsilon_{12} & \varepsilon_{13} \\ \varepsilon_{21} & \varepsilon_{22} - \varepsilon_m & \varepsilon_{23} \\ \varepsilon_{33} & \varepsilon_{32} & \varepsilon_{33} - \varepsilon_m \end{bmatrix}, \qquad (6.11)$$

where $\varepsilon_m = \dfrac{\varepsilon_{11} + \varepsilon_{22} + \varepsilon_{33}}{3}$, which is an invariant quantity (i.e., does not depend as a function of orientations). The hydrostatic or mean strain component does not cause any shape change; it changes the volume only, whereas deviatoric strain component changes the shape. Deviatoric strain tensor can be obtained by subtracting the hydrostatic strain tensor from the total strain tensor.

6.2.3 Transformation of axes

In many instances, a state of stress or state of strain may need to be detected in a set of axes different from the original axes. Transformation matrix allows states of stress or strain in the coordinates system (X_1, X_2, X_3) to be evaluated relative to a new coordinate system (X_1', X_2', X_3'). Such analysis helps to determine the state of stress in a specific plane of interest and/or to determine the maximum value of normal/shear stress in a body subjected to an external state of stress. Such a situation is shown schematically in Fig. 6.9. Each component is defined by the cosine of the angle (θ_{ij}) between X_i' (primed axis) and X_j (unprimed axis), known as direction cosines (Table 6.1). Note that θ_{ij} is positive in the counter clockwise direction as we rotate from the unprimed axes to primed axes.

Now following the angle table shown in Table 6.1 (watch the subscripts), the transformation matrix containing nine elements representing nine direction cosines can be constructed as shown below:

$$a_{ij} = \begin{bmatrix} a_{11} & a_{12} & a_{13} \\ a_{21} & a_{22} & a_{23} \\ a_{31} & a_{32} & a_{33} \end{bmatrix} = \begin{bmatrix} \cos(X_1'X_1) & \cos(X_1'X_2) & \cos(X_1'X_3) \\ \cos(X_2'X_1) & \cos(X_2'X_2) & \cos(X_2'X_3) \\ \cos(X_{31}'X_1) & \cos(X_{31}'X_2) & \cos(X_{31}'X_3) \end{bmatrix}. \qquad (6.12)$$

where a_{ij} is cosine of angle θ_{ij}. The general equation of transformation of Cauchy's stress tensor σ from (X_1, X_2, X_3) coordinate system to σ' in (X_1', X_2', X_3') coordinate system can now be written as:

$$\sigma' = a.\sigma.a_T \qquad (6.13a)$$

6.2 Development of elasticity theory

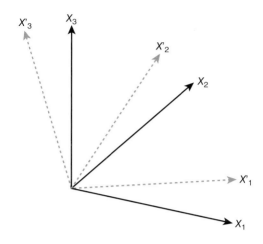

Figure 6.9
X_1-X_2-X_3 and X_1'-X_2'-X_3'—two orthogonal coordinate systems (the first one is the original system and the second one is the transformed axis system).

Table 6.1 Transformation angle designations.

θ_{ij}	Coordinate axes	X_1	X_2	X_3
	$X1'$	$\angle X_1'\ X_1$	$\angle X_1'\ X_2$	$\angle X_1'\ X_3$
	$X2'$	$\angle X_2'\ X_1$	$\angle X_2'\ X_2$	$\angle X_2'\ X_3$
	$X3'$	$\angle X_3'\ X_1$	$\angle X_3'\ X_2$	$\angle X_3'\ X_3$

which in index-based (tensorial) notation can be written as:

$$\sigma'_{ij} = \sum_{k=1}^{3} \sum_{l=1}^{3} a_{ik} a_{jl}\, \sigma_{kl} \tag{6.13b}$$

In Eq. (6.13b), i and j are treated as dummy suffixes, while k and l are considered free suffixes. The normal stresses and shear stresses can now be written in the transformed coordinate system, beginning with a normal stress taking $i = 1$ and $j = 1$:

$$\sigma'_{11} = (a_{11}a_{11}\sigma_{11} + a_{11}a_{12}\sigma_{12} + a_{11}a_{13}\sigma_{13}) + (a_{12}a_{11}\sigma_{21} + a_{12}a_{12}\sigma_{22} + a_{12}a_{13}\sigma_{13}) + \\ (a_{13}a_{11}\sigma_{31} + a_{13}a_{12}\sigma_{32} + a_{13}a_{13}\sigma_{33}); \tag{6.14a}$$

Or in a simplified form:

$$\sigma'_{11} = a_{11}^2\sigma_{11} + a_{12}^2\sigma_{22} + a_{13}^2\sigma_{33} + 2a_{11}a_{12}\sigma_{12} + 2a_{12}a_{13}\sigma_{23} + 2a_{11}a_{13}\sigma_{13} \tag{6.14b}$$

Other normal stresses (σ_{22}' and σ_{33}') in the transformed axes can also be obtained similarly. On the other hand, shear stress such as σ'_{12} is written as shown below taking $i = 1$ and $j = 2$:

$$\sigma'_{12} = (a_{11}a_{21}\sigma_{11} + a_{11}a_{22}\sigma_{12} + a_{11}a_{23}\sigma_{13}) + (a_{12}a_{21}\sigma_{21} + a_{12}a_{22}\sigma_{22} + a_{12}a_{23}\sigma_{23}) + \\ (a_{13}a_{21}\sigma_{31} + a_{13}a_{22}\sigma_{32} + a_{13}a_{23}\sigma_{33}); \tag{6.15a}$$

Or in rearranged form:

$$\sigma'_{12} = (a_{11}a_{21}\sigma_{11} + a_{12}a_{22}\sigma_{22} + a_{13}a_{23}\sigma_{33}) + (a_{11}a_{22} + a_{12}a_{21})\sigma_{12} + \\ (a_{12}a_{23} + a_{13}a_{22})\sigma_{23} + (a_{11}a_{23} + a_{21}\sigma_{13})\sigma_{31}. \tag{6.15b}$$

Similarly, expressions for all other shear stresses can be derived and are left to the reader's discretion.

Now what was introduced in basic engineering mechanics concerning plane stress is revisited in terms of whether the expression can be derived using the stress tensor concept presented above.

What is plane stress? When stress along one of the primary axes is zero, it is called a plane stress state. In certain situations, plane stress is encountered. For example, a thin sheet loaded in the plane of the sheet may have negligible or zero stress in the vertical direction. Another example could be a hollow thin-walled cylinder with internal pressurization. In this case, it may have axial stress and hoop (tangential stress) but negligible stress in the radial direction, which would lead to a plane state of stress kind of situation. Hence, by considering a 2D state of stress (assuming all stresses in the third direction, i.e., X_3 or X_3', zero) an equation for plane state of stress can be written from Eq. (6.14b):

$$\sigma'_{11} = a_{11}^2 \sigma_{11} + 2a_{11}a_{12}\sigma_{12} + a_{12}^2 \sigma_{22} \tag{6.16}$$

Now the direction cosines must be determined in terms of the angle of rotation (θ) through which X_1-X_2 axis is transformed to X_1'-X_2' as shown on the left side of Fig. 6.10. The right side of Fig. 6.10 shows the angles necessary to develop the corresponding transformation matrix a_{ij}. However, note that a_{ij} for the plane state of stress will have only four elements in the transformation matrix shown in Eq. (6.17).

$$a_{ij} = \begin{bmatrix} a_{11} & a_{12} \\ a_{21} & a_{22} \end{bmatrix} = \begin{bmatrix} \cos\theta & \cos(90°-\theta) \\ \cos(90°+\theta) & \cos\theta \end{bmatrix} = \begin{bmatrix} \cos\theta & \sin\theta \\ -\sin\theta & \cos\theta \end{bmatrix} \tag{6.17}$$

Now the appropriate direction cosines as found in Eq. (6.17) can be substituted from Eq. (6.16) to obtain:

$$\sigma'_{11} = \cos^2\theta \sigma_{11} + 2\cos\theta\sin\theta\sigma_{12} + \sin^2\theta\sigma_{22}$$

$$\text{Or, } \sigma'_{11} = \sigma_{11}\cos^2\theta + \sigma_{22}\sin^2\theta + 2\sigma_{12}\cos\theta\sin\theta \tag{6.18a}$$

Similarly, the following equation can be derived as was the above equation. Another way of getting the following equation is to replace θ with $90° + \theta$, as σ'_{11} and σ'_{12} are 90° rotated from each other.

$$\sigma'_{22} = \sigma_{11}\sin^2\theta + \sigma_{22}\cos^2\theta - 2\sigma_{12}\cos\theta\sin\theta \tag{6.18b}$$

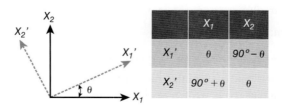

Figure 6.10
The transformation of stress axes from X_1-X_2 to X_1'-X_2' (angle of rotation θ).

6.2 Development of elasticity theory

Now the shear stress component in the transformed axis can be obtained from Eq. (6.15) using Eq. (6.17).

$$\sigma'_{12} = [(a_{11}a_{21}\sigma_{11} + a_{12}a_{22}\sigma_{22} + (a_{11}a_{22} + a_{12}a_{21})\sigma_{12}]$$
$$\sigma'_{12} = (\cos\theta)(-\sin\theta)\sigma_{11} + (\sin\theta)(\cos\theta)\sigma_{22} + (\cos\theta)(\cos\theta)\sigma_{12} + (\sin\theta)(-\sin\theta)\sigma_{12}$$
Or, $\sigma'_{12} = \sigma_{12}(\cos^2\theta - \sin^2\theta) + (\sigma_{22} - \sigma_{11})\sin\theta\cos\theta$ \hfill (6.18c)

The set of Eqs. (6.18a–c) can be also expressed in the following forms using trigonometric identities [$\sin^2\theta = (1 - \cos 2\theta)/2$; $\cos^2\theta = (1 + \cos 2\theta)/2$; $2\sin\theta\cos\theta = \sin 2\theta$; $\cos^2\theta - \sin^2\theta = \cos 2\theta$]:

$$\sigma'_{11} = \frac{\sigma_{11} + \sigma_{22}}{2} + \frac{\sigma_{11} - \sigma_{22}}{2}\cos 2\theta + \sigma_{12}\sin 2\theta \tag{6.19a}$$

$$\sigma'_{22} = \frac{\sigma_{11} + \sigma_{22}}{2} - \frac{\sigma_{11} - \sigma_{22}}{2}\cos 2\theta - \sigma_{12}\sin 2\theta \tag{6.19b}$$

$$\sigma'_{12} = -\frac{\sigma_{11} - \sigma_{22}}{2}\sin 2\theta + \sigma_{12}\cos 2\theta. \tag{6.19c}$$

Adding Eqs. (6.19a) and (6.19b) results in $\sigma'_{11} + \sigma'_{22} = \sigma_{11} + \sigma_{22}$. This is an important relation. Note that the sum of the normal stresses in the transformed axes is equal to the sum of normal stresses in the original axes, and the sum is not dependent on a function of the rotation angle (θ). Thus the sum of normal stresses is called the first invariant (I_1). Imagine that there would be a set of orientation of axes when only normal stresses act on the plane.

As long as strains are small (which is typically the case for elastic strains), they can also be treated as tensor quantities in the same way stress is treated, and the corresponding normal and shear stresses in the transformed axes can be evaluated using the following relation:

$$\varepsilon'_{ij} = a_{ik}a_{jl}\sigma_{kl}. \tag{6.20}$$

6.2.4 Simple Hooke's law (various moduli—Young's modulus, shear modulus and bulk modulus; and Poisson's ratio)

Elastic deformation can be expressed in terms of stress as a function of strain. In a generalized form, two elastic constants are generally needed. Consider a complex state of stress on a body. Poisson's effect is known to account for the transverse contraction of a specimen that undergoes tensile deformation along a longitudinal direction. Poisson's ratio (ν) is basically defined as the ratio of the transverse strain to the longitudinal strain. Poisson's ratio for metallic materials varies between 0.2 and 0.4, most hovering around 0.3. The theoretical limit of Poisson's ratio is between -1 and 0.5. Temperature change has little effect on Poisson's ratio. Materials with negative Poisson's ratio represent an interesting example of materials that can stretch in one direction while expanding in other directions. Researchers are looking into using this unique characteristic to come up with real-world applications. Typical room temperature values of Poisson's ratio of various engineering materials are given in Table 6.2.

Poisson's ratio is first defined here mathematically before delving further into an analysis of strain as a result of a bit more complicated situation. Say a uniaxial tensile stress is acting on a body in the x-direction. Because of that, the body will elongate in the x-direction. Because of the Poisson's effect, the body will contract in the transverse direction. Assume that those transverse directions are the

Table 6.2 A list of elastic constants: Poisson's ratio (v), elastic modulus (E), and shear modulus (G) for different metallic materials

Material	v	E (GPa)	G (GPa)
Aluminum alloys	0.31	72.4	27.5
Plain carbon steel	0.33	200	75.8
Titanium	0.31	117	44.8
Tungsten	0.27	411	157

other two orthogonal axes, y and z, and the material has isotropic properties. So, Poisson's ratio can be written as:

$$v = -\frac{\varepsilon_{yy}}{\varepsilon_{xx}} = -\frac{\varepsilon_{zz}}{\varepsilon_{xx}}. \tag{6.21}$$

This means that while $\varepsilon_{xx} = \sigma_{xx}/E$, the two other transverse strains, $\varepsilon_{yy} = \varepsilon_{zz} = -v\varepsilon_{xx} = -v(\sigma_{xx}/E)$. For an isotropic material, the following mathematical treatment can be applied. By considering isotropy assumption, the elastic constants like elastic modulus and Poisson's ratio do not change depending on the orientation of the reference axes. By superposition of the components of strain in the x, y, and z directions, the following net normal strains will be produced.

$$\varepsilon_{xx} = \frac{1}{E}[\sigma_{xx} - v(\sigma_{yy} + \sigma_{zz})], \text{ i.e., } \varepsilon_{11} = \frac{1}{E}[\sigma_{11} - v(\sigma_{22} + \sigma_{33})], \tag{6.22a}$$

$$\varepsilon_{yy} = \frac{1}{E}[\sigma_{yy} - v(\sigma_{zz} + \sigma_{xx})], \text{ i.e., } \varepsilon_{22} = \frac{1}{E}[\sigma_{22} - v(\sigma_{33} + \sigma_{11})], \tag{6.22b}$$

$$\varepsilon_{zz} = \frac{1}{E}[\sigma_{zz} - v(\sigma_{xx} + \sigma_{yy})], \text{ i.e., } \varepsilon_{33} = \frac{1}{E}[\sigma_{33} - v(\sigma_{11} + \sigma_{22})]. \tag{6.22c}$$

For small strains, engineering strains and stresses in the above equations can also be used to write these equations as long as ε_{ii} is less than 0.005.

Shear stress and strain relations can be given in Hooke's law relations in shear:

$$\tau_{xy} = G\gamma_{xy} = G(2\varepsilon_{xy}); \tau_{yz} = G\gamma_{yz} = G(2\varepsilon_{yz}); \tau_{xz} = G\gamma_{xz} = G(2\varepsilon_{xz}); \tag{6.23}$$

So, the proportionality constant, G, is known as the *modulus of rigidity* or *shear modulus*. The relationship among the three elastic constants, namely E, G, and v is given by

$$E = 2G(1+v). \tag{6.24}$$

Recall that this equation is usually solved in the Mechanics of Materials classes.

Another elastic constant that is less widely used is called *bulk modulus* or *volumetric modulus (K)* and is defined here. It is the ratio of the mean stress (σ_m) and the dilatation (or volume strain).

$$K = \frac{\sigma_m}{\delta_V} = -\frac{\sigma_H}{\delta_V} \tag{6.25}$$

6.2 Development of elasticity theory

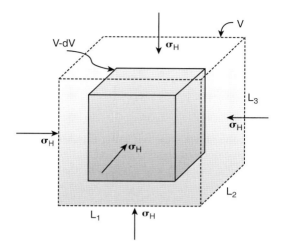

Figure 6.11

A schematic description of a cuboidal body with dimensions L_1, L_2, and L_3.

where $\sigma_m = \dfrac{1}{3}(\sigma_{11} + \sigma_{22} + \sigma_{33})$, and σ_H is known as hydrostatic stress, and δ_V is the corresponding dilatation (or volume strain). Note that historically hydrostatic stress is defined such that compression is a positive pressure. K is also defined as the reciprocal of compressibility, traditionally symbolized by β.

We now turn to deriving the relationship between K and E. To accomplish that, assume a body as shown in Fig. 6.11 with dimensions in three orthogonal primary directions, 1, 2, and 3, as L_1, L_2, and L_3, respectively.

Suppose a body undergoes changes in the dimensions owing to the application of a state of stress along three primary axes by dL_1, dL_2 and dL_3, leading to a volume change of dV in the original volume of V ($L_1 L_2 L_3$). The corresponding linear strains are ε_{11}, ε_{22}, and ε_{33}, which are very small strain values. The volume strain can then be written as

$$\delta_V = \frac{dV}{V} = \frac{\{(1+\varepsilon_{11})(1+\varepsilon_{22})(1+\varepsilon_{33})V\} - V}{V} = \frac{\left\{\left(1+\dfrac{dL_1}{L_1}\right)\left(1+\dfrac{dL_2}{L_2}\right)\left(1+\dfrac{dL_3}{L_3}\right)V\right\} - V}{V}$$

$$\text{Or, } \delta_V = \left(\frac{dL_1}{L_1}\right) + \left(\frac{dL_2}{L_2}\right) + \left(\frac{dL_3}{L_3}\right) \approx (\varepsilon_{11}) + (\varepsilon_{22}) + (\varepsilon_{22}) \tag{6.26}$$

In this case, the product of infinitesimal deformation is a further smaller quantity. Thus all product terms can be neglected. Since elastic deformation generally involves small strains, the volumetric strain is essentially given by the summation of the normal strains.

If Eqs. (6.22a–c) are added, we obtain

$$(\varepsilon_{11}) + (\varepsilon_{22}) + (\varepsilon_{33}) = \frac{1-2v}{E}(\sigma_{11} + \sigma_{22} + \sigma_{33}) = \frac{(1-2v)}{E} \cdot 3\sigma_m \tag{6.27}$$

Or, $K = \dfrac{\sigma_m}{\delta_V} = \dfrac{\sigma_m}{\dfrac{(1-2\nu)}{E}(3\sigma_m)} = \dfrac{E}{3(1-2\nu)}$ (6.28)

Another useful relationship is the equation relating all three moduli, E, G, and K, which is given below:

$$E = \dfrac{9KG}{3K+G}.$$ (6.29)

Readers are advised to derive the above equation using Eqs. (6.24) and (6.28). Other relations are also useful in certain analyses. One derived elastic constant, called Lamé's constant (λ), is a function of two fundamental elastic constants, ν and E, and is given by the following relationship,

$$\lambda = \dfrac{\nu E}{(1+\nu)(1-2\nu)}.$$ (6.30)

6.2.5 Resilience

Resilience is the ability of a material to store energy when deformed elastically and to recover the energy upon unloading. From a classical Newtonian definition, energy is the force multiplied by the distance over which it acts. For the case of an isotropic, linearly elastic body, the average elastic strain energy stored is one-half the product of force (P) and distance (δ), as the force and deformation increase monotonically, linearly, from 0 to δ. Thus the elastic strain energy can be given by

$$U = \tfrac{1}{2}P\delta = \tfrac{1}{2}(\sigma_x A)(\varepsilon_x x) = \tfrac{1}{2}(\sigma_x \varepsilon_x)(Ax),$$ (6.31)

where σ_x is the normal stress along the x-direction, ε_x is the corresponding strain, A is the cross-sectional area of the body onto which the stress acts, and x is the dimension along the x-direction. Note that Ax is nothing but the volume (V) of the material. Thus the elastic strain energy per unit volume, U_o (aka strain energy density), is given by half the product of stress and strain. In other words, this quantity can be expressed by the area under the linear stress-strain curve in the elastic region, that is, the hashed triangular region (Fig. 6.12). An important note here is that the strains generated along other directions (i.e., y and z) due to the application of stress in the x-direction are not part of the expression in Eq. (6.31) because no applied stress is acting along those directions. In cases where the strain energy per unit volume differs from point to point, the total strain energy is given by the following integral:

$$U = \int U_o \, dV$$ (6.32)

By applying Hooke's law (in this case, $\sigma_{11} = E\varepsilon_{11}$), the expression of strain energy density for the normal stress state can now be written as

$$U_o = \tfrac{1}{2}\sigma_{11}\varepsilon_{11} = \tfrac{1}{2}E\varepsilon_{11}^2 = \dfrac{\sigma_{11}^2}{2E}.$$ (6.33)

Similarly, the corresponding equations for a pure state of shear can be written as:

$$U_o = \tfrac{1}{2}\sigma_{12}\gamma_{12} = \tfrac{1}{2}G\gamma_{12}^2 = \dfrac{\sigma_{12}^2}{2G},$$ (6.34)

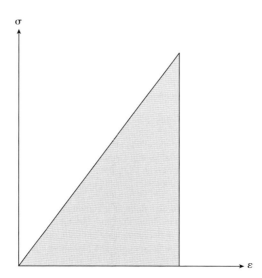

Figure 6.12
The elastic strain energy per unit volume is given by hashed triangular region (area under the curve) by integrating the stress-strain curve from the initial point to the elastic limit.

where σ_{12} and γ_{12} are the shear stress and shear strain, respectively, and $\sigma_{12} = G\gamma_{12}$.

The elastic strain energy under the 3D state of stress is given by

$$U_o = \frac{1}{2E}\{\sigma_{11}^2 + \sigma_{22}^2 + \sigma_{33}^2 - 2\nu(\sigma_{11}\sigma_{22} + \sigma_{22}\sigma_{33} + \sigma_{33}\sigma_{11})\} + \frac{1}{2G}(\sigma_{12}^2 + \sigma_{23}^2 + \sigma_{31}^2). \tag{6.35}$$

Eq. (6.35) can also be expressed solely in terms of strains. We encourage the readers to try to derive those equations above from the fundamental principles that have been discussed thus far in this chapter. Also, differentiating strain energy with respect to strain giving stress and vice versa is a powerful tool in the field of elasticity analysis. Discussion on such theory (e.g., Castigliano's theorem) is outside the scope of this book.

In practice, a term called "*modulus of resilience*" (U_R) is used. It is nothing but the elastic strain energy per unit volume required to stress the material from zero stress to close to yield stress. The expression in Eq. (6.36) is important for applications where a material should not undergo permanent deformation under loading, and mechanical springs are good examples for which the ideal material should have higher yield stress and lower elastic modulus.

$$U_R = \frac{1}{2}\sigma_o \varepsilon_o = \frac{1}{2}\sigma_o \frac{\sigma_o}{E} = \frac{\sigma_o^2}{2E} \tag{6.36}$$

Table 6.3 lists the moduli of resilience for a spring steel and a rubber-based material along with their elastic moduli and yield strength values.

Table 6.3 Elastic modulus (E), yield strength (σ_o), and the corresponding moduli of resilience (U_R).

Material	E (GPa)	σ_o (MPa)	U_R (kPa)
High carbon spring steel	207	965	2249
Rubber	0.001	2.1	2205

Concept Reinforcement – Storage and transfer of elastic energy: To remember the importance of elastic energy, think of all the sports where a ball is hit with an equipment. These sports can range from tennis, where a player is responding to a moving ball, to golf, where a stationary ball is being hit. In all these cases, the ball and the equipment are going through the elastic deflection. Try to map out the entire sequence of event. Consider how many design elements are involved. For tennis, the racquet is made lighter so that the player can swing it at higher velocity. The string tension is adjusted to change the elastic response. The racquet frame is designed for specific elastic stiffness. When the ball collides with the racquet, the ball face is compressed, and the string and frame are elastically deflected. The direction of ball travel is physically reversed by the moving racquet and ball is resident at the racquet for a finite duration. During the time ball is in contact with the racquet, the compressed face of the ball start to recover, and at the same time the string and frame elastically recover. These add velocity component to the swinging racquet. The ball leaves the racquet surface at a higher velocity than the racquet head velocity. The elastic energy release rate dictates the net impact and the final velocity of the ball. The player's ability to swing a club at a particular speed must be matched with the elastic properties of the racquet for the best outcome. Using this example, can you now work out the mechanics of a golf club hitting the ball? There is an interesting example about the Nike golf balls. The company designed a particular golf ball for Tiger Woods. An average player was not able to get the distance advantage claimed. Can you figure out why? Hint: What is the relationship between club head speed and compressibility of the golf ball? How would this impact the performance of the ball with player's skill?

6.2.6 Anisotropy in elastic property

Before delving into the details of the topic here, familiarity with various types of crystal structures and knowledge of how to index them through the Miller indexing method are assumed. The directionality of elastic properties such as elastic modulus (E) in single crystals is profound. However, in the polycrystalline state they appear to become less anisotropic because of averaging of orientations unless there is a strong preferred crystallographic orientation (texture) in the material.

In a generalized case, Hooke's law can be expressed in two ways as shown below:

$$\varepsilon_{ij} = S_{ijkl}\, \sigma_{kl} \tag{6.37a}$$

and

$$\sigma_{ij} = C_{ijkl}\, \varepsilon_{kl} \tag{6.37b}$$

In the equations above, S_{ijkl} is the compliance tensor and C_{ijkl} is the elastic stiffness tensor. Elastic modulus was noted previously to be a tensor of rank 4. Thus S_{ijkl} and C_{ijkl} are both tensors with rank 4. For example, these tensors will have $3^4 = 81$ components to express the relation between the stress and strain components. Remember that i, j, k, or l can have values 1, 2, or 3. Given that σ_{ij} and ε_{ij} are symmetric matrices, $S_{ijkl} = S_{jikl}$. The problem is simplified substantially, and only 36 of the 81 components are required to describe the most complex material. Thus S_{ijkl} can be expressed in matrix form and a 6 × 6 is usually used.

The conventional way of describing stiffness and compliance tensor is to use 2 subscripts instead of 4. Such a method is called *contracted notation* where S_{ijkl} basically becomes S_{mn}. Table 6.4 explains the meaning of the *contracted notation*, which is also illustrated in Fig. 6.13.

Table 6.4 Contracted notations (relating to subscripts).

ij or kl	11	22	33	32/23	13/31	12/21
m, n	1	2	3	4	5	6

$S_{mn} = S_{ijkl}$ if m and n are 1, 2, 3
$S_{mn} = 2 S_{ijkl}$ if m or n are 4, 5, 6
$S_{mn} = 4 S_{ijkl}$ if both m and n are 4, 5, 6

$$\begin{vmatrix} \sigma_{11} & \sigma_{12} & \sigma_{13} \\ \sigma_{21} & \sigma_{22} & \sigma_{23} \\ \sigma_{31} & \sigma_{32} & \sigma_{33} \end{vmatrix} \quad \begin{vmatrix} \sigma_1 & \sigma_6 & \sigma_5 \\ & \sigma_2 & \sigma_4 \\ & & \sigma_3 \end{vmatrix}$$

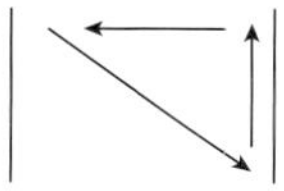

$$\begin{vmatrix} \varepsilon_{11} & \gamma_{12} & \gamma_{13} \\ \gamma_{21} & \varepsilon_{22} & \gamma_{23} \\ \gamma_{31} & \gamma_{32} & \varepsilon_{33} \end{vmatrix} \quad \begin{vmatrix} e_1 & e_6 & e_5 \\ & e_2 & e_4 \\ & & e_3 \end{vmatrix}$$

$$[\sigma_1 \quad \sigma_2 \quad \sigma_3 \quad \sigma_4 \quad \sigma_5 \quad \sigma_6]$$

$$\begin{vmatrix} S_{11} & S_{12} & S_{13} & S_{14} & S_{15} & S_{16} \\ S_{21} & S_{22} & S_{23} & S_{24} & S_{25} & S_{26} \\ S_{31} & S_{32} & S_{33} & S_{34} & S_{35} & S_{36} \\ S_{41} & S_{42} & S_{43} & S_{44} & S_{45} & S_{46} \\ S_{51} & S_{52} & S_{53} & S_{54} & S_{55} & S_{56} \\ S_{61} & S_{62} & S_{63} & S_{64} & S_{65} & S_{66} \end{vmatrix} \begin{vmatrix} e_1 \\ e_2 \\ e_3 \\ e_4 \\ e_5 \\ e_6 \end{vmatrix}$$

Figure 6.13
The use of contracted notations.

Note that the shear strains above are basically engineering shear strains. For understanding of this analysis, the matrices shown below can be followed. Remember that all contracted notations use engineering strains.

$$\gamma_{13} = e_5 = \sigma_1 S_{51} + \sigma_2 S_{52} + \sigma_3 S_{53} + \sigma_4 S_{54} + \sigma_5 S_{55} + \sigma_6 S_{56}$$

$$\text{But } \varepsilon_{13} = \frac{e_5}{2} = \frac{\gamma_{13}}{2} \tag{6.38}$$

On the other hand, Hooke's law becomes

$$e_1 = \sigma_1 S_{11} + \sigma_2 S_{12} + \sigma_3 S_{13} + \sigma_4 S_{14} + \sigma_5 S_{15} + \sigma_6 S_{16} \tag{6.39}$$

These equations show that for an anisotrcpic elastic body, normal strain in the body can have contributions from both normal stresses and shear stresses, and vice versa. Compliance S_{16} relates a shear stress σ_{12} to a tensile strain e_{11}.

Regarding the stiffness matrix, C_{ijkl} is not necessarily equal to S_{ijkl}. As for the contracted notation, $C_{mn} = C_{ijkl}$ for all values. For example, $C_{1131} = C_{15}$.

Interestingly, even though we begin with 36 compliance components required, in reality we can work with many fewer independent constants with an anisotropic, linear solid: actually 21. For higher crystal symmetry conditions, the number of independent constants is reduced further. For example, S_{mn} for the cubic system would look like the following matrix:

$$\begin{bmatrix} S_{11} & S_{12} & S_{13} & 0 & 0 & 0 \\ S_{21} & S_{22} & S_{23} & 0 & 0 & 0 \\ S_{31} & S_{32} & S_{33} & 0 & 0 & 0 \\ 0 & 0 & 0 & S_{44} & 0 & 0 \\ 0 & 0 & 0 & 0 & S_{55} & 0 \\ 0 & 0 & 0 & 0 & 0 & S_{66} \end{bmatrix}$$

However, cubic symmetry requires that $S_{11} = S_{22} = S_{33}$, $S_{44} = S_{55} = S_{66}$, and $S_{12} = S_{13} = S_{23} = S_{21} = S_{31} = S_{32}$. Thus for the cubic system, only three independent compliance constants (or stiffness constants) are required (Compliance: S_{11}, S_{12}, and S_{44}; Stiffness: C_{11}, C_{12}, and C_{44}).

$$\frac{1}{E_{uvw}} = S_{11} + (2S_{12} - 2S_{11} + S_{44}) \left[\frac{v^2w^2 + w^2u^2 + u^2v^2}{(u^2 + v^2 + w^2)} \right]. \tag{6.40}$$

This gives the Young's modulus for any given [uvw] crystallographic direction. If $2S_{12} - 2S_{11} + S_{44} = 0$, E_{uvw} would be the same for all values of [uvw]. That means the material will be isotropic with respect to its property, elastic modulus.

We can define the relative anisotropy of cubic crystals by the following expression:

$$\frac{2(S_{11} - S_{12})}{S_{44}}.$$

In cubic system, the stiffness terms can be expressed in terms of S_{11}, S_{12}, and S_{44}.

$$C_{11} = \frac{S_{11} + S_{12}}{(S_{11} - S_{12})(S_{11} + 2S_{12})} \tag{6.41a}$$

$$C_{12} = \frac{-S_{12}}{(S_{11} - S_{12})(S_{11} + 2S_{12})} \tag{6.41b}$$

$$C_{44} = \frac{1}{S_{44}} \tag{6.41c}$$

Table 6.5 includes elastic stiffness and compliance values at room temperature for a variety of materials with cubic lattice structures.

6.2.7 Elastic behavior of a polycrystal as a single crystal average

Here we try to average the elastic moduli of all the possible orientations of crystals present in the polycrystal following two approaches: (1) Voigt and (2) Reuss.

Table 6.5 Room temperature compliance and stiffness values for various materials with cubic lattices.

Material	C_{11} (GPa)	C_{12} (GPa)	C_{44} (GPa)	S_{11} (GPa)	S_{12} (10^{-3} GPa^{-1})	S_{44} (10^{-3} GPa^{-1})
Aluminum	108.0	61.0	28.0	15.7	−5.70	35.1
Copper	168.0	121.0	75.4	15.0	−6.30	13.3
Gold	186.0	157.0	42.0	23.3	−10.7	24.0
Nickel	247.0	147.0	125.0	7.3	−2.7	8.0
Silver	124.0	93.4	46.0	22.9	−9.8	22.0
Iron	237.0	141.0	115.0	8.0	−2.8	8.6
Molybdenum	460.0	176.0	110.0	2.8	−0.78	9.1
Niobium	246.0	134.0	28.7	6.6	−2.3	34.8
Tantalum	267.0	151.0	82.5	6.9	−2.6	12.2
Sodium Chloride	49.0	12.0	13.0	23.0	−4.7	79.0

1. The Voigt average method assumes uniform local strains; that is, the $e_{11}{}^A$ (normal strain in grain A) is equal to $e_{11}{}^B$ (normal strain in grain B) with B randomly oriented with respect to grain A.

$$E_V = \frac{(F-G+3H)(F+2G)}{(2F+3G+H)}$$

where, $F = \frac{1}{3}(C_{11}+C_{22}+C_{33})$

$G = \frac{1}{3}(C_{12}+C_{23}+C_{13})$ (6.42a)

$H = \frac{1}{3}(C_{44}+C_{55}+C_{66})$

For cubic materials this reduces to

$$E_V = \frac{(C_{11}-C_{12}+3C_{44})(C_{11}+2C_{12})}{(2C_{11}+3C_{12}+C_{44})} \quad (6.42b)$$

Note that the Voigt average *overestimates* the elastic modulus and is considered to give an upper bound value for a random polycrystal.

2. The Reuss average method assumes uniform local stress in each grain and randomly averaging the compliance. The average elastic modulus by this method is given by

$$\frac{1}{E_R} = \frac{1}{5}(3F^* + 2G^* + H^*)$$

Where, $F^* = \frac{1}{3}(S_{11}+S_{22}+S_{33})$

$G^* = \frac{1}{3}(S_{12}+S_{23}+S_{13})$ (6.43a)

$H^* = \frac{1}{3}(S_{44}+S_{55}+S_{66})$

For cubic materials, the Reuss average reduces to

$$\frac{1}{E_R} = \frac{1}{5}(3S_{11} + 2S_{12} + S_{44}) \tag{6.43b}$$

Interestingly, the Reuss method *underestimates* the elastic modulus and sets a lower bound.
Therefore to get a realistic value of the average elastic modulus, the average of E_V and E_R can be determined using the following:

$$\text{Arithmetic mean: } E_{av} = \tfrac{1}{2}(E_V + E_R) \tag{6.44a}$$

$$\text{Geometric mean: } E_{av} = \sqrt{E_V} \cdot \sqrt{E_R} \tag{6.44b}$$

Table 6.6 includes room temperature elastic modulus values calculated from compliance values and compared with the predicted and experimental values for four cubic materials. An easy guess from the table is which material is most isotropic with respect to its elastic modulus. The difference between elastic moduli along [100] direction and [111] directions for tungsten is zero, whereas nickel, copper, and alpha-iron show significant difference between those values.

6.3 Design of high stiffness composite materials

Nowadays composites have become ubiquitous: They are all or in part found in tennis rackets, racing bicycle frames, Formula 1 racing car bodies, and even in Boeing 787 Dreamliner planes (**Chapter 1**). The Boeing 787 represented a paradigm shift for the company. This is the first Boeing plane to shift from aluminum body to composite body. Clearly, the use of composites has become very wide in modern civilization (aerospace, automotive, sports, electronics, and many more). They are more common than we generally think! The two broad groups of composites are natural composites and engineered composites. As the term implies, natural composites are found in nature and are made by natural processes. Examples of such composites include wood, bone, and seashells, to name a few. On the other hand, engineered composites are manmade. Composites have been important since the early stages of civilization. Residential huts were made of clay reinforced with straws; that is a composite! Addition of straws in the clay made it more resistant to fracture. Composites consist of combinations of two or more materials and lead to better combinations of properties than the properties of individual constituents.

Consider a brief discussion of an instance of using composites to benefit a real-world application: a vaulting pole. Over the years, the winning pole vaulting heights in Olympics have become

Table 6.6 Predicted and experimental values (all in GPa) of elastic moduli of copper, α-iron, nickel, and tungsten.

Material	E_{100}	E_{111}	E_V	E_R	E_{av} (arithmetic mean)	E_{av} (Geometric mean)	E_{exp}
Copper (FCC)	66.9	191.7	195.2	109.8	127.5	126.3	128
α-Fe (BCC)	132.3	283.3	227.1	194.5	210.8	210.2	208
Ni (FCC)	129	304.9	239.5	197.5	218.4	217.4	207
Tungsten (BCC)	385	385	372.4	384.6	378.5	378.5	411

6.3 Design of high stiffness composite materials

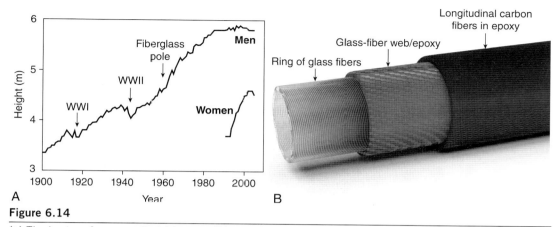

Figure 6.14

(a) The best performance (height crossed) in men's pole vaulting as a function of year.[5] (b) The composite vaulting pole with the three layers of different fibers used to maximize stiffness while minimizing twisting during use.[6]

progressively higher and higher (Fig. 6.14 [a]). In the 1896 Athens Olympics, William Hoyt scaled over 3.3 m height to win the gold medal using a bamboo pole in the pole-vault event. However, nowadays 3.3 m height would not even fetch vaulter qualification for the Olympics. (For the 2016 Rio Olympics, men's qualifying height was 5.65 m, and women's was 4.50 m.) Even though aluminum poles were introduced in the 1930s, poles made out of steel became more popular in the 1940s and 1950s. Although the development of glass fiber–based vaulting poles started in the 1940s, design and development were perfected in the 1960s. That is when performance in pole vaulting saw a sea change, with the transition from relatively rigid steel or bamboo to highly flexible fiberglass poles. Between 1961 and 1964, that is, just over 3 years, the world record in pole vaulting increased by almost 48 cm (Fig. 6.14 [a]). At present, the pole material transitioned further into highly engineered composite poles and resulted in the great Olympic records of today, the most recent one being close to 6 m (i.e., 5 97 m) in men's pole vaulting. The most recent men's pole vaulting record of 6.19 m was set recently by Armand Duplantis of Sweden.

The material used in high-performance vaulting poles is currently carbon-fiber composites. The poles have a three-layer design (Fig. 6.14 [b]). By adding the minimal-twisting requirement, the carbon-fiber composite becomes very attractive to optimize performance. An outer layer of high-strength carbon fiber provides high stiffness and an intermediate webbing of fibers together with an inner layer of wound-glass fiber contribute to twisting resistance. The glass fiber consists of 80% longitudinal and 20% radial fibers. The resulting vaulting pole is light, with optimum combination

---

[5]Linthorne N. Design and materials in athletics. In: Subic A, ed. *Materials in Sports Equipment*. Vol. 2. Woodhead Publishing; 2007:296-320.
[6]Adapted from Easterling KE. *Advanced Materials for Sports Equipment*. Chapman and Hall; 1993.

of stiffness and flexibility. From a science viewpoint, the best material for vaulting pole maximizes the following ratio:

$$\frac{(\text{Specific Strength})^2}{(\text{Specific Stiffness})} = \frac{\left(\frac{\sigma}{\rho}\right)^2}{\left(\frac{E}{\rho}\right)} = \frac{\sigma^2}{\rho E} \quad (6.45)$$

where σ is the maximum allowable bending stress, ρ is density, and E is elastic modulus. The carbon-fiber-reinforced plastic should give the best performance, followed closely by glass-fiber-reinforced plastic.

Composites can be categorized in different ways (Fig. 6.15). They can be broadly classified as particulate-reinforced, fiber-reinforced, and structural composites. Particulate composites are of two types: large particle and dispersion-strengthened. Fiber-reinforced composites can be two types based on the continuity of the reinforcements, that is, continuous aligned composites and discontinuous (short fiber) composites. Discontinuous composites can be further divided into two types, based on whether the fiber distribution is aligned or random. Another way of classifying composites is based on the matrix type: metal matrix composites (MMC), for example, Al/SiC, Al/Al$_2$O$_3$; polymer matrix composites (PMC), for example, fiberglass polyester; and ceramic matrix composites (CMC), for example, reinforced concrete, SiC/Si-Al-O-N.

6.3.1 Particulate-reinforced composites

Large-particulate composites are reinforced with particles where the matrix transfers a sizable portion of the applied load to the particles. This approach is guided by continuum mechanics. Concrete is one example of large-particulate composite. Here the mechanical behavior is dependent on strong bonding at the particle-matrix interface. On the other hand, dispersion-strengthened composites contain particle diameters much smaller (~10–100 nm) than large-particulate composites. Dispersion-strengthened composites are often regarded as the class of material akin to dispersion-strengthened alloys. One example is oxide dispersion-strengthened alloys that incorporate small, hard ceramic particles in the metallic matrix. Here dislocation-dominated particle strengthening is important in retaining high

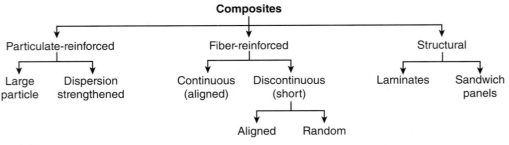

Figure 6.15

Various classes of composites.

6.3 Design of high stiffness composite materials

strength even at higher homologous temperatures. The relevant mechanism will be discussed in **Chapter 8**. One special case of particulate-reinforced composites is cermet (Ceramic-Metal), which contains sizable amounts of ceramic phases and a "minority" metallic phase. One important example of cermet is tungsten carbide–cobalt (WC-Co). Here cobalt is used as a soft/ductile binder phase with hard and stiff WC phase. Fig. 6.16 (a) shows the microstructures of WC-Co cermets containing different WC

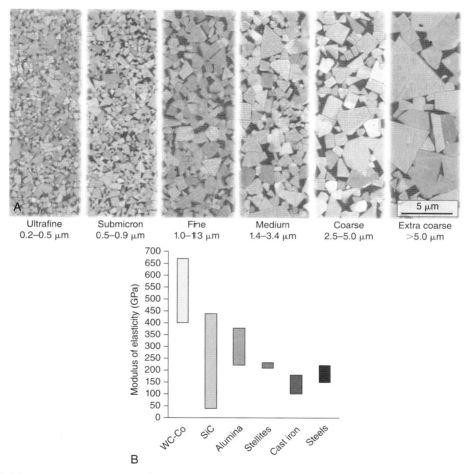

Figure 6.16

(a) Representative microstructures of tungsten carbide–cobalt (WC-Co) cermets with different WC particulate sizes. The black phases in the microstructures represent the Co binder phase. (b) Typical range of WC-Co cermets in comparison to various engineering materials.[7]

---

[7]Garcia J, Cipres VC, Blomqvist A, Kaplan B. Cemented carbide microstructures: a review. *Int J Refract Metals Hard Mater*. 2019;80:40-68.

particulate sizes. WC-Co cermets are used as the tooling material for high-temperature use with considerable hot hardness and wear resistance. By choosing suitable characteristics of constituent materials, compositions, and processing parameters, a wide range of mechanical properties can be achieved. For example, among other properties for tooling applications, WC-Co cermets are high-stiffness materials compared to other ceramics and metallic alloys, as illustrated in Fig. 6.16 (b).

We are interested in knowing what the elastic modulus of the composite would be if the moduli of the matrix and reinforcement are known. The rule of mixture equations can be used to predict elastic modulus in such cases on the assumptions.

6.3.1.1 Constant strain approach

This approach assumes that total strain in the composite (ε_c) is equal to strain in the particulates (ε_p) as well as in the matrix material (ε_m); that is,

$$\varepsilon_c = \varepsilon_p = \varepsilon_m \tag{6.46}$$

$$\text{Total force, } F = \sigma A = \sigma_p A_p + \sigma_m A_m \tag{6.47}$$

where σ is the true stress and A is the cross-sectional area. The subscript "p" refers to "particle" and "m" refers to "matrix."

So, applying Hooke's law to individual phases and the composite in Eq. (6.47), we obtain the following:

$$E_c \varepsilon_c A = E_p \varepsilon_p A_p + E_m \varepsilon_m A_m \tag{6.48a}$$

where E_c represents the composite modulus.

$$\text{Or, } E_c = E_p(A_p/A) + E_m(A_m/A) = E_p V_p + E_m V_m. \tag{6.48b}$$

In the above equation, V_p and V_m refer to the volume fraction of particles and the matrix, respectively, and $V_p + V_m = 1$. This presents the upper bound values of the composite modulus (Fig. 6.17).

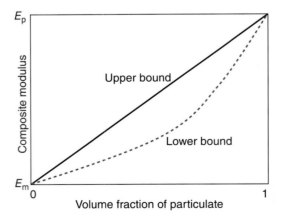

Figure 6.17

Elastic modulus of particulate composites: upper bound and lower bound predictions.

6.3.1.2 Constant stress approach

The equal stress approach was first promulgated by Reuss and assumes that $\sigma = \sigma_p = \sigma_m$. Thus overall strain in the composite is given by the following relation:

$$\varepsilon_c = \varepsilon_p V_p + \varepsilon_m V_m \tag{6.49}$$

Therefore using Hooke's law for the composite and the constituent phases obtains

$$\frac{\sigma}{E_c} = \frac{\sigma_p V_p}{E_p} + \frac{\sigma_m V_m}{E_m}$$

$$Or, \frac{1}{E_c} = \frac{V_p}{E_p} + \frac{V_m}{E_m} \tag{6.50}$$

Eq. (6.50) provides a lower bound of elastic modulus for particulate composites, as previously illustrated in Fig. 6.17.

6.3.2 Fiber-reinforced composites

Fiber-reinforced composites are a large group of composites reinforced with a variety of fibers and with different distributions. The broad categories of fiber reinforcements are (1) whiskers (strong single crystals with small diameter); (2) fibers (normally made of polymers or ceramics); and (3) wires (metals/alloys with relatively large diameter). Table 6.7 below highlights important characteristics of some types of fibers used as reinforcements in composites. Higher elastic modulus fibers as reinforcement leads to higher elastic modulus of the composite.

In fiber-reinforced composites, the fibers may have three kinds of distributions: (1) continuous aligned, (2) discontinuous and aligned, and (3) discontinuous and random. The elastic properties of the fiber-reinforced composites are discussed here. A simple way to describe elastic modulus of

Table 6.7 Glass fiber compositions and properties.[8]

Fiber	Density (g/cm^3)	Young's modulus (GPa)	Tensile strength (GPa)	Failure strain (%)	Thermal expansion coefficient (10^{-6} K^{-1})	Thermal conductivity (Wm^{-1}K^{-1})
Boron monofilament	2.6	400	4.0	1.0	5.0	38
High modulus (HM) carbon	1.95	380 (axial) 12 (radial)	2.4	0.6	0.7 (axial) 10 (radial)	105 (axial)
E-glass	2.56	76	2.0	2.6	4.9	13
Kevlar™ (aramid)	1.45	Axial (130) radial (10)	3.0	2.3	−6 (axial) 54 (radial)	0.04 (axial)
SiC whisker	3.2	450	5.5	1.2	4.0	100

---

[8]Hull D, Clyne TW. *An Introduction to Composite Materials, Cambridge Solid State Science Series*. 2nd ed. Cambridge University Press; 1996.

fiber-reinforced composites is for the ones that have aligned continuous fibers parallel to the axis of loading. The fibers used as reinforcements have greater elastic modulus and tensile strength than the matrix itself, so the expectation is that the composite modulus will be higher than that of the matrix. Consider first that the situation is as shown in Fig. 6.18 (a), where load P is being applied to the cylindrical fiber-reinforced composite with aligned continuous fibers parallel to the direction of loading. Here the assumption is that under load P the fibers and the matrix undergo strains (ε_f and ε_m, respectively) the same as the total strain (ε_c); i.e., $\varepsilon_c = \varepsilon_f = \varepsilon_m$. This type of equal strain treatment is known as the Voigt approach, which is very similar to what is done for finding the upper bound value of elastic modulus in the composite. A scanning electron microscopy image of a fracture surface of a SiC fiber-reinforced Cu-Cr-Zr alloy composite is shown in Fig. 6.18 (b).

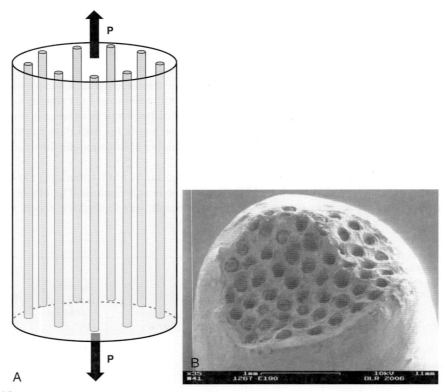

Figure 6.18

(a) A schematic of a fiber-reinforced with aligned continuous fiber under loading parallel to the fiber orientations; (b) a scanning electron microscopy image of the fracture surface of a SiC-reinforced Cu-Cr-Zr matrix composite. Note that some SiC fibers are still visible in the fracture surface.[9]

---

[9]Peters PWM, Hemptenmacher J, Schurmann H. The fibre-matrix interface and its influence on mechanical and physical properties of Cu-MMC. *Compos Sci Technol.* 2010;70:1321-1329.

Considering the cross-sectional area being composed of fiber area fraction (A_f) and matrix area fraction (A_m) and corresponding stresses in the fiber and matrix being σ_f and σ_m, respectively, we can rewrite the Eq. (6.49) as:

$$P = \sigma_c A_c = \sigma_f A_f + \sigma_m A_m,$$

$$\text{Or, } \sigma_c = (\sigma_f A_f + \sigma_m A_m)/A_c = (A_f/A_c)\sigma_f + (A_m/A_c)\sigma_m \quad (6.51)$$

where σ_c and A_c are the stress and cross-sectional area of the composite. respectively.

Note that (A_f/A_c) can also be written as ($A_f L/A_c L$) or (volume of fibers divided by volume of composite), where L is the length of the composite and the fibers. In turn, the volume fraction of fibers (V_f) can actually be given by (A_f/A_c). Similarly, the volume fraction of the matrix phase, $V_m = (A_m/A_c)$. Thus Eq. (6.51) can be written in the following form:

$$\sigma_c = V_f \sigma_f + V_m \sigma_m. \quad (6.52)$$

Using the procedure followed for Eq. (6.48), the relationship for fiber-reinforced composites becomes

$$E_{c(l)} = V_f E_f + V_m E_m = V_f E_f + (1 - V_f) E_m \quad (6.53)$$

An additional subscript "l" is used to include in E_c, to signify the loading condition parallel to the fiber orientations (longitudinal).

However, when the fiber is still elastically deforming, the matrix can undergo plastic deformation. In that case, the slope of the stress-strain curve ($d\sigma_m/d\varepsilon_m$) can be used to replace "E_m." Because $V_m (d\sigma_m/d\varepsilon_m)$ is much smaller compared to the first term in Eq. (6.53) we can essentially write

$$E_{c(l)} = V_f E_f. \quad (6.54)$$

Aligned fiber-reinforced composites are anisotropic in nature. So, consider the situation where the elastic modulus is to be calculated in the transverse direction of the fiber orientation. That is, the loading is occurring perpendicular to the fiber orientation. In that case the constant stress approach is more pertinent.

The measurement of transverse elastic modulus is generally considered difficult and does not conform to the modulus values predicted by Eq. (6.50). In the transverse case, the matrix has to experience more stress than the longitudinal case. This leads to high strain localization at the fiber-matrix interfaces, with an inhomogeneous strain field and hence inhomogeneous stress field. This condition leads to plastic deformation in the matrix, debonding at the fiber-matrix interfaces and microcracking. Basically, Eq. (6.50) provides a lower bound of transverse elastic modulus. In this regard, another, more accurate description of this quantity was developed by Halpin and Tsai based on increased fiber load bearing. The expression is as follows:

$$E_{c(t)} = \frac{E_m(1 + \xi \eta V_f)}{(1 - \eta V_f)} \quad (6.55)$$

$$\text{where } \eta = \frac{\left(\dfrac{E_f}{E_m} - 1\right)}{\left(\dfrac{E_f}{E_m} + \xi\right)}.$$

All terms in Eq. (6.55) have already been defined previously except for ξ, which is an adjustable parameter, typically close to 1.

Depending on the angle of loading relative to fiber orientation, different elastic moduli can be obtained. The elastic analysis involved would be worked out in such cases. For randomly oriented fibers, the orientation dependence of the loading direction should disappear. However, simple averaging of the moduli in all fiber orientations does not lead to a good estimate. Anyway, a useful engineering estimate of elastic modulus using certain weight factors can be obtained from the following relation:

$$E_c = (0.625)E_{c(t)} + (0.375)E_{c(l)}. \tag{6.56}$$

Here we note another variety of composites with discontinuous and randomly oriented fiber composites. To estimate elastic modulus, $E_{c(dr)}$, of this type of composite, a "rule of mixture" equation as shown below is used with a little tweak. Note that "d" and "r" in the subscripts of the term represent "discontinuous" and "random," respectively.

$$E_{c(dr)} = E_m V_m + \kappa E_f V_f. \tag{6.57}$$

In the above equation, κ is a fiber efficiency factor that generally varies between 0.1 and 0.6 depending on the fiber volume fraction and the relative ratio of elastic moduli of fibers and the matrix.

The composite strengthening aspects will be further discussed in detail in **Chapter 8**.

6.4 Key chapter takeaways

This chapter treated elastic constants as a manifestation of interatomic forces present in crystalline materials. A tensorial treatment of stress analysis was introduced to determine the state of stress in transformed stress axes. Then definitions of stress and strain were introduced, and their relationships with respect to their elastic behavior were discussed in both simplified and multiaxial conditions. An understanding of elastic compliance and stiffness was presented to explain anisotropy of elastic properties. Lastly, the elastic behavior of composites focusing on the particulate-reinforced and fiber-reinforced composites were discussed. Readers should be able to calculate the elastic modulus of the composites from the individual elastic moduli of the reinforcement and matrix phases and the corresponding phase volume fractions.

6.5 Exercises

1. A test bar of a metal with 12.83 mm in diameter and 50 mm gage length is loaded elastically with 156 kN tensile load and is stretched by 0.356 mm. Its diameter is 12.80 mm under load. What is the shear modulus of the metal? State any assumption(s) you made.

2. The 2D state of stress (with respect to x-y axes) is given as follows: $\sigma_{xx} = 50$ ksi, $\sigma_{yy} = 15$ ksi, and $\tau_{xy} = -7$ ksi. Note: ksi is 1000 lb per square inch; 1 ksi = 6.895 MPa.

    a) Show the free body diagram of the given stress state.

    b) Find the principal normal stresses.

c) Calculate the maximum shear stress.

d) Also, calculate the orientation of the principal planes.

3. Given the stress state: $\begin{bmatrix} 20 & 20 & 10 \\ 20 & 30 & 40 \\ 10 & 40 & 20 \end{bmatrix}$ MPa. (a) Calculate the sum of the principal normal stresses. (Hint: it does not need calculation of individual principal stresses). (b) Find the hydrostatic and deviatoric parts of the stress state.

4. (a) Generally, metallic materials appear to have Poisson's ratio close to 0.3. However, experimentally, a certain material does not change in volume when subjected to an elastic state of stress. Calculate its Poisson's ratio. (b) A few materials exhibit negative Poisson's ratio. Rationalize the expected physical behavior.

5. The stress state $\sigma_{ij} = \begin{bmatrix} 5 & 6 & 9 \\ 6 & 10 & 8 \\ 9 & 8 & 5 \end{bmatrix}$ MPa acts on an isotropic solid with shear modulus $(G) = 6$ GPa and Poisson's ratio $(\nu) = 0.33$, and yield strength (σ_o) is 12 MPa.

a) Draw a free body diagram of the given state of stress.

b) Evaluate the principal normal stresses.

c) Suppose the initial coordinates reference system is rotated to an arbitrary angle anticlockwise. If the normal stress in the x-direction becomes 6 MPa, then what would be the normal stress in the y-direction?

d) Find the hydrostatic and deviatoric parts of the initial state of stress.

e) Develop the corresponding strain tensor for the original state of stress. State any assumption you made.

f) What is the value of the volume strain?

g) Calculate the elastic strain energy per unit volume of the body under application of the original state of stress.

h) Determine the hydrostatic strain tensor and deviatoric strain tensor from the overall strain tensor obtained in part (e).

6. a) To create a composite, whiskers of 3.2 g/cm³ are used with a polymer matrix of 1.1 g/cm³. What is the mass percentage (or weight%) of the composite if the whisker volume fraction is 0.2?

b) If titanium matrix (density 4.43 g/cm³) is used in place of the polymer matrix, what would be the mass percentage of whiskers in the composite?

7. A unidirectional continuous boron-filaments-reinforced epoxy composite has been procured by your company for a structural application. The elastic modulus of boron $(E_f) = 380$ GPa; the elastic modulus of the epoxy matrix $(E_m) = 2$ GPa.

a) What is the elastic modulus of the composite parallel to the direction of filaments if the volume fraction of the filaments (V_f) is 0.3?

b) What is the elastic modulus of the composite transverse to the filament orientation for the same filament volume fraction as in (a)?

8. Predict the greatest value (upper bound) of the elastic modulus of a WC-Co (particulate-reinforced composite, also known as cermet) cutting tool material with 55% WC (by volume). Also, calculate the smallest value (lower bound) of the elastic modulus that is possible. Given: Elastic moduli of WC and Co are 668 GPa and 209 GPa, respectively.

9. **Thought Experiment:** Assume a metal has Young's modulus of 50 GPa and contains a cubic particle of a ceramic phase with Young's modulus of 400 GPa. The cubic particle has axis length of 10 μm, and the orientation is such that one of the cubic axes is aligned with the loading direction. Draw elastic displacement of the side of particle that is parallel to the loading direction as a function of stress. If the lattice spacing of the matrix is 2.5 Å, at what elastic stress does the mismatch in longitudinal strain in the ceramic phase and the neighboring matrix become equal to 2.5 Å? Can you think of ways to relax this stress? *In Future: Compare your current answer to the answer you will be able to give after reading Chapters 7 and 8.*

10. Derive Eq. (6.29) using the relationships for E, G, and K.

11. A test bar of a metal with 12.83 mm in diameter and 50.8 mm gage length is loaded elastically with 160 kN tensile load and is stretched by 0.356 mm. Its diameter is 12.80 mm under load. What is the elastic modulus of the metal?

12. Given the stress state: $\begin{bmatrix} 20 & 20 & 10 \\ 20 & 30 & 40 \\ 10 & 40 & 20 \end{bmatrix}$ MPa. (a) Find the principal stresses. (b) Calculate the hydrostatic part of the stress state (or mean stress tensor). (c) Find the deviatoric part of the stress state (stress deviator tensor).

13. Determine the principal normal stresses for the following state of stress: $\begin{bmatrix} 0 & -50 & 0 \\ -50 & 10 & 0 \\ 0 & 0 & -75 \end{bmatrix}$ MPa.

14. For the stress state (with respect to x-y axes): $\sigma_{xx} = 50$ ksi, $\sigma_{yy} = 5$ ksi, and $\tau_{xy} = -8$ ksi. (a) Show the free body diagram of the given stress state. (b) Find the principal stresses. (c) Calculate the maximum shear stress. (d) Also, calculate the orientation of the principal planes.

15. Calculate a safe internal pressure for a 125 mm diameter pipe of a 18-8 stainless steel. A safety factor of 3.8 below the yield stress is required for it to operate safely in service. Nominal dimensions of the pipe are as follows: Outer diameter 125 mm, and wall thickness 1.3 mm. Calculate the safe internal pressure. State your assumptions in solving this problem. Take the yield strength of 18-8 stainless steel as 172 MPa.

16. The stress state, $\sigma_{ij} = \begin{bmatrix} 20 & 20 & 10 \\ 20 & 30 & 40 \\ 10 & 40 & 10 \end{bmatrix}$ MPa, acts on an isotropic solid that has a shear modulus (G) of 6 GPa and Poisson's ratio (ν) of 0.33, and the yield strength (σ_o) of 12 MPa.

a) Evaluate the principal normal stresses.

b) Suppose the initial coordinate reference system is rotated to such an arbitrary angle anticlockwise so that the normal stress in the x-direction becomes 10 MPa and 20 MPa in the z-direction. Then what would be the normal stress in the y-direction?

c) Find the hydrostatic and deviatoric parts of the initial state of stress.

d) Develop the corresponding strain tensor (i.e., in the 3 × 3 matrix form) for the original state of stress. State any assumption you made.

e) What is the value of the volume strain?

f) Calculate the elastic strain energy per unit volume of the body under application of the original state of stress.

g) Find out the hydrostatic strain tensor and deviatoric strain tensor from the overall strain tensor obtained in part (iv).

17. Strain gages mounted on the outer surface of a pressure vessel read the elastic strain of 0.003 in the longitudinal direction and 0.005 in the circumferential direction. (a) Compute the stresses in these two principal directions if the elastic modulus (E) is 210 GPa and Poisson ration ν is 0.33. (b) What is the error if the Poisson effect is not taken into account?

18. If the elastic modulus of aluminum is 70 GPa and SiC is 45 GPa, the upper limit of shear modulus for an Al-16vol%SiC metal matrix composite is 85 GPa according to the simple rule of mixture. True or False? (Give adequate reasons supporting your choice.)

Further readings

Cook RD, Young WC. *Advanced Mechanics of Materials*. McMillan; 1985.
George E. *Dieter, Mechanical Metallurgy*. McGraw-Hill; 2001.
Hull D, Clyne TW. *An Introduction to Composite Materials*. 2nd ed. Cambridge University Press; 1996.
Linthorne N. Design and materials in athletics. In: Subic A, ed. *Materials in Sorts Equipment*. Vol. 2. Woodhead Publishing; 2007.
Thomas H. *Courtney, Mechanical Behavior of Materials*. McGraw-Hill; 2000.
William F. *Hosford, Mechanical Behavior of Materials*. 2nd ed. Cambridge University Press; 2010.

CHAPTER 7

General dislocation theory for crystalline materials

Chapter outline

7.1 Crystals and defects 127
    7.1.1 Theoretical strength of a crystal 129
    7.1.2 Miller and Miller-Bravais indices 130
7.2 Types of dislocations 132
7.3 Dislocation movement 135
7.4 Dislocation reactions 136
7.5 Dislocation generation 139
    7.5.1 Intragranular dislocation sources 139
    7.5.2 Interfacial dislocation sources 141
7.6 Geometrically necessary dislocations – how do they differ from statistical dislocations? 142
7.7 Key chapter takeaways 143
7.8 Exercises 143

Learning objectives

- Basic concepts of dislocations.
- The importance of separating the energetics and kinetics aspects of deformation micromechanisms.
- Plasticity-induced twinning and transformation – role of partial dislocations.

Dislocations are the basis for all the plastic behavior except for diffusional flow–based creep mechanisms. Chapter 6 was based on stiffness, while Chapters 8–11 are based on plastic flow of material. The first level of fundamental is just to remember that any plastic deformation of material can only happen through movement of (a) vacancies or (b) dislocations. So, the lighthearted way to remember this fact is to think of vacancies and dislocations as the "agents of deformation." The next fundamental is to consider that each deformation micromechanism involves two aspects: (a) energetics and (b) kinetics. As we establish the fundamentals of dislocations in this chapter, we want to master both these aspects.

7.1 Crystals and defects

As students are very well aware from the fundamental materials science and engineering course, the building unit of any materials is "atom." Arrangement of atoms defines the material. The first distinction we want to make is based on the nature of atomic arrangement. If the arrangement of atoms is periodic and has certain elements of symmetry, they are called crystalline materials. Polymers and glasses are noncrystalline materials. This is a very important distinction as the framework of plastic deformation of crystalline material is based on dislocations and definition of dislocations in a crystal is based on the periodic arrangement of atoms. If an atom is missing in the lattice, that site is referred to as "vacancy." The movement of vacancies is a primary mechanism for diffusion of atoms, and as we will see in the creep chapter (Chapter 11), coordinated movement of vacancies can lead to plastic deformation. The equilibrium concentration of vacancies (n_{eq}) in a crystal is temperature dependent and expressed as

$$\frac{n_{eq}}{N} = \exp\left(-\frac{\Delta H_f}{kT}\right) \tag{7.1}$$

where N is the total number of atoms, ΔH_f is energy of formation for vacancy in a perfect lattice, k is Boltzmann's constant, and T is temperature in K. At 0 K the concentration of vacancies in any metal is 0. Let us consider the value of $\Delta H_f = 0.70$ ev/vacancy for aluminum. Using Eq. (7.1), the equilibrium concentration of vacancies at 300 K and 900 K are 1.45×10^{-12} and 1.12×10^{-4}, respectively. This shows the importance of vacancies at high temperatures. Later in this chapter, we will consider the impact of this on the mobility of dislocations at high temperatures. Vacancies are also referred to as point defects and "zero" dimensional defects.

We are interested in three basic crystal structures – face-centered cubic (FCC), body-centered cubic (BCC), and hexagonal close packed (HCP). Fig. 7.1 shows these crystal structures along with a part of periodic table with elements highlighted. Most of the elements, particularly commonly used ones, fall among these three basic categories. Based on the simple concepts described in this chapter, the students

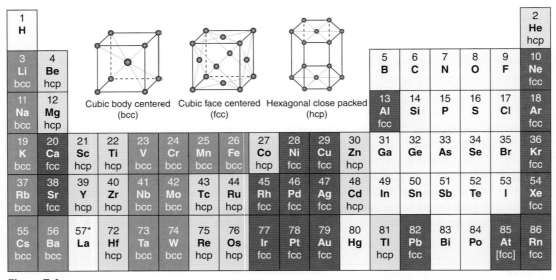

Figure 7.1

Schematic of three basic crystal structures and a portion of the periodic table highlighting elements with one of these simple crystal structures. Note that ignores polymorphism; i.e., several elements change crystal structure with pressure and temperature.

should be able to apply the concepts to more complicated crystal structures as needed for their particular interests. These crystal structures would be the basis for discussion of the dislocations which are also referred to as line defects and "one" dimensional defects. Before we discuss the dislocations, let us discuss the need for such a framework for discussion of deformation of crystalline materials.

7.1.1 Theoretical strength of a crystal

Before the development of concept of dislocation, strength of a crystal was estimated by using calculated force required to shear a crystal. Fig. 7.2 shows planes A and B shearing over planes C and D. We are interested in looking at the force variation with shear strain (γ). Energetically the configurations in (a) and (c) are equal. During shearing, when the process reaches the unstable configuration shown in Fig. 7.2(b), the applied shear force will carry it to the configuration shown in (c). Note that all the atoms are in ideal positions, same as before the shear process. So, at the end of the shear process, the crystal is defect free. Importantly, the atoms marked in (a) are shifted by one atom diameter, which is equivalent to the Burgers vector. Recall that d_{hkl} is the interplanar spacing of hkl planes. Just be careful that you pick the correct planar spacing. For example, shear of (100) planes in FCC crystal requires $d_{200} = a/2 = h$, where a is the lattice spacing. The theoretical strength of the material does depend on the selection of plane. The shear stress will vary with displacement x and a sine function is typically used (Fig. 7.2(b1)). The variation of shear stress with distance x to an equivalent lattice position (defined as b, the Burgers vector) is given as

$$\tau = \tau_{max} \sin\left(\frac{2\pi x}{b}\right) \tag{7.2}$$

Substitute $x = h\gamma$:

$$\tau = \tau_{max} \sin\left(\frac{2\pi \gamma h}{b}\right) \tag{7.3}$$

Note that $d\tau/d\gamma$ = shear modulus = G. Taking derivative of Eq. (7.3) with respect to strain, we get

$$\frac{d\tau}{d\gamma} = \tau_{max} \cos\left(\frac{2\pi \gamma h}{b}\right)\frac{2\pi h}{b} = G \tag{7.4}$$

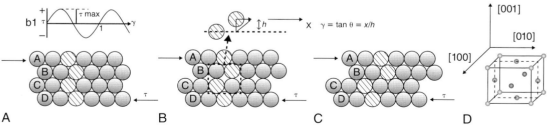

Figure 7.2

A rigid shear model of atomic sliding to produce strain. From the unstable position in (B), the strain can be calculated by the ratio of vertical to horizontal displacement.

As $\gamma \to 0$, $\cos(2\pi\gamma h/b) \to 1$, which reduces Eq. (7.4) to

$$\tau_{max}\left(\frac{2\pi h}{b}\right) = G$$

$$\tau_{max} = \frac{Gb}{2\pi h} \tag{7.5}$$

The lower bound on the theoretical strength is given by the smallest b and the largest h. The value of $|b|_{smallest} = 2 R_{atom}$ <close-packed>, where R_{atom} = radius of atom. Note that $|b|_{smallest}$ is in the close-packed direction and the largest h is usually between close-packed planes. This is also the definition for the slip system that we will refer to in **Chapter 8** quite a bit: combination of *close-packed planes and close-packed direction*.

The main takeaway from this analysis is that the theoretical or ideal shear strength based on atomic sliding is of the order of $\sim G/6$. Note that typical values of the shear modulus of pure metals range between 20 and 80 GPa and this will imply that the strength should range from ~ 3 to 14 GPa. The experimental shear strength of pure metals ranges from 1 to 30 MPa. ***Clearly the difference between theoretical and experimental is as large as three orders of magnitude***. This highlights the need for a dislocation-based framework for discussion of plasticity in metallic crystalline materials.

7.1.2 Miller and Miller-Bravais indices

At this stage of the book, it is critical to recall the basic discussion of Miller indices covered during the introductory materials science and engineering course. As we consider dislocations, basic crystallography is important and will be needed to define the slip system and related plasticity, definition of *close-packed planes and close-packed direction*, and relationship among those. The Miller-Bravais indices are applicable to hexagonal crystals and use four digits.

Let us start with a cubic system. The Miller indices convention for a set of directions is <uvw> and a set of planes is {hkl}. Fig. 7.3 shows BCC and FCC crystals to illustrate a couple of examples

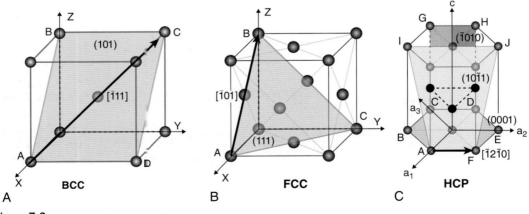

Figure 7.3

Examples of close-packed planes and close-packed directions in (a) BCC, (b) FCC, and (c) HCP crystals.

of slip systems. Consider the BCC crystal. The family of close-packed directions is represented by <111>, which includes all combinations of [111], [$\bar{1}$11], [1$\bar{1}$1], and [11$\bar{1}$]. Note that the reciprocal directions are not considered, i.e., [111] is considered equivalent to [$\bar{1}\bar{1}\bar{1}$]. Similarly, the family of close-packed planes is represented by {110}, which includes (110), (101), (011), (1$\bar{1}$0), ($\bar{1}$01), and (0$\bar{1}$1). If a plane passes through origin, shift it by one unit step while keeping it parallel to the original plane. Indices of all the parallel planes and parallel directions are same. For the <111>{110} slip system, it gives us four directions and six planes. However, note that only two of these directions lie on a particular plane. To check which direction lies on a particular plane, we have to determine if the dot product is equal to 0. The dot product of [uvw]·(hkl) is equal to $(u \cdot h) + (v \cdot k) + (w \cdot l)$. So, only [$\bar{1}$11] and [1$\bar{1}$1] lie on the (110) plane. It is important to check this criterion when writing specific slip system. This results in six non-parallel planes and two directions per plane, a total of 12 slip systems.

Fig. 7.3(c) shows the HCP crystal with the Miller-Bravais planes and direction. The Miller-Bravais system uses four digits in place of three. The direction family becomes <uvtw> and the plane family becomes {hkil}. Key thing to check when changing from the three-digit system to four-digit system is the rules $t = -(u + v)$ and $i = -(h + k)$. Note that the close-packed direction in HCP system is <11$\bar{2}$0>. AF in Fig. 7.3(c) denotes one close-packed direction. Note that the three example planes drawn in this figure share this direction. The ABCDEF plane is referred to as the basal plane. There are two non-basal planes drawn in this example. The CGHD is the prismatic plane and the AIJF is a pyramidal plane. There is one unique basal plane, while there are three prismatic planes and six different pyramidal planes. The slip activities in HCP crystal depend on the c/a ratio of the crystal. For example, the c/a ratios of some common elements are Cd – 1.886, Zn – 1.856, Mg – 1.624, alpha-Zr – 1.590, Ti – 1.588, and Be – 1.586. The ideal c/a ratio is 1.633. If one only considers the basal slip plane, then there are three slip systems. A minimum of five independent slip systems are required to accommodate discontinuities during deformation of a polycrystal. Note that the basal slip systems do not provide adequate number of slip systems. This means that the ductility of an HCP alloy will depend on activation of non-slip systems including twins, which is discussed later.

It is also important to consider the spacing between the planes and the angles between the planes. For the cubic system, the spacing between the planes in a cubic lattice of size "a" is given by

$$d_{hkl} = \frac{a}{\sqrt{h^2 + k^2 + l^2}} \tag{7.6}$$

The angle between two directions $[h_1k_1l_1]$ and $[h_2k_2l_2]$ in a cubic system is given by

$$\cos\theta = \frac{h_1h_2 + k_1k_2 + l_1l_2}{\sqrt{h_1^2 + k_1^2 + l_1^2} + \sqrt{h_2^2 + k_2^2 + l_2^2}} \tag{7.7}$$

Using Eq. (7.7), the angle between (111) and (1$\bar{1}\bar{1}$) is

$$\cos\theta = \frac{-1+1+1}{\sqrt{1+1+1} + \sqrt{1+1+1}} = \frac{1}{3}.$$

This gives the value of θ as 70.53 degrees, which also means that the angle (111) and (1$\bar{1}\bar{1}$) is 109.47 degrees. The four {111} planes in the cubic system form a tetrahedron. Keep this in mind when we discuss dislocation-dislocation interaction later in this chapter.

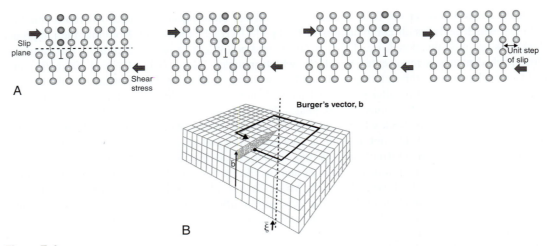

Figure 7.4
Schematic of (a) edge and (b) screw dislocations. Note that the closure vector is the unit step required to take atoms from one lattice position to the next in the direction of motion and this is referred to as the Burgers vector.

7.2 Types of dislocations

There are two basic dislocations that you were introduced to in any introductory materials science and engineering course: (a) edge dislocation and (b) screw dislocation. Students are encouraged to refer to a book by Hull and Bacon[1] for information beyond this chapter. Fig. 7.4 shows the schematic of these two types of dislocations. It is important to visualize that the shear of the crystal using one of these dislocation motions leads to shear strain. Try to follow the displacements of atoms around the dislocation core. In a pristine defect-free crystal, all the atoms will be at the ideal position and there will not be any distortion in the lattice. However, the atoms around a dislocation core are shifted from their ideal positions. This is referred to as the strain field in the lattice associated with the presence of a dislocation. The concept of this strain field or atomic displacement associated with the dislocations is important and is used for imaging of dislocations. As mentioned earlier, the Burgers vector (b) defines the magnitude and direction of displacement associated with the dislocation. Its magnitude can be determined as a closure step. The symbol ξ is used to mark the direction of a dislocation. For pure edge dislocation, the dot product $\xi \times b = 0$. However, ξ is parallel to b for the screw dislocation, so $\xi \cdot b = 0$. It is important to note that screw dislocations can slip on more than one plane, i.e., cross-slip onto a separate close-packed plane. This provides additional freedom for screw dislocations to overcome obstacles. The geometrical characteristics of dislocations are summarized in Table 7.1.

In a real crystal, the dislocations are rarely pure edge or pure screw. Most of the dislocations are curved and therefore have a mixed character. Fig. 7.5 shows schematic of a mixed dislocation changing

---

[1] Hull D, Bacon DJ. *Introduction to Dislocations*. Butterworth-Heinemann; 2011.

7.2 Types of dislocations

Table 7.1 Key properties of dislocations

Properties	Type of dislocation	
	Edge	Screw
Orientation of Burgers vector	Perpendicular to dislocation	Parallel to dislocation
Slip direction	Parallel to Burgers vector	Parallel to Burgers vector

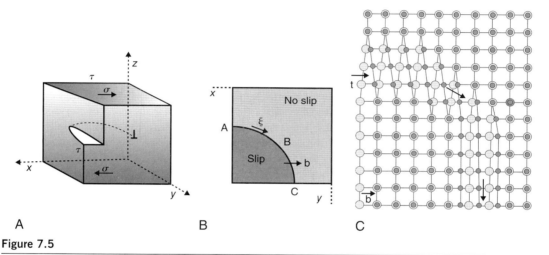

A B C

Figure 7.5

(a) and (b) Schematic of mixed dislocation showing change of dislocation character from screw to edge along the dislocation line. (c) Atomic configuration involved with a mixed dislocation.

its character from screw to edge. The atomic displacement associated with such a dislocation core is also shown to help visualization. As you go through this course and coming chapters you will see reference to dislocation loop. As you consider the geometrical characteristics of the dislocation, you will understand that the nature of dislocation changes along the dislocation loop. As we will discuss, the dislocation line energy changes with the character of dislocation, as does the mobility. So, when we think of the large-scale deformation of crystals, it is important to consider the types of dislocations that are created and contribute to the overall strain.

At this stage, consider the type of common sources used for imaging or characterization of materials: (a) light – used in optical microscopy; (b) x-rays – used for diffraction studies of crystals; and (c) electrons – used in scanning and transmission electron microscopy. Both x-rays and electrons are diffracted by atomic positions and that leads to shift in the corresponding diffraction peaks. When a dislocation is imaged in the transmission electron microscope, the contrast in the image (presence of lines) comes from diffraction of electrons from the strain field associated with the dislocation core. Any time atoms are displaced from their ideal position, their energy changes. The combined effect of all atoms displaced around the core constitutes the dislocation core energy. Integrating this elastic energy associated with all the atoms around the core would then give the dislocation line energy per unit length.

The displacement is uniform around the core of a screw dislocation, whereas the stresses above the glide plane of an edge dislocation are compressive and below the glide plane is tensile. The dislocation line energy per unit length (E/L) is given by (Hull and Bacon)

$$\text{Screw dislocation} \quad -\frac{E}{L} = \frac{Gb^2}{4\pi}\ln\left(\frac{R}{r_o}\right). \tag{7.7a}$$

$$\text{Edge dislocation} \quad -\frac{E}{L} = \frac{Gb^2}{4\pi(1-\nu)}\ln\left(\frac{R}{r_o}\right). \tag{7.7b}$$

where R is the cutoff radius for the strain field, r_o is the core radius of the dislocation, and ν is Poisson's ratio. It is generally difficult to define R and r_o and various textbooks give different numbers. For our purposes and using this in the overall context of discussion of strengthening mechanisms and plasticity, a better approach would be to define the range of values for R and r_o and then consider the impact of such choices. Starting with r_o, we are defining the core of the dislocation where the atomic displacement would be highest or distortion in atomic arrangement would be largest. The minimum value of r_o can be b. In many textbooks, the value of r_o is taken as $5b$. So, the range for r_o is b–$5b$. Switching to R, as mentioned earlier, it denotes the cutoff radius for the strain field. The values for this in various publications can range from $50b$ to the size of the single crystal containing that dislocation. We are interested in defining the physical basis for R.

At what distance from the dislocation core does the elastic strain or stress field associated with the dislocation go to zero? It is unreasonable to use the size of the single crystal as we need to consider the **dislocation density** in a crystal to create a physical basis for this understanding. Even in a well-grown single crystal of large dimension, the experimentally determined dislocation density, ρ, is 10^{10} m^{-2}. Note that the unit of dislocation density comes from line length per unit volume, m/m³. Such dislocation density results from crystal growth accidents. The spacing between dislocations is given by $\frac{1}{\sqrt{\rho}}$. For this single crystal, the dislocation spacing would then be 10^{-5} m, i.e., 10 µm. This then would be the upper limit for the value of R for grains that are larger than this. In thermomechanically processed metals and alloys, the value of dislocation density in annealed condition is considered to be 10^{12} m^{-2}. That would give the spacing between dislocations as 1 µm. The upper limit of dislocation density in a highly deformed crystal is considered 10^{17} m^{-2}, meaning 3.2×10^{-9} m. A typical value of the Burgers vector is taken as 2.5×10^{-10} m. So, the value of dislocation spacing would go down to ~$13b$ in highly deformed crystal. The value of R for such a deformed material cannot be then greater than ~$7b$. This is quite a small number and the dislocations will enter different configurations to lower the overall energy, i.e., formation of dislocation network. With such high dislocation density, you can visualize that each dislocation core strain field starts interacting with the core strain field of neighbouring dislocations. Getting back to the value of R, a reasonable number is $50b$. Putting this value of R and r_o as b in Eq. (7.7a), we get the E/L for screw dislocation as ~$\frac{Gb^2}{\pi}$. The Poisson's ratio is approximately 0.33 for a number of metals. Substituting this value results in the dislocation line energy per unit length for edge dislocation as ~$\frac{Gb^2}{2}$. This is the generalized dislocation line energy that we will consider throughout the book; and this is also easy to remember as a generic value.

7.3 Dislocation movement

We have covered the range of dislocation density in a crystal and basic geometrical properties of dislocations along with the relationship with slip system. We need to get some basic understanding of dislocation movement and how it contributes to strain and strain hardening. To begin with, two fundamental relationships are referred to as Orowan equations and relate movement of dislocation to shear strain (γ) and shear strain rate ($\dot{\gamma}$). These are:

Orowan Equations:
$$\dot{\gamma} = \rho_m b v \tag{7.8a}$$
$$(\gamma) = \rho_m b v \tag{7.8b}$$

where x is the average distance traveled by mobile dislocations (ρ_m) and v is the average velocity of mobile dislocations. Generally two keywords are used to refer to mobile (glissile) and immobile (sessile) dislocations. What makes a dislocation glissile? For a dislocation to be glissile, the dislocation line direction and Burgers vector must lie on the slip system. Fig. 7.6 shows a couple of examples of dislocation-dislocation interaction. In both cases, one stationary dislocation interacts with a moving edge dislocation, and in the process it creates a step in the stationary dislocation. In Fig. 7.6(a), the

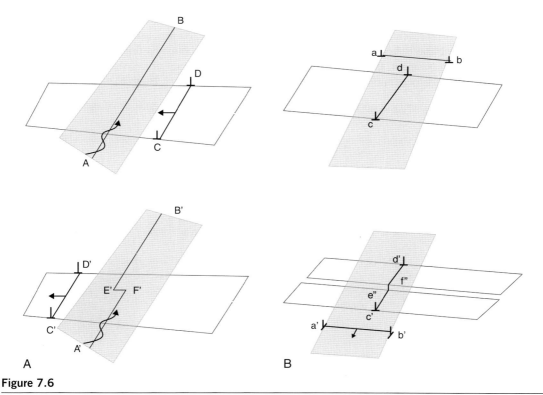

Figure 7.6

Illustration of dislocation-dislocation interactions. (a) An edge dislocation creates a kink E'F' in the screw dislocation segment. (b) An edge dislocation creates an out-of-plane jog e"f" on a segment of edge dislocation.

edge dislocation represents the extra half plane that glides on the slip plane and shifts the top half of the slip plane containing the screw dislocation. This creates a segment E'F' that lies in the same plane as the screw dislocation. This segment is called a "kink" as it lies in the same slip plane, and it has edge character as the Burgers vector is perpendicular. Because it lies in the same plane, the kink can glide. Now consider the edge dislocation "ab" interacting with the edge dislocation "cd" in Fig. 7.6(b). As this dislocation glides through, it creates a step in the c'd' dislocation that is out-of-plane. This segment is referred to as "jog" and does not lie in the same slip plane. Such segments become immobile and become a pinning point for the dislocation c'd'. Think back to the dislocation density increase due to plastic deformation. As the dislocations increase during deformation, they interact with each other as they glide on different planes. These interactions create a number of different kinks and jogs that alter the mobility of dislocations. Burgers vector of interacting dislocations defines the resultant Burger's vector of the kink or jog that forms from such interaction.

7.4 Dislocation reactions

For this section, we will consider the FCC lattice to illustrate the splitting of dislocations into partials and impact of such movements. Fig. 7.7 shows the FCC lattice with coloring of atoms in subsequent (111) planes. The stacking of these (111) planes in A, B, and C layers leads to ABCABCABCABC sequence which represents the perfect crystal. Note the difference between B and C positions. The HCP crystal is represented by ABABABABABAB sequence. Now consider the slip of B plane on the top of A plane. If the B atoms go to B position, that is displacement by a full Burger vector b. When you visualize this movement, think of the fact that all the (111) planes above that glide plane experience the same displacement and the atomic layer sequence is preserved. Burgers vector for FCC crystal belongs to $\frac{a}{2}[110]$ family, where "a" is the lattice parameter. The magnitude of the Burgers vector is calculated using

$$|b| = \frac{a}{2}\sqrt{u^2 + v^2 + w^2} \qquad (7.9)$$

Note that the distance between B and C positions or C and A positions is a lot closer than the full Burger's vector. However, if C atom goes to A position, then the sequence of planes will change. For example, the first four planes of the {ABCA}BCABC sequence will go to {ABAB}CABC sequence,

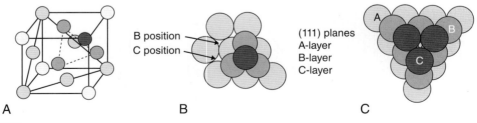

Figure 7.7

Illustration of (111) planes in FCC lattice and definition of various atomic positions used for stacking sequence.

7.4 Dislocation reactions

in which the first four stacking layers are of an HCP sequence. In other words, the movement of C atom to A position created a "stacking fault." This vector being smaller is called a partial dislocation. It is represented by the Burgers vector family of $\frac{a}{6}$ [112] and referred to as Shockley partials. Recall that the dislocation energy was proportional to Gb^2. Such dissociation of dislocations into partials will only occur if the reaction is energetically favorable. The dissociation of a full dislocation into two partials can be written as

$$b_1 \to b_2 + b_3 \qquad (7.10)$$

Frank's rule states that for the dissociation to be favorable,

$$b_1^2 > b_2^2 + b_3^2 \qquad (7.11)$$

An example of a dissociation reaction is

$$\frac{a}{2}<110> \to \frac{a}{6}<211> + \frac{a}{6}<12\bar{1}> \qquad (7.12)$$

To evaluate the energy change using Frank's rule, we need to calculate the magnitude of the vector. Also, we need to learn how to write and check the dissociation reaction. The magnitude of $\frac{a}{2}<uvw>$ is $\frac{a}{2}\sqrt{(u^2 + v^2 + w^2)}$. So, checking for Frank's rule, the dissociation reaction in Eq. (7.12) can be expressed in the form of Eq. (7.11),

$$\left(\frac{a\sqrt{1^2+1^2+0^2}}{2}\right)^2 \to \left(\frac{a\sqrt{2^2+1^2+1^2}}{6}\right)^2 + \left(\frac{a\sqrt{1^2+2^2+\bar{1}^2}}{6}\right)^2$$

which after simplification reduces to $\frac{a^2}{2} \to \frac{a^2}{3}$. Clearly, this dissociation reaction is favorable as it results in lower energy. While figuring out the dissociation reaction, it is important to check that the vectors add up (it is a simple algebraic sum) and that the dissociated vectors lie on the same glide plane. The visualization is helped by the use of Thompson tetrahedra as shown in Fig. 7.8. Take the (111) plane marked by the triangle ABC. This is one of four independent {111} planes. Each of these planes has three <110> directions, making the 12 possible independent slip systems. The directions in this figure are marked with [uvw> to aid the visualization. Take the <0$\bar{1}$1] perfect dislocation. It belongs to two planes (you can check by doing dot product of [uvw] and (hkl); it should be zero). Considering the (111) plane, it will dissociate into [$\bar{1}$12] and [$\bar{1}$2$\bar{1}$]. Note the change in the sign of the second vector as we are going from C to A (CA → Cδ + δA). Again, the dot products of the partials and the (111) plane are zero.

An important implication of dissociation of dislocations is formation of **stacking faults**. Stacking faults are the regions bounded by the partials and locally resemble h.c.p. structure in the overall f.c.c. crystal. The reduction in energy of a set of dislocations from dissociation (e.g. consider Eq. (7.12)) goes toward the stacking fault. This then dictates the distance between the partials, d, which is given by Hull and Bacon.[2]

$$d = \frac{Gb^2}{4\pi\gamma_{SF}} \qquad (7.13)$$

---

[2]Hull D, Bacon DJ. *Introduction to Dislocations*. Butterworth-Heinemann; 2011.

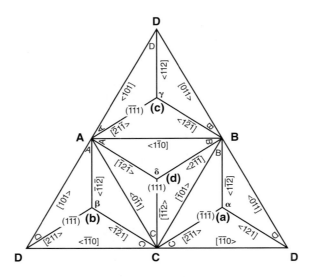

Figure 7.8

Schematic of Thompson tetrahedra showing the dissociation of $\frac{1}{2}\langle 110\rangle$ perfect dislocations into Shockley partials. It is important to ensure that the dissociated dislocations lie on the same glide plane. Students are encouraged to make a copy of this figure and fold the cutout to visualize the four {111} planes and three <110> directions that make up the 12 possible slip systems for the f.c.c. lattice.

where γ_{SF} is the stacking fault energy (SFE). SFE is a very important aspect of alloy design. Not only does the overall behavior of dislocation dynamics depend on this, but there are important derivative aspects that emerge from dissociated dislocations. Two deformation-induced phenomena are "deformation-induced twinning" and "deformation-induced transformation." These are more commonly referred to as twinning-induced plasticity (TWIP) and transformation-induced plasticity (TRIP). These are very significant for steels and many new alloys are emerging under the category of "complex concentrated alloys" (CCAs). The naming of TWIP and TRIP is based on the fact that formation of deformation twins or phase transformation during plastic deformation enhances the strain hardening and leads to enhanced overall ductility. Note that deformation twins are different from annealing twins. While annealing twins also create obstacle for dislocation motion, we are more interested in the deformation twins. The density of which evolves during deformation and spacing between them decreases, resulting in enhanced strain hardening. Another way to think about these is that deformation leads to formation of twins and transformation of crystal structure. Although the details are out of scope for the current chapter, it is important to understand that most theories regarding TWIP and TRIP consider stacking faults as the origin. The key to observation of these phenomena lies in the range of SFE values, and therefore students should understand the general range of SFE values and effect of alloying on SFE values. Let us consider common FCC elements and alloys. Aluminum is considered a high SFE element (160–200 mJ m^{-2}). On the other hand, pure copper has SFE of 70–80 mJ m^{-2} and addition of Zn or Al can lower it to 3–10 mJ m^{-2}. Stainless steel has SFE values of 30–40 mJ m^{-2}. In general, if the SFE is lower than ~40 mJ m^{-2}, the alloy can exhibit twinning and steels or CCAs can exhibit transformation from γ-phase (FCC) to ε-phase (HCP) when SFE is lower than ~20 mJ m^{-2}.

Concept visualization – Partial dislocations leading to twinning and transformation: To visualize the twinning and transformation because of the movement of partial dislocations, consider movement of partial dislocations on each successive plane and alternate planes as shown below. We start the process of visualization by writing the lattice plane sequence of ABCABCAB-CABCABC ..., which represents stacking of the closest packed (111) planes in FCC. Remember that the sequence of the closest packed (0001) basal planes in HCP is ABABABABABABABAB .... Movement of the partial dislocations on the successive planes results in formation of twin planes. On the other hand, if the partials move on alternate planes, the FCC structure on the left side in (b) transforms to HCP structure on the right side in (b).

7.5 Dislocation generation

To sustain plastic deformation, dislocations are generated; they move through the crystal or grain and get annihilated. Therefore, the generation of dislocation is an important step. In section **7.2**, we considered that a well-annealed metal or alloy can have a starting dislocation density of 10^{12} m^{-2}. On application of external stress, the pre-existing dislocations can move and start the deformation process but these are not adequate to sustain large-scale plasticity. In **Chapter 4**, we looked at a few examples of microstructures in common alloys. As we discuss a limited number of mechanisms of dislocation generation, it is important to keep in mind those microstructural examples to understand the context.

7.5.1 Intragranular dislocation sources

The first consideration in the generation of new dislocations is the location of such dislocation sources. A classic dislocation generation mechanism is the Frank-Read source depicted in Fig. 7.9. The externally applied stress leads to a force (τb) on the dislocation segment DD'. Under this applied force, the dislocation segment forms a loop, and after pinching, one part comes back to the original segment L. Throughout the process, the dislocation line tension (T) is balanced by the applied force. This force balance can be written as

$$\tau b L = 2T \sin\theta \quad (7.14)$$

where θ is the angle of dislocation segment at the original node with respect to the original line direction. The applied stress needs to be increased till $\theta = 90°$ when the shear stress τ_{max} is reached,

140 Chapter 7 General dislocation theory for crystalline materials

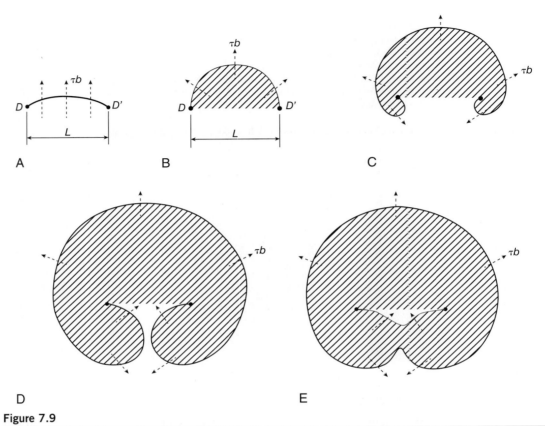

Figure 7.9

Schematic illustration of operation of a Frank-Read source.

and after that the configuration becomes unstable and drives itself through rest of the loop formation. So, the stress required to operate a Frank-Read source with dislocation segment of length L is

$$\tau_{max}^{F-R} = \frac{2T}{bL} = \frac{Gb}{L} \tag{7.15}$$

Concept drill – Applicability of Frank-Read source in actual engineering materials: In **Chapter 4** we looked at microstructures in engineering materials. As we learn basic concepts of dislocations, it is good to check the context under which they can be applied. This can force us to think of situations where we need to modify the basic concepts or develop new concepts to explain the experimental observations.

The simple discussion of an intragranular Frank-Read source and the stress required to operate is good to raise several conceptual questions. Let us think of the following:
- If we consider the F-R dislocation loop size to be 3L, then the grain size for operating such a mechanism at the minimum must be 3L! For large grained materials, it is easy to visualize operation of such a mechanism. However, most of the high-strength alloys have fine grain size; review Figs. 4.1–4.6 in Chapter 4. Students can create a plot of τ_{max}^{F-R} vs. grain size using Eq. (7.15) by taking L as 1/3 of the grain size. As researchers work on ultrafine grained materials (grain size <1 μm) and nanocrystalline materials (grain size <100 nm), this discussion becomes particularly relevant.

- Think of the character of dislocation segments through the process. If the starting segment DD' is an edge dislocation, then the Burgers vector is perpendicular to this segment. As the loop forms, different parts of the dislocation loop will have character of edge to screw.
- How would this process work for a dissociated dislocation in a low SFE alloy? Will the maximum stress still be at the point of $\theta = 90°$ or at the point of pinching (a step between Figs. 7.9(d) and 7.9(e)) where the dissociated dislocation will need to constrict? This type of thought experiments can help in visualization of the dislocation process in a microstructure that is more relevant to you.
- What if the material has lath microstructure or lamellar microstructure? The spacing between interfaces for a given slip plane depends on the orientation. In most cases, the spacing will be too small for F-R type source.

Answers to some of these thoughts or questions can lead us to look for alternative intragranular mechanisms. Cross-slip of dislocations and interaction with other dislocations can create nodes that are immobile (sessile) and this can lead to segments that can operate in Frank-Read mode or as a spiral mechanism (much like a sprinkler head). For many cases, we will conclude that intragranular dislocation sources are not viable and we will need to switch to sources at the interfaces (a generic term covering both grain boundaries and boundaries between phases as in pearlitic microstructure in steels).

7.5.2 Interfacial dislocation sources

The simplest of interfacial dislocation sources are from grain boundaries. Li[3] proposed generation of edge dislocations from grain boundary ledges as an alternative to Frank-Read source. This was an attempt to provide alternative explanation for grain boundary strengthening (Hall-Petch relationship) that you will learn in **Chapter 8**.

As we go over the basics in this chapter, the central idea is to recap some of the fundamentals that would be needed for Chapters 8-11 which are based on plastic deformation. The grain boundaries and other interfaces are very important in static and dynamic plastic deformation. The historical shift in the description of grain boundaries happened in the 1960s, when the understanding of the structure of grain boundaries started emerging. Before this period, grain boundaries were considered a region without any crystalline structure. So, first thing that you should grasp is that the interfaces have structure. While the details are beyond the scope of this chapter and book, it is important to learn that interfaces can be classified in terms of their intrinsic structure. The structure of the grain boundaries depends on the misorientation between the two crystals. The overall misorientation can be split in terms of "tilt" and "twist." The crystal structure of the individual grains extends right up to the interface. Based on the numbers of atoms at the interface that can belong to either lattice (common sites), the interface structure can be described using these "coincidence site lattice" (CSL). The energy of a grain boundary depends on this misorientation, and at certain particular tilt angle, twist angle, or combination of tilt and twist angles, the CSL values are unique – referred to as special S boundaries. The S boundaries have low energy configuration because of high lattice registry across the interface. In addition to the lattice registry, other geometrical features like ledges help in the dislocation nucleation process.

Consider the nucleation of an edge dislocation from the grain boundary. As we have seen before, the edge dislocation is an extra half-plane. So, if a dislocation is nucleated from a very plane grain boundary, it leaves a step in that grain boundary plane. So, the energy associated with grain boundary surface energy increases in addition to the dislocation line energy. If the grain boundary already contains a ledge, it is geometrically simpler to just nucleate the dislocation from there. Another thing to consider is the movement of this newly dislocated dislocation. If a segment has been nucleated, the segment can move across the grain by bulging out, keeping the ends pinned at the nucleating site. In

---

[3] Li JCM. Petch relation and grain boundary sources. *Trans Metall Soc AIME*. 1963;227:239-247.

this case, the radius of curvature keeps changing as the dislocation moves. Alternatively, the nodes can move by keeping the original radius of curvature. The complication in this is that the node movement requires local rearrangement at the interface boundary.

Considering that most real microstructures of engineering materials contain fine grain size, the generation of dislocations from grain boundaries is very important to consider. It is also necessary to keep the details of micromechanisms involved in the context of the microstructural detail. At this stage, the students should start connecting their thoughts on mechanisms of deformation with the microstructure of specific material. As you cover **Chapters 8–11**, keep this thought process going.

7.6 Geometrically necessary dislocations – how do they differ from statistical dislocations?

Before closing this chapter, let us consider an important geometrical aspect of deformation. Ashby (1970) proposed the distinction between "geometrically necessary" dislocations (GNDs) from the "statistically stored" dislocations which accumulate in pure crystals during straining.[4] Note that Eq. (7.8) referred to total dislocation density.

Even the single-phase polycrystals of pure metals are plastically non-homogeneous. Let us look at this concept and then extend it to multiphase material discussion. Consider a group of randomly oriented grains in a polycrystalline material. During uniaxial loading of this group of grains, as in the tensile test, each grain starts to first deform elastically. In **Chapter 6** we noted that the elastic constants are crystal direction dependent. So, each grain elastically deforms by different amount in any particular direction. This creates strain mismatch among the grains in this group. During the elastic loading the magnitude of strain is small. Whether the mismatch is large enough to require nucleation of dislocation obviously depends on the magnitude. A simple way to think about this is to calculate the elastic strain mismatch and see if the magnitude is near the Burgers vector. In that case nucleation of a dislocation would reduce the local strain.

As the deformation proceeds to the plastic stage, different slip systems activate in individual grains. Note that a grain will undergo shape change directly as a result of activation of a set of parallel slip planes with lowest critical resolved shear stress (CRSS). Because we assumed that each grain is oriented randomly, the active slip system orientation will also differ. If we take a FCC crystal structure metal, the active slip system is {111}<110>. Activation of slip on {111} planes will result in elongation of grains in different directions, although the tensile axis for all of them is in one direction. This leads to inhomogeneity in shape change and incompatibility among neighboring grains. To accommodate this geometrical incompatibility and maintain continuity with neighbors, GNDs have to nucleate from the interfacial regions. It is easy to visualize that such a deformation results in inhomogeneous deformation. So, while we mathematically expressed a uniform dislocation density in Eq. (7.8) to describe overall strain and strain rate, microscopic level inhomogeneity evolves as the deformation proceeds. How this inhomogeneity is accommodated dictates the continuation of plasticity and onset of failure processes.

Engineering microstructures often contain multiple phases of various length scales. So, this discussion of inhomogeneous plasticity in multiple grains can now be extended to a generic discussion of multiphase materials. Let us consider a material with two phases, as in pearlitic microstructure. One phase is ferrite (α-Fe) and the other is cementite (Fe_3C). Ferrite is relatively weaker phase and will deform more readily. However, both phases are geometrically constrained in terms of maintaining continuity. If for this discussion

---

[4]Ashby MF. The deformation of plastically non-homogeneous materials. *Philos Mag.* 1970;21(170):399-424.

we assume that cementite is non-deformable, it is easy to visualize that the ferrite phase will deform to accommodate the overall strain imposed on the material. Within the ferrite phase, there will be strain gradient, varying from near zero at the ferrite/cementite interface to a larger number inside the ferrite phase. These local variations in the dislocation density also lead to development of internal stresses or back stress that can oppose the deformation, a form of work hardening. If we generalize this discussion to any two-phase material, then the dependence of the overall process on the length scale of non-deforming particles is quite obvious. Conceptually, very small particles will not cause significant strain mismatch, whereas larger particles will result in quite significant mismatch. This has very important implications for deformation and failure, some intended and some unintended. Intended use of this concept includes design of microstructures to enhance work hardening or extend overall plasticity. An interesting example of "unintended" consequences includes the influence of constituent particles. For example, aluminum alloys contain Fe as an impurity element. The way composition of engineering alloys is specified, the intended alloying elements are given a range and impurity elements are mentioned as maximum allowed value. Fe in aluminum alloys forms constituent particles during solidification and typical size range of these particles is 20–100 μm. These particles are significantly harder than the matrix and non-deformable. They lead to reduction in fracture toughness and fatigue life of high-performance aluminum alloys. As we go through the rest of the book, we need to understand not only the intended deformation of materials but also unintended consequences. Dislocations are the agents of deformation and cause the final failure of materials. Understanding the way we can apply fundamentals to real engineering microstructures can allow us to address questions like: ***Why do fail-safe designed components fail?*** Was the design flawed, or did the unintended microstructural feature lead to failure? If an inclusion gets trapped in an alloy during casting and processing, what is its impact?

7.7 Key chapter takeaways

We learned that at a very fundamental level, we just need to consider three steps: (a) generation of dislocations, (b) movement of dislocations, and (c) annihilation of dislocations. However, the specifics of these get complicated quite fast when we consider "real" microstructural features in engineering materials. First, we acknowledged that even a very well-annealed material starts with significant dislocation density. In that case, the generation of additional dislocations will occur after the preexisting dislocations begin to move. As we generate more dislocations and move them, dislocation-dislocation interactions lead to mobile and immobile dislocation segments. That results in work hardening, which will be discussed further in the next chapter. We also learned that alloys with lower SFE have dissociated dislocations or partial dislocations. Movement of such dislocations in a particular manner can lead to formation of deformation-induced twins and transformation. As we will see more in the next chapters, activation of deformation-induced twinning and transformation can have profound influence on the structural performance of alloys.

7.8 Exercises

1. What is the typical range of values for ratio of theoretical strength to experimental strength?
2. What are the favored slip systems in FCC, BCC, and HCP crystals?
3. What are the values of the ideal c/a ratio for a HCP metal?

Chapter 7 General dislocation theory for crystalline materials

4. Using Thomson tetrahedra, show an example of dislocation dissociation to partial dislocations.

5. How to check if a dislocation line lies in a particular plane?

6. Fill the following (perpendicular or parallel):

   (a) The Burgers vector of a dislocation is _____ to the dislocation line for screw dislocation.

   (b) The Burgers vector of a dislocation is _____ to the dislocation line for edge dislocation.

7. The stacking sequence for (111) planes in FCC crystals can be represented as ABCABCABC. What happens to this sequence, when you have a

   (a) Twin? _____

   (b) Stacking fault? _____

8. What is the range of dislocation density of a well-annealed FCC metal?

9. (a) Define a slip system. (b) Give the conventional slip systems of FCC, BCC, and HCP metals. Show at least one slip system in respective unit cells with clear sketches and identify the Miller-Bravais indices of the corresponding slip plane and direction.

10. (a) Discuss the role of twinning in plastic deformation. (b) Compare mechanical twins and annealing twins. (c) Both slip and twinning are the mechanisms by which plastic deformation occurs in crystals. However, they are very different. Compare the two processes in a tabular form.

11. A group of edge dislocations present in a bulk specimen of mass "M" has a total length of "L." The physical density of the material is ρ. All the dislocations are present in parallel slip planes and they have Burgers vector of b. However, it is noted that only 50% of the dislocations are mobile in nature and the rest are sessile.

    (a) What would be the shear strain rate if the mobile dislocations move at an average velocity of "V"?

    (b) To create a shear strain of "γ," how much time should elapse from the start of the dislocation movement? (Hint: Consider using Orowan's equation.)

12. Consider a single crystal of copper (FCC) of 8 mm diameter. The lattice constant of copper at room temperature is 0.3615 nm. Assume that movement of perfect dislocations through the crystal produces plastic strain. The mobile dislocation density is 10^8 m^{-2}. What is the shear strain rate produced by dislocations which move at a velocity of 240 m s^{-1}?

13. We have learned how to calculate the magnitude of Burgers vector of a dislocation from its lattice parameter and slip direction in a cubic crystal. What would be the Burgers vector of a perfect dislocation in a simple cubic metal like alpha-polonium (α-Po) with a lattice parameter of 0.335 nm?

14. Why is cross-slip easier in aluminum (FCC) but difficult in austenitic stainless steel (FCC)?

15. (a) State Frank's rule. (b) Show that the Lomer-Cottrell reaction follows Frank's rule.

7.8 Exercises

16. Which of the following is a valid slip system(s) for an aluminum single crystal under room temperature plastic deformation? Explain your answers properly.

   (a) (111), [110]

   (b) (11$\bar{1}$), [00$\bar{1}$]

   (c) (101), [11$\bar{1}$]

   (d) (100), [0$\bar{1}$1]

17. A symmetrical tilt boundary was formed due to recovery process in aluminum (lattice constant: 0.405 nm). The mean dislocation spacing was found to be 3 nm via a TEM study. (a) If perfect dislocations are assumed to be forming the tilt boundary, what is the angle (in degree) of the boundary? Show your calculation. (b) At what distance do you think the dislocation model in aluminum will break down? Estimate the transition angle when the low angle boundary will become high angle boundary.

18. The total dislocation density of an iron crystal is 10^{13} m/m^3.

   (a) Calculate the total amount of energy per m^3 of the crystal due to the presence of those dislocations. Assume the geometrical factor (α) is 1.

   (b) What would the rise in temperature be if only 6% of the strain energy of the dislocations calculated in part (a) could be released as heat?

   Given: Lattice constant of alpha-Fe – 0.286 nm; crystal structure – FCC; density – 7.2 g/cm^3; atomic weight – 55.85 g/mol; specific heat – 0.415 cal/g.°C; elastic modulus (E) – 200 GPa; and Poisson's ratio – 0.3.

19. Consider the following dislocation reactions and write the product dislocation Burgers vector for each reaction and also state whether the reactions below are energetically feasible (Frank's rule).

$$\frac{a}{2}[1\bar{1}0] + \frac{a}{2}[110] = ... \quad \text{(i)}$$

$$\frac{a}{2}[101] + \frac{a}{2}[01\bar{1}] = ... \quad \text{(ii)}$$

$$\frac{a}{2}[1\bar{1}0] - \frac{a}{2}[101] = ... \quad \text{(iii)}$$

20. Differentiate between (a) Frank partial and Shockley partial; (b) dislocation jog and dislocation kink.

21. Consider one of the dislocation dissociation reactions that leads to the creation of stacking fault in an FCC crystal:

$$\frac{a}{2}[110] = \frac{a}{6}[21\bar{1}] + \frac{a}{6}[121]$$

Assume that dislocation energy per unit length is given by Gb^2 (where G and b carry the usual meaning) irrespective of their type.

   (a) Derive the total decrease in energy per unit length of the perfect dislocation in terms of a (lattice constant) and G (shear modulus). Neglect stacking fault energy.

   (b) On which type of {111} plane must all the three dislocations lie? Explain.

22. Why is dislocation climb called non-conservative motion, whereas dislocation glide is considered conservative motion?

23. (a) Describe how a typical Frank-Read (F-R) source works.

   (b) Assume that you have two FCC crystals. Crystal-1 has twice the shear modulus of crystal-2 but the lattice constant of crystal-1 is half that of crystal-2. Which crystal would activate F-R source at a lower shear stress if a pre-existing perfect dislocation of the same length is the F-R source? If not lower, then what? State any assumptions involved.

   (c) If the grain diameter of a material is very small (in the extreme nanocrystalline range), the F-R sources may not be able to operate at all. Why? Think of a scenario of how that can be possible.

24. In a lithium fluoride crystal subjected to an applied stress of 10.8 MPa, the dislocation velocity at $-23.9°C$ is 6×10^{-3} cm/s and that at $-45.7°C$ the velocity is 10^{-6} cm/s. The data suggest an Arrhenius-type relationship between the dislocation velocity and temperature as given by the following equation:

$$v = A \exp\left(-\frac{Q}{RT}\right)$$

where v is the dislocation velocity, A is a constant of proportionality, Q is an effective activation energy in J/mol, T is the temperature in absolute scale, and R is the universal gas constant (8.314 J/mol K). Determine Q and A from the given data and above equation.

Qualify the following statements as "True" or "False." Give adequate reasons for supporting your choice.

25. Rigid body translation of a crystal would produce a theoretical shear strength of a material several orders of magnitude smaller than the experimentally observed one.   True   False

26. Slip systems in BCC metals are of {111}<100> type.   True   False

27. An edge dislocation cannot glide on its slip plane; it must climb.   True   False

28. Aluminum has a higher stacking fault energy (~200 mJ/m²) compared to that of copper (~80 mJ/m²). That is why cross-slip of screw dislocations in copper will be more difficult than in aluminum.   True   False

29. Single crystals of HCP metals like cadmium (Cd) exhibit extensive "laminar flow" regime during strain hardening.   True   False

30. If "n" number of dislocations gliding on a slip plane reaches the surface of a single crystal, the final slip step width will most possibly be $(2n + 1).b$ where b is the magnitude of the Burgers vector.   True   False

31. Dislocation strain energy is directly proportional to the magnitude of the Burgers vector (b).   True   False

32. The value of Poisson's ratio changes drastically with temperature.   True   False

33. Dislocation velocity is not dependent on applied stress.   True   False
34. Cold worked material contains very low dislocation density.   True   False
35. Geometrically necessary dislocations are created at the grain boundaries upon plastic deformation.   True   False
36. Stacking fault energy is inversely proportional to staking fault width and thus has important influence on the cross-slip process.   True   False

Further readings

Ashby MF. The deformation of plastically non-homogeneous materials. *Philos Mag.* 1970;21(170):399-424.
Hull D, Bacon DJ. *Introduction to Dislocations.* Butterworth-Heinemann; 2011.
Li JCM. Petch relation and grain boundary sources. *Trans Metall Soc AIME.* 1963;227:239-247.

CHAPTER 8

Yielding and work hardening: strength limiting design

Chapter outline

8.1 Strengthening mechanisms in engineering materials 150
    8.1.1 Strain hardening/work hardening 150
        8.1.1.1 Stress-strain curve of single crystals *150*
        8.1.1.2 Single crystal to polycrystal *152*
    8.1.2 Grain boundary strengthening 157
        8.1.2.1 Dislocation pileup model of grain boundary strengthening *158*
        8.1.2.2 Li's general model for Hall-Petch relationship *161*
    8.1.3 Solid solution strengthening 161
    8.1.4 Fine particle strengthening 167
        8.1.4.1 Particle cutting or shearing mechanism *167*
        8.1.4.2 Particle hardening via strong barriers *174*
        8.1.4.3 Transition from particle cutting to Orowan looping *179*
        8.1.4.4 Some further topics on fine particle strengthening *179*
    8.1.5 Superposition of strengthening mechanisms 187
8.2 Large-scale plasticity—dislocation generation, storage, and arrangement 188
8.3 Failure mechanisms 188
    8.3.1 Low-temperature tensile fracture 191
        8.3.1.1 Ductile fracture *191*
        8.3.1.2 Cleavage fracture *198*
8.4 Microstructural distribution and consequent effects 198
8.5 Effect of multiaxial loading on yielding 202
    8.5.1 von Mises criterion 203
    8.5.2 Tresca's yield criterion 204
    8.5.3 Yield locus 204
8.6 Principles and examples of strength limiting design 206
8.7 Key chapter takeaways 207
8.8 Exercises 208

Learning objectives

- Learn about various dislocation-based strengthening mechanisms.
- Discuss each strengthening mechanism in detail based on dislocation theory.
- Understand low-temperature fracture mechanisms and their relation to yielding and plastic deformation.

Chapter 6 addressed the elastic behavior of engineering materials, which is basically a measure of the interatomic forces at work between atoms. However, plasticity, which is manifested in the nonlinear part of the stress-strain curve following the (linear) elastic regime, is much influenced by the presence of crystal defects. Without the knowledge of crystal defects, specifically dislocations, understanding the essential nature of plasticity is destined to be futile. The important distinction between elasticity and plasticity must be remembered as we move forward. Hopefully the reader has already mastered the required knowledge of dislocations in the previous chapter (**Chapter 7**) on dislocation theory.

8.1 Strengthening mechanisms in engineering materials

Most engineering materials are by nature crystalline. We have already seen how dislocations greatly affect crystal plasticity. Strengthening or hardening of crystals can be achieved by impeding easy motion of dislocations. So, in simple terms strength of a crystal is inversely related to dislocation mobility. Dislocation movement can be obstructed by various factors that lead to the strengthening effect. Depending on the type of factors at play, strengthening mechanisms could be different and multiple strengthening mechanisms can be incorporated simultaneously. The fundamentals you learn in this chapter can allow you to design alloys and engineer their microstructure to obtain desired strengthening mechanisms.

8.1.1 Strain hardening/work hardening

> *"It is sometimes said that the turbulent flow of fluids is the most difficult remaining problem in classical physics. Not so. Work hardening is worse."*
>
> **Sir Alan H. Cottrell**[1]

Here we first introduce the topic of strain hardening for single crystals and then for polycrystals. Strain hardening is an important strengthening mechanism. Often materials that cannot be hardened by other means need to be strain hardened. Strain hardening increases the yield strength of materials by increasing dislocation density. However, a strain-hardened material exhibits decreased ductility. This tradeoff of strength and ductility is an important aspect of engineering materials, where balance of properties is critical. As we move through the remaining chapters in this book, the students should particularly pay attention to influence of microstructure on various properties.

8.1.1.1 Stress-strain curve of single crystals

Strain hardening occurs in single crystals in a unique way. This understanding is essential before wading into the discussion of strain hardening in polycrystals. A classic crystal structure to discuss strain hardening behavior of single crystals is face centered cubic (FCC). For an FCC crystal, a typical stress-strain curve shows three stages of deformation, as originally suggested by Seeger[2] and illustrated in Fig. 8.1. Note that the rate of strain hardening is given by the first derivative of the stress-strain curve, that is, $\theta = \dfrac{d\tau}{d\gamma}$, where τ is the shear stress and γ is the shear strain.

---

[1]Cottrell AH. Commentary. A brief view of work hardening. In: Nabarro FRN, Duesbery MS, eds. *Dislocations in Solids, Volume 11*. Elsevier; 2002.
[2]Seeger A. *Dislocations and Mechanical Properties of Crystals*. John Wiley & Sons; 1957.

8.1 Strengthening mechanisms in engineering materials

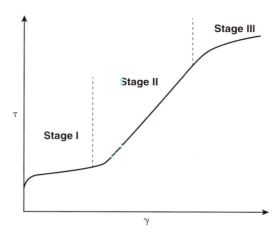

Figure 8.1
A generalized flow curve of a FCC single crystal.

The first stage, or Stage I, is called "easy glide" regime, where dislocations move on a single set of parallel slip planes. Here, dislocation stress fields interact with stress fields of the neighboring dislocations, but the obstacles provided by dislocation stress fields are weak. The slip system with the highest Schmid factor is activated, and many dislocations escape from the surface and produce surface slip steps. That is why the rate of strain hardening noted in the first stage is little or negligible ($\theta \approx G/10{,}000$; G is the shear modulus). Thus the slip process involved in such plastic deformation is sometimes termed as laminar flow.

As deformation progresses, plastic deformation enters Stage II, with the stress-strain curve attains a steeper slope ($\theta \approx G/300$) than in Stage I (note that the use of G/300 is not universal; some textbooks have it as G/200; the point is simply that Stage II slope is significantly greater than Stage I, by orders of magnitude). The strain hardening rate becomes more or less constant (linear hardening region). However, this stage is relatively independent of temperature and strain rate. Dislocations start gliding on multiple slip systems. Those dislocations encounter sessile dislocations, such as Lomer-Cottrell barriers (see **Chapter 7**). Given that this stage does not depend on temperature, dislocation pileup is the main reason for the enhanced strain hardening.

Due to extensive slip, dislocation density increases and is accompanied by increasing flow stress. In this region, flow stress follows the theory proposed by Taylor. Eq. 8.1 relates flow stress ($\Delta \tau$) to dislocation structure for this strengthening mechanism:

$$\Delta \tau = \alpha G b \sqrt{\rho_d} \qquad (8.1)$$

where α is a numerical constant (typically 0.3–0.6), G the shear modulus, b the Burgers vector, and ρ_d the dislocation density.

Stage III is characterized by decreasing strain hardening rate because of dynamic recovery. Thus this stage is also known as parabolic hardening stage. This decrease takes place because of the cross-slip of piled-up dislocations after attaining a high level of stress. Strain hardening that takes place at this stage is generally related to the intersection of dislocations and dislocation forests. This stage is temperature dependent.

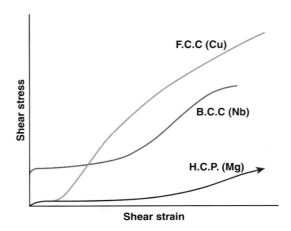

Figure 8.2

Schematic illustration of stress-strain curves of single crystals with three different crystal structures.[3]

Higher temperature generally shortens Stage I and Stage II as slip activity on secondary slip planes and cross-slip is enhanced. For FCC metals, a decrease in stacking fault energy (SFE) affects cross-slip activity, which increases stress required to initiate transition from Stage II to Stage III. For instance, the low SFE of alpha-brass with FCC lattice structure stretches Stage II to higher stress levels than would be expected for pure copper. Similarly, aluminum with a high SFE exhibits very limited Stage II while showing an extended Stage III region, as cross-slip takes place easily. Generally, the extent of Stage I or the easy glide region involves slip in a single set of slip systems which is affected by material purity, low temperature, the apparent absence of surface oxide films, and stress axis orientation for simple slip. Most hexagonal closed packed (HCP) metals have limited number of slip systems and exhibit more pronounced, easy-glide regimes than other metals. Fig. 8.2 illustrates the nature of stress-strain curves for Cu (FCC), Nb (BCC), and Mg (HCP).

8.1.1.2 Single crystal to polycrystal

Now consider what the strain hardening behavior for polycrystalline materials would look like. In fact, the situation is a bit more complicated because of multiple slip phenomena that ensue due to mutual interference of neighboring grains, all of which lead to considerable strain hardening. The behavior of polycrystalline material deviates from its single crystalline counterpart because of the presence of grain boundaries. In single crystals, Schmid's law analysis gives the resolved shear stress on a slip plane as:

$$\tau = \sigma \cos\theta \cos\phi = \sigma/M \tag{8.2}$$

where M is the reciprocal of Schmid factor and is treated as an orientation factor. θ is the angle between the loading direction and the glide direction, whereas ϕ is the angle between the loading direction and the vector normal to the glide plane. Recall that the slip system is combination of closest packed plane and the closest packed direction in this plane. The Schmid factor then defines the ease of glide based on orientation. For example, if the closed packed plane is oriented 45° to the loading direction, it is favorably oriented for slip as the maximum value of Schmid factor, 0.5, can be observed. This is also referred as "easy" slip orientation. On the other hand, when the closed

---

[3]Smallman RE, Ngan AHW. *Modern Physical Metallurgy*. 8th ed. Butterworth-Heinemann; 2014.

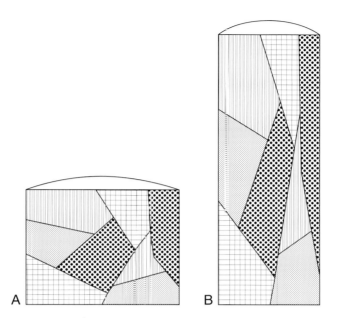

Figure 8.3

The schematic representation of a polycrystalline aggregate used in Taylor-Elam experiment: (a) before deformation and (b) after 125% strain.

packed plane is parallel or perpendicular to the loading direction, then the Schmid factor will be 0 and such crytal orientation is referred as "hard" orientation for slip. This implies that in a polycrystalline material this M value will change from one grain to another due to the difference between their orientations. Generally, an average orientation factor ($\bar{M}$) is defined. Taylor was the first to determine the value of this mean orientation factor in FCC polycrystalline materials, and accordingly, it is known as Taylor factor.

The Taylor-Elam experiment designed to calculate σ-ε response of polycrystal was based on τ-γ data. Aluminum crystals, both single crystal and polycrystals, were used. Each crystal undergoes the same shape change as the test specimen. If the test specimen elongates 125%, each grain elongates by 125% by multiple slip systems (Fig. 8.3). The hypothesis is that a system under fixed stress will arrange itself to do as little slip deformation as possible. We need to pick five slip systems that minimize the sum of the shear strains, that is, $\sum_{i=1}^{5} \gamma_i$, where γ_i is the strain on i-th slip system.

The following relations can be obtained from the assumptions and can be written as:

$$\sigma = \bar{M}\tau \tag{8.3}$$

and

$$\varepsilon = \frac{\gamma}{\bar{M}} \tag{8.4}$$

From Eqs. (8.3) and (8.4), the strain hardening rate in a polycrystalline material, $\frac{d\sigma}{d\varepsilon}$, is related to the strain hardening rate in a single crystal, $\frac{d\tau}{d\gamma}$, by

$$\frac{d\sigma}{d\varepsilon} = \bar{M}^2 \frac{d\tau}{d\gamma} \tag{8.5}$$

For materials with FCC lattice and random texture (i.e., no preferred orientations), $\bar{M} = 3.06$. For BCC lattice with similar texture, $\bar{M} = 2.90$. For example, using Eq. (8.5), the strain hardening rate in polycrystalline FCC metals is almost 9.4 times that of the single crystal for FCC. So, here an important point is that the strain hardening rate in polycrystals is many times larger than that of its single crystal counterpart. In practice, fine-grained materials actually work harden more than their coarse-grain counterparts. As grain size decreases, plastic deformation becomes more homogeneous. Due to the constraints imposed by grain boundaries, slip occurs on several systems even at low strains. As grain size is reduced, more effects of grain boundaries are felt at the grain interior. Fine grain size reduces slip line length and thus diminishes stress concentration and delays flaw nucleation. These also help in translating into improvement in the uniform portion of total ductility with decreasing grain size. However, when the grain size crosses into the nanocrystalline range, it does not work harden the same way because of the preponderance of grain boundaries. For nanocrystalline metals, the grains are too small to contain multiple dislocations and the lack of dislocation storage changes the work hardening and associated uniform elongation. In contrast, single crystal ductility can sometimes be more than polycrystalline ductility given the lack of interference coming from the neighboring grains that open up flaws. The foregoing discussion provides a glimpse of the difficulty in predicting ductility of materials quantitatively in a unified way.

The Taylor factor may vary depending on the textures and can be calculated as the weighted fraction of various orientations, which can be expressed mathematically as

$$\bar{M}_{texture} = \sum f_i M_i \tag{8.6}$$

where f_i is the fraction of crystals with Taylor factor M_i. Suppose in a polycrystalline material 80% of crystals are of <111> orientation (with M value of 3.674) and the remaining 20% are of <100> orientation (with M value of 2.449). The average Taylor factor can be calculated using Eq. (8.6) for such a textured microstructure as $\bar{M} = 0.8\,(3.674) + 0.2\,(2.449) = 3.429$. The problem is better understood by referring to Fig. 8.4, which shows the constant minimum M {110}<111> contours for slip system as a function of tensile axis orientation for BCC lattice superimposed on standard stereographic triangle. The figure is divided into regions A to E, demarcated by dashed lines. Each region corresponds to specific set of active slip systems as given in Fig. 8.4. Interestingly, each region contains more than five active slip systems, which means that multiple combinations of five or more slip systems could be active at a certain point in the stereographic triangle that have the same minimum. Hence, the final texture developed due to lattice reorientation is decided by the collective contribution of reorientation of individual active slip systems. Bishop and Hill[4] (1951) gave a more complete discussion of grain size effects and Taylor factor. These principles also point to a different strengthening mechanism known as texture strengthening. This is a vast subject by its own merit. So, we will keep our attention to the dislocation-assisted strengthening mechanisms here.

---

[4] Bishop JFW, Hill R. A theory of the plastic distortion of a polycrystalline aggregate under combined stresses. *The Philosophical Magazine*. 1951;42:414-427.

8.1 Strengthening mechanisms in engineering materials

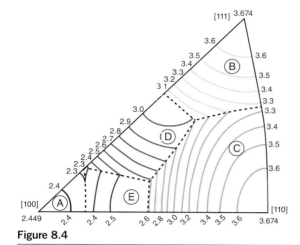

Figure 8.4

The distribution of Taylor factor (M) in BCC lattice as a function of tensile axis orientation for {110}⟨111⟩ slip system, superimposed on a stereographic triangle along with the slip systems active in each region.[5]

Plastic deformation carried out in a temperature range, and over a time interval where strain hardening is not relieved is referred to as cold working. Heavily cold-worked materials tend to contain dislocation density (ρ) of 10^{16} m^{-2} with a mean dislocation spacing of 10^{-8} m or 10 nm. However, well-annealed crystals can contain ρ of 10^{10} to 10^{12} m^{-2}; that is, a mean dislocation spacing of ($1/\sqrt{\rho}$) 10^{-6} to 10^{-5} m, that is, in the range of 1 to 10 μm. Note that this can now be used to visualize the separation between the dislocations with respect to grain size.

In the initial stages of plastic deformation, slip is limited mostly to primary slip planes, and the dislocations tend to form coplanar arrays. However, as deformation proceeds, cross-slip takes place and dislocation-dislocation interaction processes start to operate. The cold-worked structure then forms high dislocation density regions or tangles that soon develop into tangled networks. Thus the characteristic microstructure of the cold-worked state is a cellular substructure. Fig. 8.5 shows the presence of dislocation cell structure in 10% cold-worked, high-purity aluminum. With increasing strain, cell size decreases in the lower strain regime and soon reaches a fixed size, which implies that as deformation proceeds, dislocations sweep across the cells and join the tangles in the cell walls, thus making the interior of the cells relatively devoid of dislocations. The exact nature of the cold-worked microstructure depends on material, strain, strain rate, and deformation temperature. Cell structure development, however, is less pronounced for low temperature and high strain rate deformation and in materials with low SFE.

Strain hardening is an important mechanism by which single-phase ductile metals/alloys can be strengthened easily. With increasing strain, mechanical property of the material changes. For example, yield strength and ultimate tensile strength increase, whereas measures of ductility such as percentage elongation to fracture and percentage reduction in area suffer, as shown in the schematic plots of Fig. 8.6. Generally, the strain hardening rate is lower in HCP metals compared to FCC metals and decreases with increasing temperature. The final strength of the cold-worked alloy is always greater

---

[5] Chin GY, Mammel WL, Dolan MT. Taylors theory of texture for axisymmetric flow in body-centered cubic metals. *Trans Metall Soc AIME*. 1967;239:1854.

Chapter 8 Yielding and work hardening: strength limiting design

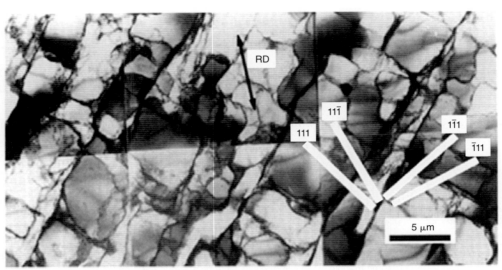

Figure 8.5

A transmission electron microscopy image of cell block structures in a 10% cold-rolled high-purity aluminum. RD is the rolling direction, and the dashed lines represent traces of {111} planes.[6]

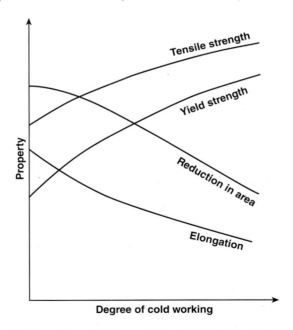

Figure 8.6

General trend of variation in strength and ductility with degree of cold working.

---

[6]Liu Q, Hansen N. Geometrically necessary boundaries and incidental dislocation boundaries formed during cold deformation. *Scripta Metall Mater*. 1995;32:1289.

than that of the pure metal cold-worked to the same extent. Other effects on physical properties because of cold working might be a slight decrease in density, electrical conductivity decrease due to the increase in electron-scattering centers, and thermal expansion coefficient increase. Furthermore, chemical reactivity usually increases, and thus corrosion resistance decreases.

A high rate of strain hardening implies mutual obstruction of dislocations gliding on intersecting slip systems. This can arise from three sources: (1) interaction of the stress fields of dislocations, (2) interactions that produce sessile locks, and (3) the interpenetration of one slip system by another (like cutting trees in a forest) that results in the formation of jogs (short segments of dislocations that don't lie on the glide plane). Another complexity of strain hardening is the number of dislocations that take part in the process. Researchers have been performing computational simulations to capture collective (ensemble) dislocation behavior to better understand the strain hardening behavior of crystals.

The basic equation relating flow stress increment ($\Delta\sigma$) due to this kind of strengthening mechanism in the polycrystalline material can also be given by the following equation:

$$\Delta\sigma = \alpha \bar{M} G b \sqrt{\rho_d} \tag{8.7}$$

All terms in the above equation have already been defined in Eq. (8.1) except for the introduction of Taylor factor ($\bar{M}$). We now also know about Taylor factor from our preceding discussion. For example, yield strength of an annealed (O-temper) AA 3003 Al alloy (Al—1–1.5 wt.% Mn alloy) can be increased to 165 MPa (H-temper) by cold working. Cold working is used for wrought aluminum alloys as the major strengthening mechanism if they are not responsive to precipitation heat treatment to enhance strength.

If the cold-worked material is heated, it would undergo recovery, recrystallization, and grain growth, which are collectively called "*annealing phenomena*." However, those topics are not discussed here, as they are generally covered in detail in standard physical metallurgy courses.

8.1.2 Grain boundary strengthening

Cold working and how the presence of grain boundaries influence the strain hardening behavior of polycrystalline materials were elaborated in Section 8.1.1. Nonetheless, grain boundaries (both high angle and low angle grain boundaries) as well as twin boundaries have a direct effect on hardening. They strengthen the crystalline material by acting as obstacles to easy dislocation motion. This strengthening mechanism is discussed now.

The classical grain boundary strengthening mechanism was proposed by E.O. Hall in 1951 while discussing the effect of grain size on yielding behavior, and N.J. Petch in 1953 further refined the concept. Thus grain boundary strengthening became known also as Hall-Petch strengthening. The concept is based on a dislocation pileup model giving yield strength of material as a function of grain size and is expressed by the following equation:

$$\sigma_y = \sigma_i + k_y d^{-1/2} \tag{8.8}$$

where σ_y is yield strength, σ_i is intrinsic strength of the grain (i.e., solid solution strengthening or Peierls-Nabarro stress to move dislocations or a combination of other mechanisms), k_y is the Hall-Petch grain boundary strengthening parameter or Hall-Petch coefficient (or unpinning constant), and d is the mean grain diameter. To express the dependence of hardness, fracture strength, and the like, grain size can be used in Hall-Petch type of equations. In addition, grain size can be substituted by martensitic lath width, interlamellar width of pearlite, or interdendritic spacing depending on the type of microstructure. Basically, the closer the spacing of barriers to dislocation motion, the higher the strength. However, a complete understanding of the mechanisms behind the Hall-Petch equation still eludes materials researchers. The two types of models discussed in the following sections are based on dislocation pileup and dislocation source.

8.1.2.1 Dislocation pileup model of grain boundary strengthening

First consider a grain that has a dislocation source near the center of the grain (Fig. 8.7). The dislocation source (such as Frank-Read source) is activated, and the dislocations move on the slip plane until they get obstructed at the grain boundary. If the dislocations are stuck in this "*traffic jam*" without being able to escape, dislocation pileups are created. Unless stress reaches a high value, dislocations cannot break through the grain boundary. Note that the slip plane orientation in the neighboring grain is different, so slip transfer from one grain to the next involves complex process at the interface, that is, the boundary. The number of dislocations in a pileup can be expressed by the following equation:

$$n = \frac{\beta' \tau_s d}{Gb} \quad (8.9)$$

where τ_s is the mean resolved shear stress on the active slip system, d is the grain diameter, G is the shear modulus, b is the Burgers vector, and β' is a constant, typically less than 1 (it is about 0.6 for edge dislocation and about 0.8 for screw dislocation provided the source is at the center of the grain), depending on the dislocation type and source. In fact, stress concentration is generated at the tip of the

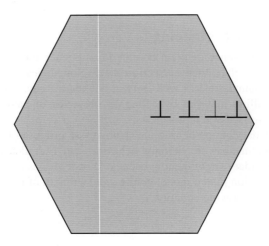

Figure 8.7

A schematic of a dislocation pileup configuration at the grain boundary of a grain. (Note that the actual model needs to have more dislocations than those shown for illustration above.)

8.1 Strengthening mechanisms in engineering materials

pileup. This stress must exceed a critical shear stress (τ_c) before the slip can transmit past the grain boundary and can be expressed by the following relation:

$$\tau_c = n\tau_s = \frac{\beta'\tau_s^2 d}{Gb} \tag{8.10}$$

However, τ_s is resolved shear stress expressed as applied shear stress (τ) less friction stress (τ_i). Depending on the microstructure, τ_i could be the P-N stress, solid solution strengthening, or any other type of strengthening that creates an obstruction to the dislocations until they reach the grain boundary. τ_s can then be replaced by ($\tau - \tau_i$) in Eq. (8.10) to get the following expression:

$$\tau_c = n\tau_s = \frac{\beta'(\tau - \tau_i)^2 d}{Gb} \tag{8.11}$$

The above equation can be rearranged into the following form:

$$\tau = \tau_i + \left(\frac{\tau_c \beta' Gb}{\pi d}\right) d^{-1/2} = \tau_i + k_y' d^{-1/2}$$

or it can also be written by replacing yield strength (σ_y) as

$$\sigma = \sigma_i + k_y d^{-1/2} \tag{8.12}$$

where σ_i is the appropriate friction stress, and k_y is the Hall-Petch parameter or locking parameter that basically gives the grain boundary hardening contribution. The locking parameter can be obtained from the slope of the linear plot of σ versus $d^{-1/2}$. Eq. (8.12) is essentially the same relation as in Eq. (8.8) and is known as the Hall-Petch relation. A similar form of equation has also been used to describe grain size dependence of hardness, fracture stress, and fatigue strength. The Hall-Petch equation has been noted to apply well to the conventional grain size range. However, when the grain size becomes very small, the Hall-Petch relation will break down. Instead of approaching theoretical strength as would be expected from a small grain size of less than 10 nm, some studies reported strength decrease as a function of grain size with a negative Hall-Petch slope. This effect is referred to as the "inverse Hall-Petch" effect. However, one has to be very careful with this interpretation. Depending on the synthesis route used, there can be processing defects that can influence the observation. Often the conclusions were drawn from compression test or indentation test. Those are not the same as tensile deformation and a word of caution is important. The overall framing of the grain size effect is illustrated in Fig. 8.8; the lower and upper bounds of strength for a pure metal are quite clear. Single crystal of a pure metal is its weakest form as there are no strengthening mechanisms. So that forms the lower limit of the strength. At the other extreme, no metal can exceed theoretical strength, as at that stress, the crystal planes will simply separate by cleavage. So, between these two bounds we expect the interfacial strengthening to be effective. Indeed, there is significant data to support the effectiveness of this mechanism. Table 8.1 summarizes the value of k_y for three different crystal structures. Note that the value is higher for BCC and HCP metals and low for FCC metals. For example, the implication is that Hall-Petch strengthening is quite ineffective for aluminum alloys. So, while considering grain boundary strengthening, this is something to keep in mind for microstructural engineering of structural alloys. Coming back to the breakdown of Hall-Petch strengthening in the nanocrystalline region, let us

Chapter 8 Yielding and work hardening: strength limiting design

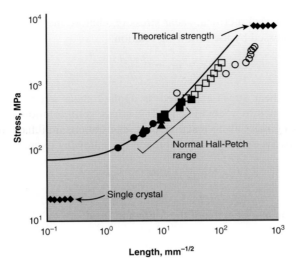

Figure 8.8

Hall-Petch plot for iron and low-carbon steel from single crystal to nanocrystals.[7]

Table 8.1 Hall-Petch parameters σ_i and k_y for BCC, FCC, and HCP structures.[8]

Material	σ_i (MPa)	k_y (MN m$^{-3/2}$)	Material	σ_i (MPa)	k_y (MN m$^{-3/2}$)
Body-centered cubic (BCC)			Copper-30% Zn	45.1	0.31
			Aluminum	15.7	0.07
Mild steel	70.6	0.74	Aluminum-3.5% Mg	49.0	0.26
Swedish iron	47.7	0.71	Silver	23.5	0.17
Fe-3% Si (−196°C)	506.0	1.54	**Hexagonal close packed (HCP)**		
Fe-18% Ni	650.1	0.22			
FeCo, ordered	50.0	0.9	Cadmium (−196°C)	17.7	0.35
FeCo, disordered	319.7	0.33	Zinc	32.4	0.22
Chromium	178.5	0.9	Magnesium	6.7	0.28
Molybdenum	107.9	1.77	Magnesium (−196°C)	14.7	0.47
Tungsten	640.3	0.79	Titanium	78.5	0.40
Vanadium	318.7	0.3	Zirconium	29.4	0.25
Niobium	68.64	0.04	Beryllium	21.6	0.41
Tantalum, with O$_2$, 0°C	186.3	0.64			
Face-centered cubic (FCC)					
Copper	25.5	0.11			
Copper-3.2% Sn	111.8	0.19			

[7]Smith TR, Armstrong RW, Hazzledine PM, Masumura RA, Pande CS, Pile-Up Based Hall-Petch Considerations at Ultra-Fine Grain Sizes, MRS Online Proceedings Library. Springer Link, Vol. 362. 1995:31-37.
[8]Adapted from Armstrong RW. In: Bunshah RF, ed. *Advances in Materials Research*. Vol. 5. Wiley Interscience; 1971:101.

consider the historical context. The classical Hall-Petch observation was based on the fact that the pileup contains a number of dislocations, more than 50. However, as we decrease the grain size, stress required to operate Frank-Reed source within the grain becomes too large. At some stage, in the grain size range of 100 to 1000 nm, the Frank-Reed source will not be viable and a dislocation pileup within the grain will not be feasible. So, the original mechanistic understanding fails at this level and a transition in micromechanisms should be expected. In addition, researchers argued that creep processes (covered in **Chapter 11**) will get activated. Although students are encouraged to wait until they learn those concepts, for closing the current discussion, a brief comment is made here. The majority view behind the interpretation of inverse Hall-Petch relationship was based on onset of grain boundary–related creep deformation mechanisms. The argument goes that below 10 nm grain size, the grain boundary area per unit volume is so high that in such materials Coble creep and grain boundary sliding get promoted. Basically the material is weakened rather than strengthened!

8.1.2.2 Li's general model for Hall-Petch relationship

Many models have attempted to explain grain boundary strengthening. One issue that came to the fore is the relative scarcity of real evidence of dislocation pileups. J.C.M. Li[9] (1963) proposed a dislocation-source model that considers grain boundary ledges as the major source of dislocations. Note that in the previous subsection, we also mentioned that as grains are refined below 1 μm (domain of ultrafine grained and nanocrystalline alloys), it is difficult to operate Frank-Reed sources within the grain. In that context as well, the generation of dislocations from grain boundaries becomes important. We know the following equation holds for flow stress given in terms of dislocation density:

$$\tau = \tau_i + \alpha G b \sqrt{\rho_d} \tag{8.13}$$

where ρ is dislocation density and α is a numerical constant (generally varying between 0.3 and 0.6). The smaller the grain size, the greater the dislocation density. Thus Li stated that ρ is actually an inverse function of d. Thus

$$\tau = \tau_i + \frac{k}{\sqrt{d}} \tag{8.14}$$

which is of the same form as the Hall-Petch equation.

8.1.3 Solid solution strengthening

Both interstitial and substitutional atoms create distortion in the solvent lattice. Dislocations are also associated with stress and strain fields of their own. When these stress fields created by the solute interact with the approaching dislocations, in most cases solid solution hardening results. In a handful of cases, however, solid solution softening may result depending on the details of the interaction, with examples in Mo-Re and W-Re alloys. However, the focus of this section will remain on solid solution strengthening. Fig. 8.9 shows the representation of interstitial and substitutional solid solutions in which atoms with different color/shade represent the solute atoms and blue (lighter color in B&W version) circles the solvent atoms.

Substitutional atoms produce a spherically symmetric distortion as shown in Eq. (8.15a). Note that any transformation of the coordinate system produces the same stress matrix of the solute atom as

---

[9] Li JCM. Petch relation and grain boundary sources. In: *Transactions of the Metallurgical Society of AIME*. Vol. 227. 1963:239-247.

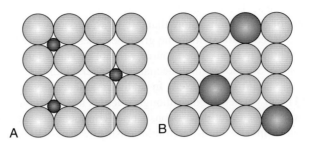

Figure 8.9

Schematic illustrations of (a) interstitial and (b) substitutional solid solutions (*orange circles* in [a] and *green circles* in [b] represent the solute atoms (darker color atoms in the B&W version)).

shown in the stress matrix below. From **Chapter 7**, we learned that edge dislocations are associated with a stress field that has both normal and shear stress components as described in Eq. (8.15b). That is why the stress fields created by substitutional atom and edge dislocation can interact.

$$(\sigma_{ij})_{sub.atom} = \begin{bmatrix} \sigma_m & 0 & 0 \\ 0 & \sigma_m & 0 \\ 0 & 0 & \sigma_m \end{bmatrix} \text{ and} \quad (8.15a)$$

$$(\sigma_{ij})_{edge} = \begin{bmatrix} \sigma_{xx} & \tau_{xy} & 0 \\ \tau_{xy} & \sigma_{yy} & 0 \\ 0 & 0 & \sigma_{zz} \end{bmatrix} \quad (8.15b)$$

On the other hand, screw dislocations at the first approximation (smooth spiral distortion) exhibit only shear stress. Thus substitutional solutes produce no interaction with screw dislocations. However, such is not the case for dissociated screw dislocations in BCC crystals.

$$\begin{vmatrix} \sigma_x & 0 & 0 \\ 0 & \sigma_y & 0 \\ 0 & 0 & \sigma_z \end{vmatrix} \begin{matrix} \sigma_x \neq \sigma_y \\ \sigma_y \neq \sigma_z \end{matrix} \quad (8.16)$$

Interstitial atoms produce a tetragonal distortion of the solvent lattice as shown above. However, here the transformation of the coordinate system produces shear components. Different interatomic displacements would create a tetragonal distortion! Transformation to a coordinate system with b as one of the axes will produce a shear on the (101) plane in the direction of b. The transformed matrix will still have a hydrostatic component, too. Thus screw dislocations interact with both interstitials and substitutional atoms. Overall, interstitials are more effective hardeners than substitutional solutes.

Solute atoms can interact with dislocations in a variety of ways as summarized by Dieter[10]:

a) Elastic interaction (i.e., size factor effect)
b) Modulus interaction

---

[10]Dieter G. *Mechanical Metallurgy*. 3rd ed. McGraw Hill;1986.

c) Long-range order interaction
d) Stacking fault interaction
e) Electrical interaction
f) Short-range order interaction

The first three interactions (i.e., [a] to [c]) are long range, and they are relatively insensitive to temperature and are active up to 0.6 to 0.7 T_m (where T_m is the melting point of the material in K). The last three (i.e., [d] to [f]) involve short-range barriers that contribute strongly to strength only at low temperatures, typically less than 0.3 T_m. However, these interactions tend to become ineffective at higher temperatures. The first two factors are the most important in solid solution strengthening.

a) **Elastic Interaction:** The atomic size difference between solute atoms and solvent atoms would stretch the lattice in different ways. This interaction has already been discussed, albeit only minimally. The stress field of solutes and the stress field of dislocations interact with each other, and the interaction would be such that local strain energy would decrease, thereby creating the hardening effect. For example, during interaction with an edge dislocation, smaller size substitutional solute atoms (relative to the solvent lattice atom size) would segregate to the compressive region of the dislocation (i.e., above the slip plane), whereas the oversize substitutional solutes would segregate to the tensile region of the stress field (below the slip plane) along with interstitial solutes. Because this interaction arises due to atomic size difference, the lattice misfit strain (ε_s) is one way of expressing elastic interaction in the following form:

$$\varepsilon_s = \frac{1}{a}\frac{da}{dc} \qquad (8.17a)$$

where a is the lattice parameter of the solvent lattice and c is the atomic fraction of the solute. Fleischer[11] further noted that this misfit strain can also be expressed by the following expression:

$$\varepsilon_s = \frac{V_{solute} - V_{solvent}}{V_{solvent}} \qquad (8.17b)$$

where V_{solute} and $V_{solvent}$ are the atomic volume of solute and solvent, respectively.

b) **Modulus Interaction:** Solute atoms generally create additional stress fields in the lattice of the surrounding matrix phase because of the difference in their shear moduli. Particularly, this type of modulus interaction occurs if the presence of a solute atom changes the local modulus of the crystal. When solutes have smaller shear moduli than the solvent (i.e., $G_{solute} < G_{solvent}$), the energy of the stress fields around dislocations will be reduced (i.e., elastic strain energy is reduced) and an attraction between the solutes and the dislocations will result. Both edge and screw dislocations are subject to this interaction. The equation describing the modulus interaction (extent of modulus mismatch) can be written as

$$\varepsilon_m = \frac{1}{G}\frac{dG}{dc} \qquad (8.18)$$

---

[11] Fleischer RL. Substitutional solution hardening. *Acta Metall.* 1963;11(3):203-209.

R.L. Fleischer in a series of papers in the early 1960s described the mathematical formulation of solid solution strengthening. An expression was given for calculating the total strain (ε_t) created by both the elastic misfit strain (ε_s) and a modulus mismatch parameter (ε_m'):

$$\varepsilon_t = \left| \varepsilon_m' - \beta \varepsilon_s \right| \tag{8.19}$$

where $\varepsilon_m' = \dfrac{\varepsilon_m}{1 + \frac{1}{2}|\varepsilon_m|}$. The parameter β is dependent on plastic deformation characteristics.

General substitutional solid solution hardening has been estimated by Fleischer as expressed in Eq. (8.20).

$$\tau = \frac{G \varepsilon_t^{3/2} c^{1/2}}{700} \tag{8.20}$$

Examples of materials showing $c^{1/2}$ dependence were provided previously. In most solid solution materials, *size and modulus effects dominate strengthening*. The modulus interaction energy can be either positive or negative depending on the sign of ε_m. The best hardening combination for substitutional solute hardening is a solute that decreases the modulus upon addition to the solvent and has a large difference in size (+) or (−).

Labusch[12] modified Fleischer's model for alloys with high solute concentrations by using an alternative mean statistical treatment of the interaction between dislocations and solute atoms. The expression was given as

$$\tau = \frac{G \varepsilon_t^{4/3} c^{2/3}}{550} \tag{8.21}$$

This equation has shown better agreement with the data from single crystals of gold-, silver-, and copper-based alloys.

Addition of carbon in solution lowers the shear modulus of iron (Fig. 8.10). But atom misfit must also be considered. An important point is that interstitials interact with both edge and screw dislocations. Substitutional elements have the greatest interaction with edge dislocations and to a lesser extent with screw dislocations. Fig. 8.11 shows yield stress increase as a function of different solute contents in alpha-Fe. This confirms that the interstitial solutes impart much more hardening compared to substitutional solutes. However, the solubility of interstitial solutes in most metals is quite limited. For instance, carbon solid solubility in alpha-Fe (BCC) is only about 0.007 wt.% at room temperature. In fact, a major portion of hardening in martensitic strengthening of steels comes from solid solution strengthening effect imparted by the trapped carbon atoms. Martensite is considered a non-equilibrium supersaturated solid solution of carbon in a distorted lattice. Martensite is so unique that it is sometimes considered to have its own strengthening mechanism, but when looked at closely, it is essentially a combination of various strengthening mechanisms, including a large density of misfit dislocations. A martensite formed in a eutectoid composition (0.77 wt.% C) steel can reach a Rockwell hardness of 65 R_C and is quite brittle.

---

[12] Labusch R. Statistical theories of solid solution hardening. *Acta Metall.* 1972;20(3):917-927.

8.1 Strengthening mechanisms in engineering materials

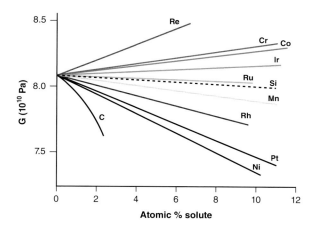

Figure 8.10

The effects of various solutes on the shear modulus (**G**) of alpha-iron at room temperature (25°C).[13]

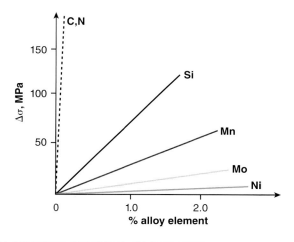

Figure 8.11

The effects of solute concentrations on strengthening in alpha-Fe. Note that interstitial solutes are more potent solid solution strengtheners than substitutional solutes such as Si, Mn, Mo, and Ni.[14]

Fig. 8.12 illustrates that solid solution hardening imparted by interstitial solutes (C in alpha-Fe and N in Nb) follows a $\Delta\tau \propto c^{1/2}$ relation. The equation provides the extent of solid solution hardening by interstitial solutes:

$$\Delta\tau = \gamma G c^{1/2} \tag{8.22}$$

where γ is the distortional strain and G is the matrix shear modulus.

---

[13]Data from G.R. Speich, cited in Met. Trans., 3:5 (1972).
[14]Adapted from Pickering FB. *Physical Metallurgy and the Design of Steels*. Applied Science Publishers; 1978.

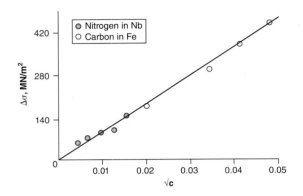

Figure 8.12

The effect of interstitial solutes (C and N) on the strengthening of Fe and Nb, respectively.[15]

Interstitial solutes in face-centered cubic lattice have a hydrostatic stress field, and hence, interactions with edge dislocations are only possible. Interstitials in HCP crystals, however, have a tetragonal distortion and hence are very effective hardeners. For example, the presence of oxygen in different amounts can embrittle titanium.

c) **Long-Range Order Interaction:** This interaction occurs only in superlattices (ordered alloys), not in conventional disordered alloys. A superlattice features a long-range periodic arrangement of dissimilar atoms, such as in Cu_3Au, where each element occupies a specific lattice site. The movement of dislocations through a superlattice causes regions of disorder called *antiphase boundary* (APB). The dislocation dissociates into two ordinary dislocations separated by an APB. As the slip proceeds, more APBs are created. Atoms across APBs have wrong neighbors, representing an increase in lattice energy. Ordered alloys with a fine domain size are stronger than the disordered state. Details are discussed in the next section in the context of the creation of APB in ordered precipitates.

d) **Stacking Fault Interaction:** This interaction is important, as solute atoms segregate preferentially to stacking faults (contained in extended dislocation). This effect is also known as *Suzuki effect* or chemical interaction. The SFE gets reduced due to the increasing concentration of solutes in the SF, and thus the separation between partial dislocations increases, making it concomitantly more difficult for partial dislocations to move.

e) **Electrical Interaction:** This strengthening interaction arises from the fact that solute atoms always have some charge due to dissimilar valences, and the charge remains localized around the solute atoms. The solute atoms can then interact with dislocations with electrical dipoles. This interaction contributes negligibly compared to the elastic and modulus interaction effects in metallic materials. It becomes significant only when the difference in valence between solute and matrix is large, and elastic misfit is small. However, this interaction can contribute significantly in materials with ionic bonding.

f) **Short-Range Order Interaction:** Solid solutions with high solute concentrations have a tendency to create regions of significant short-range order where one atom has more unlike neighbors than

---

[15]Sckopiak ZC. Flow stress of Niobium-oxygen and Niobium-Nitrogen alloys. *J Less Common Met.* 1972;26:9-17.

usual. Clusters that have more like-neighbors than usual may also have a similar strengthening effect. Additional energy will be required to move dislocations through a lattice with this kind of ordering, and hence, flow stress will be increased. The shear stress required to move a dislocation because of this kind of effect is given by[16]

$$\tau = 16\left(\frac{2}{3}\right)^{\frac{1}{2}} \frac{c(1-c)\omega a_s}{a^3} \qquad (8.23)$$

where ω is the interaction energy and a_s is the short-range order coefficient. All terms in Eq. (8.23) are temperature independent, and thus strengthening caused by this effect is temperature independent, as its contribution to overall strengthening is athermal. However, a_s is sensitive to thermal history, as it increases with decreasing annealing temperature. So, flow stress increase can be influenced by the thermal history of the alloy in question. The work of Nordheim and Grant[17] suggested that Ni-Cr alloys with Cr concentrations in the range of 20 to 25 wt.% exhibit short-range ordering. Thus alloys like Hastelloy-X and Inconel 625, each containing about 25% Cr, have likelihood of exhibiting short-range order interactions.

8.1.4 Fine particle strengthening

We will now introduce the concept of fine particle strengthening, which has been instrumental in developing high-strength, creep-resistant alloys of great engineering importance. Fine particles are generally of two types: (1) precipitates that are formed due to the precipitation process, typically through aging, and (2) particles or dispersoids that can be incorporated in the metallic matrix by a powder metallurgy type method, where hard particles in powder form are mixed with metallic powder. Other ways to include dispersoids include nucleation of particles in the melt during solidification process. In such cases, the alloying element forms an intermetallic or ceramic phase that does not solutionize at any temperature in solid state. Since the latter types of particles are not formed due to the aging process, there is no crystalline relationship with the metal matrix, and invariably the particles are incoherent in nature. Dislocation-particle interaction is generally a significantly more powerful strengthening mechanism than solid solution strengthening. Many structural materials derive their strength from the fine particles present in their microstructures, and this mode of strengthening is key for high temperature strength. This section focuses only on the low-temperature mechanical behavior of fine-particle-bearing materials (high-temperature part is covered in **Chapter 11**). Under such conditions, two main types of mechanisms can be operative—either dislocations will cut through the particles or loop around them. Thus increased stress must be applied to create yielding; in other words, the need for enhanced stress is manifested as the fine particle strengthening effect.

8.1.4.1 Particle cutting or shearing mechanism

When particles are cut, the crystallography of both the matrix and the precipitate becomes important. First, the slip plane of the precipitate may not be parallel with the matrix slip plane. Second, the magnitude of the slip may be different, that is, b_{matrix} (b_0) ≠ $b_{precipitate}$ (b_1). This may also result in a misfit dislocation,

---

[16]Stoloff NS. Alloying of nickel. In: Walter JL, Jackson MR, Sims CT, eds. *Alloying*. ASM International; 1988.
[17]Nordheim R, Grant NJ. Resistivity anomalies in the nickel chromium system as evidence of ordering reactions. *J Inst Met.* 1954;82(9):440-444.

168 Chapter 8 Yielding and work hardening: strength limiting design

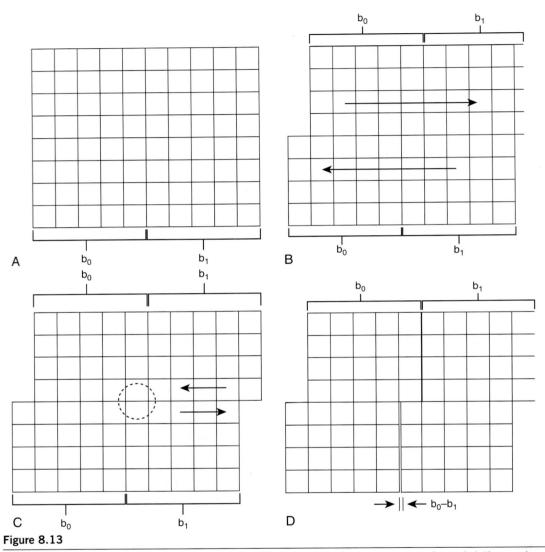

Figure 8.13

(a)–(d) Several stages involved in the creation of a misfit dislocation when slip occurs across a change in lattice spacing.

as shown by Fleisher.[18] The effect is illustrated in Fig. 8.13. Two types of dislocation loops are left at the particle/matrix interface: (1) a dislocation loop around the particle with magnitude $|b_0 - b_1|$; and (2) a dislocation loop on an interface of magnitude $|b_0|$ that results from misorientation of the slip planes.

Refer to Fig. 8.14. The segment BC that is in the precipitate may have some interesting properties. For example, if the modulus of the precipitate was less than that of the matrix, the self-energy of the

---

[18]Fleisher RL. Effects of non-uniformities on the hardening of crystals. *Acta Metall.* 1960;8: 598-604.

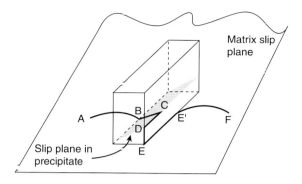

Figure 8.14

The dislocation ABCDEF passing through a particle.

dislocation segment would be less. Because the slip plane for the particle and the matrix is not the same, a jog DE is created.

We have learned in **Chapter 7** how the dislocation strain energy is proportional to self-energy per unit length or Gb^2. Thus dislocations may be attracted to the precipitates, and free energy of the system (matrix + precipitate + dislocation) can be lowered by keeping the dislocation inside the precipitate. To remove the dislocation will require an extra applied force. If the modulus of the particle is higher than that of the matrix (the more general case), an extra force is required to push the dislocation into the precipitate. Most of the models to be discussed henceforth start at the level of what extra forces act on the dislocation.

a) *Coherency hardening*

If the precipitate is coherent but the lattice parameter of the precipitate is different from that of the matrix, the region around the precipitate is strained. This will lead to repulsive or attractive interaction based on the type of mismatch. An earlier attempt by Mott and Nabarro[19] for an estimate for coherency hardening did not receive much traction because of its failure to show dependence on particle size. Gerold and Haberkorn[20] treated the coherency hardening in somewhat of a similar way as Fleischer did for solid solution strengthening and developed the relation shown in Eq. (8.24) for screw dislocations:

$$\tau_{\text{coherency}} = AG(\varepsilon_{misfit})^{3/2}\left(\frac{r}{b}\right)^{1/2}V_f^{1/2} \qquad (8.24)$$

where A is a numerical constant (for edge dislocation, A = 3, and screw dislocation, A = 1), G is the shear modulus of the matrix, b is the Burgers vector, r is the radius of second phase precipitate, V_f is the volume fraction of the second phase, and ε_{misfit} is misfit strain.

While ε_{misfit} is mostly used in the form of δ as given in Eq. (8.25a), the constrained strain does depend on the elastic properties of both the matrix and the precipitate. For a spherical coherent particle,

---

[19]Mott NF, Nabarro FRN. Dislocation theory and transient creep. *Report Bristol Conf, Strength of Solids, Phys Soc.* 1948;1-9.
[20]Gerold V, Haberkorn H. On the critical resolved shear stress of solid solutions containing coherent precipitates. *Phys Stat Solidi*. 1966;16:675.

ε_{misfit} can be related to the unconstrained misfit as shown in Eq. (8.25b) and was used in Gleiter's analysis of coherency hardening as shown in Eq. (8.26):

$$\delta = \frac{a_{ppt} - a_{mat}}{a_{mat}} \quad (8.25a)$$

$$\varepsilon_{misfit} = |\delta| \frac{3K_{ppt}}{3K_{ppt} + 2E/(1+v)} \quad (8.25b)$$

where a_{ppt} and a_{mat} are the lattice parameter of the precipitate and the matrix, respectively; v and E are the Poisson's ratio and Young's modulus of matrix, respectively; and K_{ppt} is the bulk modulus of the precipitate. Note that the misfit strain will be temperature dependent as the lattice expansion coefficient of the matrix and particle are likely to be different, thereby making this strengthening contribution temperature dependent. The expression of coherency stress proposed by Gleiter is suitable for misfit strengthening with flexible edge dislocations[21]:

$$\tau_{coherency} = 11.8 G (\varepsilon_{misfit})^{3/2} \left(\frac{r}{b}\right)^{1/2} V_f^{5/6} \quad (8.26)$$

where G is the shear modulus of the matrix, b is the Burgers vector, r is the radius of second phase (precipitate), V_f is the volume fraction of the second phase, and ε_{misfit} is constraint misfit.

Here a couple of things must be noted. First, Witt and Gerold[22] used a relationship with $\tau \propto V_f^{1/2}$, and second, many of the particle cutting models have a $(r/b)^{1/2} V_f^{1/2}$ dependence (Fig. 8.15). However, $V_f^{5/6}$ is preferred by many!

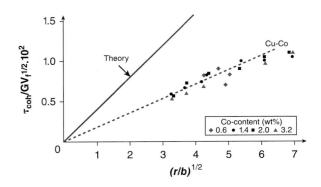

Figure 8.15
Yield stress data of Cu-Co alloys plotted according to the coherency strain model.

---

[21]Gleiter H. *Z Angew Phys*. 1967;23:108.
[22]Witt M, Gerold V. The critical resolved shear stress of Cu-Co alloys containing coherent precipitates. *Scripta Metall*. 1969;3:371-373.

b) Modulus effect

Using Eq. (8.27), another strengthening effect can be estimated due to the modulus difference between the matrix and the coherent particle:

$$\tau_{\Delta G} = r^{1/2} V_f^{1/2} \frac{\Delta G}{4\pi^2} \left(\frac{3\Delta G}{Gb}\right)^{1/2} \left[0.8 - 0.143 \ln\left(\frac{r}{b}\right)\right]^{3/2} \tag{8.27}$$

where ΔG is essentially the difference between the shear modulus of the matrix and the precipitate, that is, $|G_{matrix} - G_{ppt}|$ (mod sign makes it positive); b and G values hold the usual meaning for the matrix.

The above equation would predict void hardening, too. Consider the use of phosphorus in tungsten bulb filaments! Phosphorus becomes vapor at high temperatures. When the dislocation line reaches a void, the energy associated with the dislocation line becomes zero locally, and this leads to the attachment or strengthening. The counter-argument would be to consider radiation damage in the fuel rod of a nuclear reactor. Here also, the $r^{1/2}$ and $V_f^{1/2}$ dependence of the stress would hold.

c) Stacking fault energy

SFE difference ($\Delta\gamma_{SFE}$), which is related to the difference in the equilibrium stacking fault width between the matrix (w_m) and the particle (w_p), can also lead to a strengthening effect. For example, if the SFE of the second phase is lower, the separation of the partials becomes higher. A reference to the situation illustrated in Fig. 8.16 shows that additional force is required to collapse partial dislocations inside the particles. In the model equation below, the matrix and the precipitate must have the same crystal structure.

$$\tau_{\Delta SFE} = G_{matrix} \left(\frac{\Delta\gamma_{SFE}}{G_{matrix} b_{matrix}}\right)^{3/2} \left(\frac{r}{b_{matrix}}\right)^{1/2} V_f^{1/2} \tag{8.28}$$

where $\Delta\gamma_{SFE} = |\gamma_{SFE}^{matrix} - \gamma_{SFE}^{ppt}|$ is always positive. The other terms are either explained previously or self-explanatory.

d) Order hardening

The ordered structure of the particles affects how particle shearing can occur. Generally, Eq. (8.29) is used to express the order strengthening contribution:

$$\tau_{APB} = G_{matrix} \left(\frac{\gamma_{APB}}{G_{matrix} b_{matrix}}\right)^{3/2} \left(\frac{r}{b_{matrix}}\right)^{1/2} V_f^{1/2} \tag{8.29}$$

However, other expressions of order strengthening depend on the level of APB energy and the stage of precipitation.

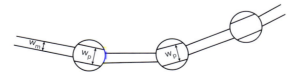

Figure 8.16

Configuration of dislocation split into partials both in the matrix and the particles. The stacking fault width in the matrix is represented as "w_m" and that in the particles is termed "w_p." Here $w_p > w_m$, that is, $\gamma_p < \gamma_m$.

Examples of the precipitation hardening system using order hardening include Ni-based superalloys containing γ FCC Ni-based solid solution matrix and γ′ (Ni$_3$X, X being Al, Ti, and similar elements, depending on alloy composition) coherent particles, and in Al-Li alloys having δ′(Al$_3$Li) particles. Figs. 8.17 (a) and (b) show the schematic microstructure and the actual SEM micrograph of a nickel-based superalloy, ASTRA 100 (Ni-13.5Al-9.0Co-6.0Cr-2.2Ta-0.6Mo-1.98W-1.27Ti, wt.%) from the work of Probstle et al.[23] The cuboidal phases are the ordered γ′ phase (L1$_2$ structure) present in the FCC γ matrix.

APB is the abbreviated form of antiphase boundary. The extra energy that is associated with these APBs (γ_{APB}) is an important parameter in strengthening by ordered precipitates. Table 8.2 lists the values of γ_{APB} for different alloys. Eq. (8.29) overestimates the shear stress required, since it is based on a single dislocation cutting the precipitates (Fig. 8.18). The second dislocation is actually attracted to the APB and repairs the damage caused by the first. A TEM image of sheared δ′-Al$_3$Li precipitates in an Al-Li alloy is shown in Fig. 8.19. This dislocation pair configuration is referred to

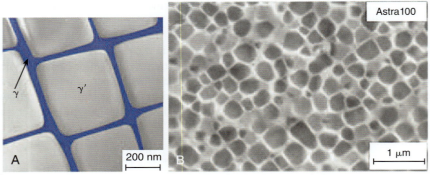

Figure 8.17
(a) A schematic microstructure of a precipitation strengthened Ni-based superalloy, and (b) a SEM micrograph of the Ni-based superalloy Astra 100.

Table 8.2 Approximate antiphase boundary energies of γ′ in various alloys.[24]

Alloy composition	APB energy (mJ/m^2)
Ni-12.7/14.0Al (at.%)	153
Ni-18.5Cr-7.5Al (at.%)	104
Ni-8.8Cr-6.2Al (at.%)	90
Fe-Cr-Ni-Al-Ti (Ti/Al=1)	240
Fe-Cr-Ni-Al-Ti (Ti/Al=8)	300
Ni-19Cr-14Co-7Mo-2Ti-2.3Al (at.%)	170–220
Ni-33Fe-16.7Cr-3.2Mo-1.6Al-1.1Ti (wt.%)	270

---

[23] Pröbstle M, Neumeier S, Feldner P, et al. Improved creep strength of nickel-base superalloys by optimized γ/γ′ partitioning behavior of solid solution strengthening elements. *Mater Sci Eng A*. 2016;676:411-420.
[24] Brown LM, Ham RK. Strengthening methods in crystals. In: Kelly A, Nicholson RB, eds. *Applied Science*. London: Applied Science Publishers; 1971:9.

8.1 Strengthening mechanisms in engineering materials

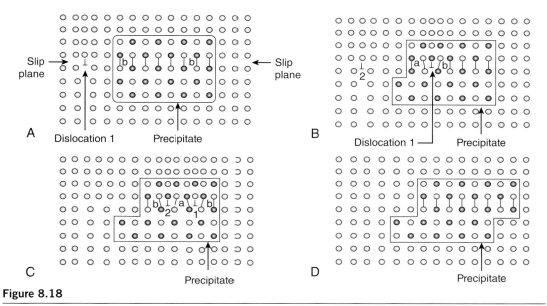

Figure 8.18

A schematic representation of an (a) ordered particle (b) being cut by an edge dislocation to create a new interface at the entry as well as the exit.[25]

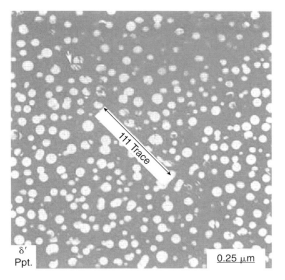

Figure 8.19

Sheared δ′ particles (note one is shown by the white arrow) along with other non-sheared particles in a dark field TEM image of an aged Al-Li alloy.[26]

---

[25] Adapted from Nembach E. *Particle Strengthening of Metals and Alloys*. John Wiley & Sons; 1997.
[26] Cassada, W. A., G. J. Shiflet, and E. A. Starke Jr. "The effect of germanium on the precipitation and deformation behavior of Al-2Li alloys." Acta metallurgica 34.3 (1986): 367-378.

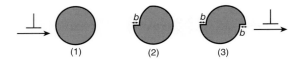

Figure 8.20
The sequence of steps involved (steps 1 through 3) in the particle cutting process leading to "surface or chemical strengthening."

as "super-dislocation." Interestingly, each of these dislocations can further dissociate into a pair of partial dislocations with a stacking fault in between them. So, APB can be sandwiched between two stacking faults. Thus this type of configuration can lead to complicated situations in ordered alloys; those aspects will not be discussed here.

e) *Surface hardening*

Note that surface area is created each time the particle is cut by dislocation. The two surface steps produced will have the same width as the magnitude of the Burgers vector of the dislocation (Fig. 8.20). The newly created surface involves the higher surface energy that is required during the particle shearing process. Thus an extra stress will be needed. The relation below gives an expression of this kind of strengthening.

$$\tau_{Surface} = G_{matrix} \left(\frac{\gamma_{p\text{-}m}}{G_{matrix} \, b_{matrix}} \right)^{3/2} \left(\frac{b}{r} \right) V_f^{1/2} \quad (8.30)$$

where $\gamma_{p\text{-}m}$ is the surface energy between matrix and particle. In this model, the $1/r$ dependence looks like the Orowan relationship. This type of strengthening is also called "chemical strengthening." Eq. (8.30) predicts a decrease in shear stress needed to cut particle as the particle size increases for a constant volume fraction of particles, which contradicts experimental observations with coherent particles. That is why this mechanism is believed to contribute little to the total strengthening effect. In fact, one should also consider an additional implication of this concept. While the surface area increases, the actual effective radius on the particle on that glide plane decreases. Note that the overall cutting stress increases with increasing radius of the particle (see the earlier relationships for particle cutting effect). So, there is a counteracting factor in this mechanism that one should also consider and depending on the specific details. The overall effect can lead to planar slip, which is usually detrimental for work hardening and fatigue properties.

8.1.4.2 Particle hardening via strong barriers

This type of strengthening occurs when the particles cannot be cut by dislocations. These particles are for non-deforming, or Orowan, barriers, and the strengthening mechanism is called Orowan bypassing or dislocation bowing mechanism. These particles are either semi-coherent or incoherent precipitates, or they could also be oxides, nitrides, and carbides that may be introduced through powder metallurgical means. Typically powder metallurgy route of introducing dispersoids with different crystal structure and lattice spacing than the matrix means that the matrix-particle interfaces will be incoherent.

The shear stress required to move a dislocation through a dispersion of second-phase particles that are "non-deforming" is given by the Orowan relationship:

$$\tau_{Orowan} = \frac{Gb}{\lambda} \quad (8.31)$$

8.1 Strengthening mechanisms in engineering materials

where τ_{Orowan} is the resolved shear stress required to move the dislocation through the dispersion of particle barriers, G is the shear modulus of the matrix, and λ is the effective distance between the Orowan barriers. Note that several refined expressions of Orowan strengthening have been worked out. Eq. (8.31) presents the most basic form.

Interparticle spacing λ cannot be measured directly but is related to geometric spacing L of the barriers. Several relations based on different interpretations of interparticle spacing have been developed. If the particles are spherical in shape and are distributed uniformly, the following mathematical expression involving the linear mean free path or interparticle spacing (λ), particle radius (r), and volume fraction (V_f) can be used.[27]

$$\lambda = \frac{4(1-V_f)r}{3V_f} \tag{8.32}$$

Dislocation moves through an array of nondeforming particles by looping around the particles (Fig. 8.21 [a]). The schematic representation of the Orowan bypassing mechanism shows that the

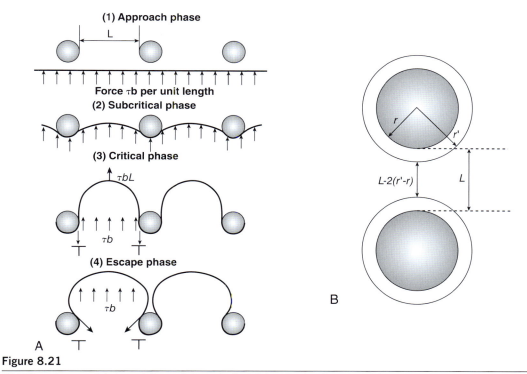

Figure 8.21
(A) Different sequential stages (phases 1 through 4) of dislocation looping around Orowan barriers, and (B) the change in the interparticle spacing (L) involving particle radius of r and the first Orowan loop radius r' after Orowan bypassing.

---

[27]Corti CW, Cotterill P, Fitzpatrick GA. The evaluation of the interparticle spacing in dispersion alloys. *Int Met Rev.* 1974;19:77-88.

mechanism consists of sequential stages: (1) approach, (2) subcritical, (3) critical, and (4) escape. Only three particles are shown here. Remember that dislocation separates the sheared crystal from the non-sheared one. Thus the crystal enclosed by r' (the radius of the dislocation loop left) has not been sheared. The effective spacing to the particle barrier would also decrease from the original interparticle spacing, that is, L in this case, to $\{L - 2(r'-r)\}$ as shown in Fig. 8.21 (b), and this would lead to work hardening since the next dislocation would encounter a closer obstacle spacing.

Actual evidence of the microstructure that has undergone Orowan bypassing mechanism is shown in Fig. 8.22. Note the Orowan loops left behind around Ni_3Si particles in the Ni-6% Si alloy single crystal. The details of the loop size left after the bypass are also critical to consider. For example, the repulsive force experienced by the dislocation at the approach side depends on the shear modulus of the particle. An understanding of such details is very helpful for design of dispersion-strengthened alloys.

a) *Derivation of Orowan equation*

We start by looking at the relationship between the curvature of a dislocation under an applied shear stress. Fig. 8.23 shows a situation where a dislocation has begun to loop around two particles. The radius of curvature (ρ) with the bowing radius of curvature is expressed as

$$\rho = \frac{T}{\tau b} \tag{8.33}$$

where τ is the applied stress, and T is the line tension of the dislocation usually taken as ½ Gb^2. T is the restoring force that tries to minimize the length of the dislocation.

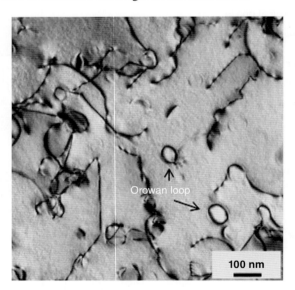

Figure 8.22

A TEM picture of Orowan loops left behind around Ni_3Si particles in the Ni-6% Si single crystal.[28]

---

[28]Sun F, Gu YF, Yan JB, Zhong ZH, Yuyama M. Dislocation motion in Ni-Fe-based superalloy during creep-rupture beyond 700°C. *Mat Lett*. 2015;159:241-244.

8.1 Strengthening mechanisms in engineering materials

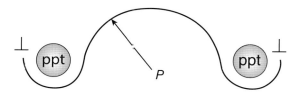

Figure 8.23
A schematic of a dislocation around two particles. For convenience the particles are labeled as "ppt," but note that it can generically represent a precipitate or a dispersoid.

The easiest case to examine is when ρ is equal to $\frac{1}{2}\lambda$, the effective particle spacing. In other words, when the dislocation bows out to form a semicircle. This gives an expression for τ_{max} via the following steps:

$$\rho = \frac{\lambda}{2} = \frac{T}{\tau_{max} b} \tag{8.34}$$

$$Or, \tau_{max} = \frac{2T}{\lambda b} = \frac{2\left(\frac{1}{2}Gb^2\right)}{\lambda b} = \frac{Gb}{\lambda}$$

The Orowan relationship may also be derived by considering a balance of forces on the dislocation line in the y-direction. From the Peach-Koehler equation, the force per unit length of dislocation in the y-direction is given by

$$\frac{F}{L} = \tau b \tag{8.35}$$

Thus from Eq. (8.35), we can write

$$+ F_y = \tau b \lambda \text{ (writing } L \text{ as } \lambda\text{)} \tag{8.36}$$

Also, from the force balance under static equilibrium,

$$-F_y = 2T \tag{8.37}$$

Summing the forces in the y direction, that is, summing Eqs. (8.36) and (8.37), we get

$$\Sigma F_y = -2T + \tau b \lambda = 0$$

Therefore, at equilibrium

$$2T = \tau b \lambda \tag{8.38}$$

Now, setting $T = \frac{1}{2}Gb^2$ in Eq. (8.38) we get

$$Gb^2 = \tau b \lambda$$
$$\tau_{Orowan} = (Gb)/\lambda$$

$T = \alpha Gb^2$, where α is a constant,
  and

$$\tau_{Orowan} = (2\alpha Gb)/\lambda$$

However, we will assume $\alpha = \frac{1}{2}$.
  Thus we can write the following equation

$$\tau_{Orowan} = Gb/\lambda \tag{8.39}$$

which is Orowan's equation described in Eq. (8.31).

b) *Modification of Orowan equation*

Ashby worked on Orowan's equation and further modified it considering an effective interparticle spacing and the effects of statistically distributed particles. This is known as Ashby-Orowan equation. The equation is of the following form:

$$\tau_{Ashby-Or} = \frac{Gb}{1.18 \times 2\pi X \left(\sqrt{\frac{\pi}{6V_f}} - 1\right)} \ln\left(\frac{X}{2b}\right) \tag{8.40}$$

where X is the real diameter of particles and other terms have their usual meanings. Fig. 8.24 shows the plots of yield strength increase as a function of particle volume fraction in an iron matrix following Orowan equation and Ashby-Orowan equation. Note that this equation does not consider any intrinsic property of particles but depends on particle geometry and volume fraction. So, if the precipitates are in needle- or plate-shaped instead of spherical, additional geometrical considerations need to be incorporated into the equation. Indeed, many computational simulations have been done to understand the impact of this. An example is illustrated in the next concept builder box. Fig. 8.24 shows clearly that the Ashby-Orowan equation predicts a reduced particle strengthening contribution compared to the Orowan equation.

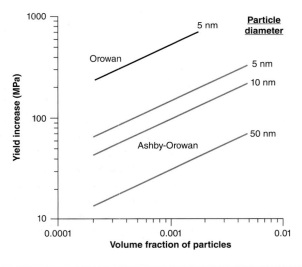

Figure 8.24

Predicted yield strength increment according to Ashby-Orowan and Orowan equations.[29]

---

[29]Gladman T. Precipitation strengthening. *Mater Sci Technol*. 1999;15:30-36.

> *Concept Builder*—What is the impact of particle size distribution and particle spatial distribution on Orowan strengthening? As mentioned in the text leading to Figure 8.24, it is important to consider the details of the barrier spacing and distribution. Note that in developing a model, "ideal" microstructure is assumed. For example, *uniform distribution of mono-dispersed particles*! However, the actual microstructure never produces ideal microstructure! Therefore in this book we covered the actual microstructure in engineering materials (**Chapter 4**) before going into the theoretical concepts of deformation. Kulkarni et al. (2004) studied the impact of the particle distribution on the magnitude of strengthening (Kulkarni AJ, Krishnamurthy K, Deshmukh SP, Mishra RS. Effect of particle size distribution on strength of precipitation-hardened alloys. *J Mater Res*. 2004;19:2765). The journal article also made a comprehensive table listing all the computational efforts till 2004. Students are encouraged to look at the paper to understand the progression of understanding from 1960s to early 2000s. It is instructive to see how ideas evolve and in hindsight some of the aspects look obvious. It is important to note that consideration of realistic microstructure leads to as much as 20% knockdown of the "ideal" theoretical Orowan strengthening estimate.
>
> A key lesson is not only that the ideal microstructure "overestimates" the theoretical values, but how we consider such calculations in an effort to compare the experimental results with theory. For example, a typical researcher first measures the experimental strength and then finds out average particle size. Then there is an attempt to compare the calculated strength with the experimental value. A number of papers report that "*our calculations are in good agreement with experiments*." NOTE: The experiments should be significantly lower than the ideal Orowan strength calculation. So, it should ***never*** match the experimental results! Indeed, we need to then remark about the efficacy of actual microstructure. Point to ponder: In processing the material, can we create a narrower distribution of microstructural features like particles? Try to find distribution of particle size in an engineering alloy, and think about the influence of the distribution on the strength and work hardening.

c) *Modeling strong dislocation barriers*

While modeling strong dislocation barriers, the following assumptions are made: (1) dispersed phase is incoherent, and (2) no volume change occurs upon precipitation. The second assumption is more critical since a particle which misfits with the matrix will have a long-range stress field and will interact with the approaching dislocation. We will assume that hardening mechanisms may be added linearly:

$$\tau_{CRSS} = \tau_{Matrix} + (Gb)/\lambda \qquad (8.41)$$

where τ_{CRSS} is the critical resolved shear stress and τ_{matrix} is the strength of just the matrix. The assumption of linear addition is not always appropriate. This is considered further later in the chapter.

8.1.4.3 *Transition from particle cutting to Orowan looping*

Note that the r dependence of τ_{Orowan} is quite different from the particle shearing stress. The weak barrier compared with the Orowan model would result in the trend as shown in Fig. 8.25. On the same figure, curves B and C represent two weak barrier models. Curve A is the Orowan limit. That is to say, the shear stress required to deform a two-phase mixture cannot exceed the Orowan stress calculated for the same V_f and r. Thus at point P, the precipitate will be looped rather than cut. As precipitate size increases, coherency of the particle disappears. The energy needed to shear the particle becomes correspondingly larger, and the energy needed to bypass the particle decreases. The intersection point of the curves representing two different precipitation strengthening mechanisms is deeply significant.

8.1.4.4 *Some further topics on fine particle strengthening*
a) *What about a polycrystal?*

So far, we obtained the shear yield strength resulting from different models of particle strengthening mechanisms. Here we present some descriptions of strengthening of an actual polycrystal where more than one mechanism needs to be considered.

Chapter 8 Yielding and work hardening: strength limiting design

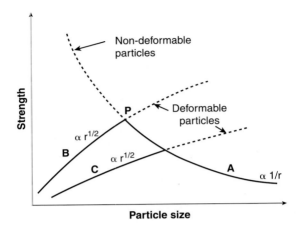

Figure 8.25
Strength due to precipitate strengthening effect as a function of particle size. Smaller particles tend to be coherent precipitate and undergo particle shearing, and larger ones follow dislocation bypassing mechanism.[30]

(1) Simple Taylor factor. M:

$$\sigma_y = M\left(\tau_{Matrix} + \frac{Gb}{\lambda}\right) \tag{8.42}$$

(2) Incorporating simple Hall-Petch grain size effect:

$$\sigma_y = M\left(\tau_{matrix} + \frac{k}{\sqrt{D}} + \frac{Gb}{\lambda}\right) \tag{8.43}$$

Consider the reported values for the k carefully since it may have been derived for $\sigma_y = \tau^* + (k/\sqrt{D})$, which may not incorporate Taylor factor.

(3) Kocks model:

$$\sigma_y = M\tau_{CRSS}\left[1 + \frac{4d}{D}\left(\frac{\sigma_B}{M\tau_{CRSS}} - 1\right)\right] \tag{8.44}$$

where $\tau_{CRSS} = \tau_{Matrix} + (Gb)/\lambda$, D is the grain diameter, and d is the width of grain boundary of strength σ_B.

b) *When does looping occur?*

The second phase particle must be able to resist the stress imposed by the dislocation loop. Fig. 8.26 shows a schematic of a particle of radius r_v and a dislocation loop of radius of curvature ρ. The dislocation loop exerts a shear stress of

$$\tau_{imposed} = \frac{Gb}{2\rho} \tag{8.45}$$

---

[30]Smallman RE, Ngan AHW. *Physical Metallurgy and Advanced Materials*. 7th ed. Butterworth-Heinemann; 2007.

8.1 Strengthening mechanisms in engineering materials

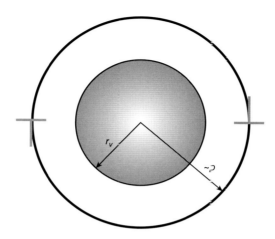

Figure 8.26

A schematic of a particle with radius r_v and a surrounding dislocation oop with a radius of curvature (ρ).

The second-phase particle must support stress $\tau_{imposed}$ without fracturing or yielding. Note that large particles are not as severely stressed because ρ is larger. Estimate $\rho = \eta r_v$, where $1 < \eta < 1.5$. Let $\tau_{imposed}$ equal shear stress of the second-phase particle.

$$\tau_{imposed} = \tau_{CRSS}^{2nd} \text{ of second phase} \tag{8.46}$$

$$\tau_{CRSS}^{2nd} = \frac{G_{matrix} \, b_{matrix}}{2\eta r_v^{2nd}}$$

A particle of critical size is needed. Assuming $\eta = 1$ (most severe case) and Eq. (8.46), the following relationship can be written

$$r_v^{2nd} > \frac{G_{matrix} \, b_{matrix}}{2\eta \tau_{CRSS}^{2nd}} \tag{8.47}$$

Now the above equation can be applied to answer how big an Al$_2$O$_3$ particle must be to withstand a dislocation loop in a nickel matrix.

For nickel: $G = 76$ GPa, $b = 2.5$ Å

For Al$_2$O$_3$: $\tau = 19$ GPa, $r_v > 5$ Å, which is quite small. However, under such a situation the second phase may fracture or interfacial decohesion may occur instead of yielding!

c) *Some case studies in fine particle strengthening*

One of the most important strengthening mechanisms is precipitation strengthening or age hardening. Alfred Wilm unknowingly discovered precipitation hardening in an Al-Cu-Mn alloy in the early 20th century. Since then, precipitation hardening has been applied successfully to a number of alloys including precipitation-hardenable aluminum alloys.

A quick review of what precipitation entails follows. Precipitation strengthening alloy systems have second phases that can go into solid solution sufficiently at elevated temperatures and that experience significant decrease of solubility at lower temperature. The Al-rich Al-Cu alloy system is considered a classical precipitation hardening system. The Al-rich end of the phase diagram is shown in Fig. 8.27

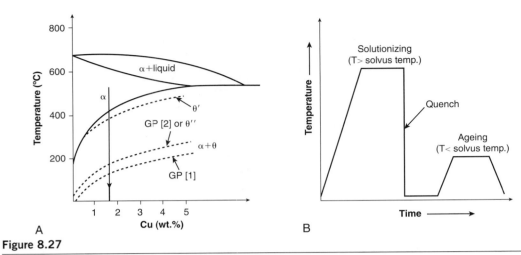

Figure 8.27

(a) Aluminum-rich end of Al-Cu binary phase diagram along with metastable solvus lines of GP zone, θ'' and θ' phases; (b) a typical precipitation hardening heat treatment cycle.

(a). A general precipitation heat treatment consists of specific steps, as shown in Fig. 8.27 (b). First, the alloy is heated to the solution heat treatment (SHT) temperature above the equilibrium solvus line. Then all metastable second phases are dissolved, and thus all Cu solutes go into solution in the solutionizing step. Following SHT, the alloy is cooled fast (generally quenched in cold water) so that it forms a supersaturated solid solution (SSSS). This solid solution contains more solute than what Al matrix would otherwise hold at room temperature. That is, SSSS is a non-equilibrium solid solution, and with higher than equilibrium concentration. Additionally, the equilibrium thermal vacancy concentration is fixed at a particular temperature. However, quenching does not provide enough time for the material to even out all the vacancies created at the SHT temperature. Hence, these extra vacancies are called "quenched-in vacancies" and are in non-equilibrium state. Since the solubility of copper in Al at room temperature is very low, there is a large thermodynamic driving force for the nucleation of precipitates. However, at room temperature the process of nucleation and growth tends to be too sluggish. The presence of enough vacancies in alloy causes the solute diffusion to be enhanced. Thus precipitation occurs at room temperature, albeit slowly. This type of aging process is known as *natural aging*. When sufficient thermal energy is put into the system, for instance, by holding the material at 150°C, precipitates will nucleate at an accelerated rate. If this process is allowed to continue long enough, equilibrium configuration is reached; this step of higher temperature aging is called *artificial aging*. Currently many different aging schemes are practiced to process different Al-based precipitation hardenable alloys. From these aging treatments temper designations emanate.

In the Al-Cu system, similar to many other precipitation-hardenable systems, transformation from SSSS to the equilibrium microstructure does not take place in a single step; rather, it occurs via sequential formation of different intermediate metastable precipitates. These intermediate metastable precipitates may differ from the equilibrium phase in composition, coherency, and shape. TEM images of the precipitates formed in Al-Cu alloys during the aging treatment are shown in Fig. 8.28.

8.1 Strengthening mechanisms in engineering materials

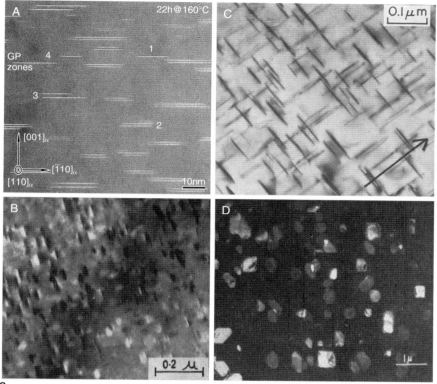

Figure 8.28

Precipitate structure in Al-Cu alloys: (a) GP zones (coherent), (b) θ'' (coherent), (c) θ' (semicoherent), and (d) θ (incoherent).[31,32,33]

In the early stages of aging, yield strength (or hardness) will increase rapidly through the nucleation of copious small coherent phases, which act as barriers to dislocation motion. These phases are called Guinier-Preston (GP) zones and basically are solutes clustering on certain crystallographic planes of the Al matrix. That is why they are not called particles or precipitates. Yield stress continues to increase as aging progresses, as the stress required for particle cutting mechanisms increases with the increase in precipitate radius. With continued aging, the GP-zones convert to transitional precipitates. In Al-Cu alloy system, it is converted to metastable coherent θ'' phase. But as precipitates grow further, the coherency strain increases, and at some stage, to relieve the misfit strain, they lose coherency; that is, a few dislocations are generated at the particle-matrix interface. The precipitate transforms to another metastable variant called θ' phase, which is semicoherent. Finally, the precipitate becomes completely incoherent with its surrounding matrix with the misfit strains having very small value. When the

---

[31] J.D Boyd, R.B Nicholson, The coarsening behaviour of θ'' and θ' precipitates in two Al-Cu alloys, Acta Metallurgica, Volume 19, Issue 12, 1971, Pages 1379-1391.; 2.

[32] Li, Z. Q., Ren, W. R., Chen, H. W., & Nie, J. F. (2023). θ'''' precipitate phase, GP zone clusters and their origin in Al-Cu alloys. Journal of Alloys and Compounds, 930, 167396.

[33] Eto, T., Sato, A., & Mori, T. (1978). Stress-oriented precipitation of GP Zones and θ' in an Al-Cu alloy. Acta Metallurgica, 26(3), 499-508.

184　Chapter 8 Yielding and work hardening: strength limiting design

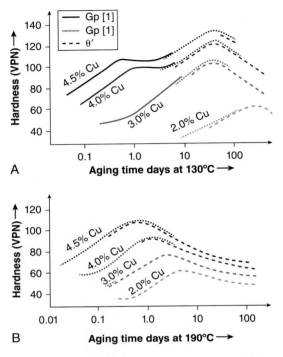

Figure 8.29

Aging curves (Vickers hardness vs. aging time) of different Al-Cu alloys at aging temperatures of (a) 130°C and (b) 190°C.

particle loses coherency, hardening due to particle cutting essentially disappears. Thus the particles serve as Orowan barriers where dislocations bypass the particles rather than cut them, and Orowan bypassing takes over (see Fig. 8.29). Recall that according to the Orowan bypassing mechanism, as interparticle spacing increases, yield stress decreases. This occurs because when precipitate growth stops and coarsening takes over (through an Ostwald ripening type mechanism wherein bigger particles become even bigger at the expense of the smaller particles), the volume fraction of precipitates is essentially kept constant. Note this distinction between initial growth period where precipitate size increases and the volume fraction of precipitates increases as well. At some point, the supersaturated matrix solute concentration reaches the equilibrium solubility at that temperature, and from this point on, the volume fraction of the precipitates becomes constant during further coarsening of the precipitates. Under such a situation following Eq. (8.32), interparticle spacing should increase with increasing particle radius. That will decrease the Orowan strengthening effect (Fig. 8.29). So, when yield stress increases because of the increase in both particle radius and volume fraction, that aging regime is known as underaging. When yield stress decreases, that aging phase is known as overaging. The intersection of the underaging and overaging curves represents peak aging, with yield stress being maximum. A reference to Fig. 8.25 may facilitate better understanding of the bell-shaped curve of the aging curves.

The beginning of the discussion on fine particle strengthening distinguished between precipitation strengthening and dispersion strengthening. If the particles are fine, stable, and incoherent, dispersion strengthening is applicable. Dispersion strengthening follows the Orowan bypassing mechanism; it is the same as the precipitation strengthening mechanism for incoherent particles. However, unlike incoherent precipitates, the dispersoid particles are generally not soluble. Dispersion strengthening is a significant strengthening mechanism used for high-temperature alloys. The Orowan bypassing mechanism at the low-temperature regime is applicable here also. One early example of dispersion strengthening system is thoria-dispersed nickel (TD-Ni). Many mechanically alloyed Ni-based (such as MA 754) and Fe-based alloys (MA 956) also used the dispersion strengthening effect imparted by yttria particles. Another example of dispersion strengthened materials is sintered aluminum powder (SAP), wherein alumina (Al_2O_3) particles are distributed in the Al matrix. All these materials fall under the high-temperature alloy category of oxide dispersion–strengthened alloys. However, the particles do not need to be always oxides—they can be borides, nitrides, and carbides if they are incorporated into the alloy matrix via powder metallurgy methods. These alloys have much greater high-temperature deformation resistance compared to their matrix. Certain aspects of high-temperature strengthening will be discussed in **Chapter 11** in greater detail.

Concept Builder—Nonequilibrium or far-for-equilibrium processing of materials for microstructural tailoring. In the last section, we discuss precipitation strengthening in Al-Cu alloys and a passing mention of dispersion strengthening in oxide dispersion strengthening alloys. As you go through this book, you want to connect the elements of the "materials systems" approach. Processing is an integral aspect of microstructural engineering. In a somewhat circular fashion, the old becomes new. Let us very briefly look at emergence of some new additive manufacturing processes and their connection to older processes. Mishra and Gupta (2023) have discussed microstructural engineering for a few far-from-equilibrium processes (Mishra RS, Gupta S. Microstructural engineering through high enthalpy states: implications for far-from-equilibrium processing of structural alloys. *Front Met Alloy.* 2023;2:1135481). For those of you who are curious to connect the fundamentals in this chapter to additive manufacturing processes, let us consider a fusion-based process and a solid-state process.

Laser powder–based additive manufacturing (LPBAM): LPBAM has emerged as a leading AM process. It has been developed over the last 20 years. Mishra and Thapliyal (2021) have reviewed design of aluminum alloys for LPBAM (Mishra RS, Thapliyal S. Design approaches for printability-performance synergy in Al alloys for laser-powder bed additive manufacturing. *Mater Des.* 2021;204:109640). One of the new alloys is Al-3Ni-1Ti-0.8Zr alloy. You will notice that the alloying elements in this alloy do not have much solid solubility. The particles in such an alloy come out during solidification. In such a case, the second-phase particles cannot be dissolved back and therefore are dispersoids. Also, such particles cannot be sheared by the dislocations, and therefore Orowan strengthening is the appropriate mechanism. Recall that for Orowan strengthening, the finer the particles, the higher the strength. The size of the particle in LPBAM can be linked to the solidification rate. The cooling rate during LPBAM is around 10^5 K/s. Note that the cooling rate is similar to rapid solidification processing rates. There were a lot of activities in 1980s to develop high-strength, high-temperature aluminum alloys using rapid solidification processing. Some of the compositions developed were Al-7Cr-6Zr, Al-8Fe-4Ce, and Al-8Fe-1V-1Si (composition in wt.%). These alloys were processed through powder metallurgical processing. The processing involved production of powders or ribbons, canning the powders/ribbons, degassing, and then extrusion/forging/rolling as the final step. The powder processing resulted in many difficulties, and these failed to become commercial success. Fast-forward to the current period of LPBAM, which uses rapid solidification *in situ* and the component is produced without the steps of conventional powder processing steps. However, the fundamental of microstructural evolution is similar to the older rapid solidification processing. The mechanical behavior aspects, therefore, can also borrow from the understanding of such alloys from 1980s and 1990s!

Friction stir–based additive manufacturing (FSAM): Friction stir processes use intense shear deformation of materials (Mishra RS, Haridas RS, Agrawal P. Friction stir-based additive manufacturing. *Sci Technol Weld Join.* 2022;27(3):141–165). Again, note that severe plastic deformation has been researched for more than 50 years in various forms. The microstructure during FSAM evolves in the solid state. A number of non-equilibrium microstructures can be generated using this process. For example, nano- and microparticles can be added during friction stir processing. Depending on the size, a number of strengthening mechanisms can be activated.

d) *Fine particle strengthening and associated work hardening behavior*

The level of strain hardening achievable in particle-bearing alloys is guided by the nature of the particular particle strengthening mechanism operating, that is, on whether particles are sheared or are bypassed by dislocations. We already know that coherent and/or deformable particles are cut through by dislocations. Hence, in this case strain hardening characteristics are quite akin to those of materials without any particles. However, dislocations continue to shear the particles. This does not result in any appreciable increase in dislocation density. It just makes it easier for the next dislocation to pass through. That is why this kind of particle cutting leads to planar slip, which is mainly localized in those areas and creates coarse planar slip. On the other hand, non-deformable or incoherent particles cannot be sheared. Dislocations must bypass these particles via the Orowan looping mechanism. Also, dislocation loops form around the particles each time a dislocation passes through. Thus the next dislocation bypasses through a shorter interparticle distance, thus increasing the strain hardening rate to much greater than that of a material without particles or the presence of deformable particles.

Fig. 8.30 shows the stress-strain curve for three materials: pure Cu, precipitation-strengthened Cu-Be, and dispersion-strengthened Cu-BeO alloy. The Cu-Be alloy contains coherent second-phase

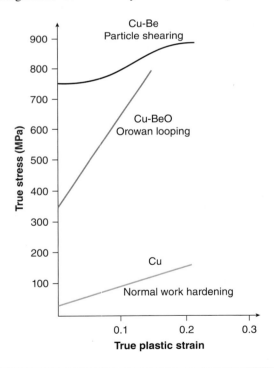

Figure 8.30

True stress—true plastic strain curves for Cu (conventional work hardening), Cu-BeO (Orowan looping), and Cu-Be (particle shearing).[34]

---

[34]Original adapted from Kelly A. Proc. of Royal Soc., A282 (1964) 63; T.H. Courtney, Mechanical Behavior of Materials, CRC Press, 2nd Edition, Waveland Press, Long Grove, IL, USA.

particles (intermediate phases of γ-CuBe phase). The precipitates are cut by the particle shearing mechanism. While this leads to a significant enhancement in strength, the strain hardening rate in the Cu-Be alloy is minimal, just a bit above pure copper. On the other hand, note the far greater work hardening rate in the Cu-BeO alloy because the alloy contains non-deformable BeO dispersoid particles which strengthened the matrix via the Orowan bypassing mechanism (because of stored dislocation loops). This results in a smaller increase in the overall strength of the Cu-BeO alloy, but flow stress catches up with that of Cu-Be alloy.

> **Points to ponder:**
> - What happens to these strengthening mechanisms at elevated temperatures?
> - What happens to ductility when strength increases?
> - What is the maximum strength of a material?
> - If element "A" has 2% solubility and element "B" has 2% solubility, can we get solid solution strengthening by adding 2% of each?
> - What factors govern the additive rules for various strengthening mechanisms?

8.1.5 Superposition of strengthening mechanisms

The discussion of dislocation-dominated strengthening mechanisms in the foregoing sections now begs the question of how to use them to predict the strength of materials. Can the strength of materials be predicted from these individual hardening mechanisms? The additivity rule through superposition approach needs to be ascertained. But the details of the approach are not clear-cut. The main complication is that interactions between the individual strengthening mechanisms could be significant. Furthermore, one additivity could be applicable to a particular material system and may not work so well for other types of systems.

Kocks and Argon[35] (1975) specifically addressed the question of superposition and presented different expressions for adding different strength contributions. Brown and Ham[36] (1971) and Ardell[37] (1985) mentioned these issues in the case of precipitation hardening. The additivity rule that first comes to mind is the linear superposition method, which involves adding all individual strength contributions, and is expressed by the following relation:

$$\tau = \tau_1 + \tau_2 + \tau_3 + \ldots \tag{8.48}$$

When the length-scales of these various contributions are not widely different, simply resorting to the linear superposition method would be inappropriate. However, when the hardening mechanisms are of obviously different length scales, linear summation could be valid. The examples of Peierls-Nabarro stress and grain boundary strengthening contributions, where the former is on the scale of atomic distances and the latter is on microstructural scale, can be added linearly.

To address this issue, another method by which these various strengthening contributions to flow stress can be added is the Pythagorean superposition. In other words, total shear strength can be expressed as the square root of the sum of the squares of the contributions from various barriers:

$$\tau = \sqrt{\tau_1^2 + \tau_2^2 + \ldots} \tag{8.49}$$

---

[35] Kocks UF, Argon AS, Ashby MF. Thermodynamics and kinetics of slip. In: Chalmers B, Christian JW, Massalski TB, eds. *Progress in Materials Science*. Vol 19. Pergamon Press, 1975.
[36] Brown, L. M., and R. K. Ham. *Strengthening Methods in Crystals*. A. Kelly, R.B. Nicholson (Eds), Applied Science, London 9 (1971).
[37] Ardell AJ. Precipitation hardening. *Metall Trans A*. 1985;16:2131-2165.

The above superposition philosophy can be applied provided the individual hardening contributions that may be added present at least two sets of discrete obstacles of equivalent strengths, but with varying barrier densities—for instance, the precipitation strengthening effect coming from two different types of precipitates.

Another type of situation could be where two sets of barriers have different strengths but the same density. A statistical treatment of this kind of situation given by Labusch[38] (1970) led to the following equation:

$$\tau = \left(\tau_1^{\frac{3}{2}} + \tau_2^{\frac{3}{2}}\right)^{2/3} \tag{8.50}$$

One issue is that most superposition models do not assume any interaction between strengthening mechanisms. But in reality, that is not the case. Nembach[39] (1992) addressed this issue. Some examples are presented to increase understanding of the intricacies of the problem. As was noted before, during Orowan by passing, dislocation lines leave dislocation loops behind, around the particles. This in turn also influences strain hardening behavior. Another example is that solutes affecting the mechanisms behind Peierls-Nabarro stress can be considered. Also, grain boundary strengthening can be generated by the dislocations being created at the grain boundary sources. The question is, how will that contribute to strain hardening?

Overall, superposition of hardening mechanisms does not involve a clear-cut method, and a number of different factors need to be considered to fully assess each specific situation. Unfortunately, no general method approach to this problem is recognized, so each case requires a careful analysis to determine the appropriate model. Nowadays, another approach uses computer simulations at a fundamental level, which is still limited to relatively small and ideal systems. These computer simulations are not yet ready to tackle the complex alloy systems.

8.2 Large-scale plasticity—dislocation generation, storage, and arrangement

While the focus in this chapter is on strengthening mechanisms and failure of alloys with a goal of relating those to strength limiting design, it is interesting to briefly discuss the situations where large-scale plasticity is important. Many alloys are processed by severe plastic deformation and examples include mechanical alloying, equal channel angular extrusion, accumulative roll bonding, wire drawing, and friction stir processing. Most of these involve high strain at low temperatures where recovery is through arrangement of dislocations leading to formation of high angle grain boundaries. Fig. 8.31 shows a sequence proposed for equal channel angular extrusion.[40]

8.3 Failure mechanisms

Some intricacies of fracture characteristics and their origin in materials are discussed here. These intricacies are quite varied which depend on factors including test/service modes (temperatures, strain

---

[38]Labusch R. A statistical theory of solid solution hardening. *Phy Stat Sol*. 1970;41:659-669.
[39]Nembach E. Synergistic effects in the superposition of strengthening mechanisms. *Acta Met Mat*. 1992;40:3325-3330.
[40]Mishra A, Kad BK, Gregori F, Meyers MA. Microstructural evolution in copper subjected to severe plastic deformation: experiments and analysis. *Acta Mater*. 2007;55:13-28.

8.3 Failure mechanisms

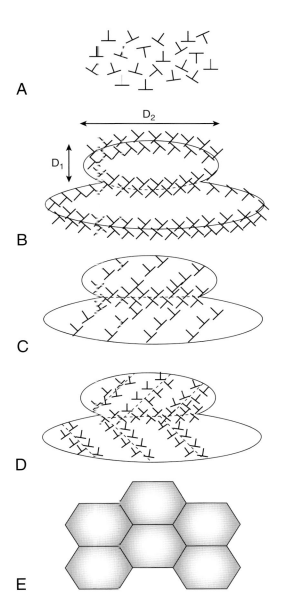

Figure 8.31
Schematic illustration of microstructural evolution during severe plastic deformation. (a) Homogeneous distribution of dislocations, (b) elongated cell formation, (c) dislocations blocked by subgrain boundaries, (d) break up of elongated subgrains, and (e) reorientation of subgrain boundaries and formation of ultrafine grain size.

rates, state of stress, chemical environment) and intrinsic materials characteristics (purity, orientation, and microstructure). The discussion here is limited to low-temperature fracture mechanisms. High-temperature fracture behavior will be discussed in **Chapter 11**.

In essence, fracture occurs when plasticity is exhausted after plastic deformation of ductile materials or gross plasticity is lacking. Thus the two main categories of fracture are ductile fracture (former condition) and brittle fracture (latter condition). We consider brittle fracture to be a catastrophic, fast-fracture event. On the contrary, ductile fracture is resistance to fast fracture. Ductile fracture involves gross plastic deformation, whereas brittle fracture does not. Crystallographic mode of fracture in ductile case is generally shear based, whereas in brittle fracture it is called cleavage. Generally, the fracture surface of the failed specimen is examined to determine the fracture mode. At low magnification, the surface appearance of the ductile fracture surface appears dull and fibrous, whereas in brittle fracture, the surface appears shiny and granular. For a detailed examination, fractographic examination is generally done using scanning electron microscopy.

There are also two types of fracture that depend on the fracture path. If the fracture path is across the grain interior parts, it is called transgranular or intragranular fracture. On the other hand, if the fracture path is along the grain boundaries, the fracture mode is termed intergranular fracture. Some examples of such fracture modes are shown in the schematic diagrams in Fig. 8.32. The most brittle fracture modes (Mode-I) are shown in Figs 8.32 (a) and (b), which represent the transgranular cleavage fracture and intergranular fracture, respectively. Mode-II fracture occurs by nucleation of cracks

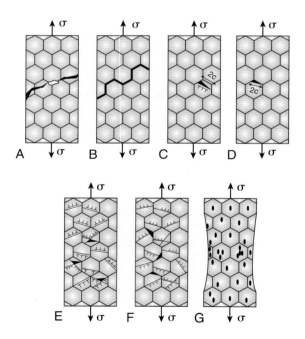

Figure 8.32

(A–G) Schematics of various fracture modes.[41]

---

[41]Adapted from Gandhi C, Ashby MF. Fracture mechanism maps for materials which cleave. *Acta Metall.* 1979;27:1565.

preceded by microscopic plastic deformation. Figs. 8.32 (c) and (d) represent Mode-II transgranular cleavage fracture and brittle intergranular fracture, respectively. Figs. 8.32 (e) and (f) represent Mode-III transgranular cleavage fracture and intergranular brittle fracture, respectively; however, in this mode of fracture gross plastic deformation precedes nucleation of cracks. Note that the various modes of brittle fracture discussed here are not in any way related to the loading modes (Mode-I to Mode-III) used with respect to fracture toughness measurement in **Chapter 9**. Lastly, Fig. 8.32 (g) represents the tensile fracture that is preceded by void nucleation, growth, and coalescence in the externally necked region of the tensile specimen.

8.3.1 Low-temperature tensile fracture

In metallic materials, the different types of low-temperature tensile fracture that can take place are summarized in the schematic diagrams of Fig. 8.33. Fig. 8.33 (a) shows the situation where tensile forces can rupture the atomic bonds without any gross deformation. Brittle fracture with relatively flat fracture surface normal to tensile stress is the result and usually occurs on the cleavage planes unless it is related to grain boundary embrittlement. Generally, this type of fracture may occur in both single and polycrystals of BCC and HCP metals. For BCC metals, at low temperature and particular orientations, grains are particularly prone to this kind of fracture. Fig. 8.33 (b) shows the situation where shear fracture occurs because of slip across a single slip plane. This type of fracture can happen in HCP single crystals oriented for single slip with the basal plane participating. In this case, fracture surface may appear flat but is caused by shear. In some cases as shown in Fig. 8.33 (c), the fracture can progress post-necking down to a point with essentially a percentage reduction in area of 100%. This type of fracture takes place in the polycrystalline form of highly ductile materials like gold and lead. For moderately ductile materials, the general form of fracture is the "cup-and-cone" type (Fig. 8.33 [d]). In this case, the tensile specimen necks down and fracture starts at the center of the necked region, extending outward by shear at the planes (at close to 45° to the tensile stress).

8.3.1.1 Ductile fracture

Void formation is important for ductile fracture. The microscale dimensions of these voids lead to their being generally known as microvoids. Microvoids are nucleated at the particle-matrix interface. Then the voids grow plastically and coalesce with neighboring microvoids by local necking. Once they coalesce, they grow outward to meet the external necking that is growing inward. This kind of fracture surface appearance at high magnification (specifically in SEM) is composed of copious depressions (called dimples). Fig. 8.34 shows such a dimpled fracture surface in a tensile-tested iron-based alloy specimen. The appearance of this type of feature generally refers to a ductile type of fracture. Each dimple represents a fracture nucleation site and is often associated with second-phase particles. Various mathematical theories describe the microvoid nucleation and growth phenomenon. One common fracture type in ductile materials is the cup-and-cone type fracture, which is schematically depicted in Fig. 8.35.

a) *The Brown-Embury analysis for tensile ductility*

This analysis (schematically shown in Fig. 8.36) assumes that a void forms at the second-phase particle as soon as tensile stress is applied. As the metal is subjected to tensile strain, voids elongate parallel to applied tensile stress. Failure is defined when the length of the elongated void is equal to the particle spacing. This criterion is based on a plastic slip line field that is 45° to the tensile axis (see Fig. 8.36 (c)). Until then, the material is plastically constrained.

192 **Chapter 8** Yielding and work hardening: strength limiting design

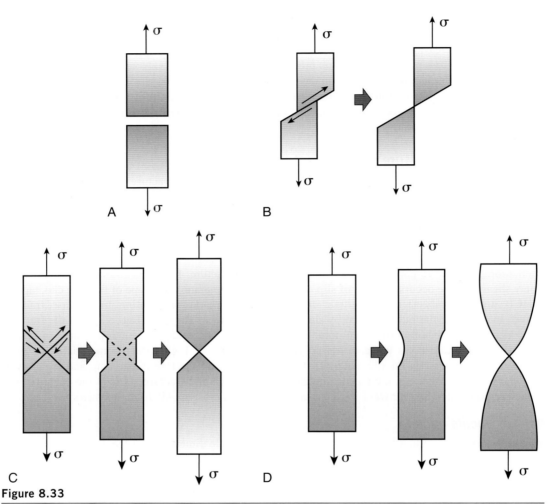

Figure 8.33

Various low-temperature tensile fracture modes: (a) brittle fracture, (b) shear fracture, (c) cup-cone type fracture, and (d) perfectly ductile fracture.

The strain to failure by void growth (ε_g) is defined by the following equation:

$$\varepsilon_g = \frac{1}{2}\ln\left[\left(\frac{\pi}{6V_f}\right)^{1/2} - \left(\frac{2}{3}\right)^{1/2}\right] \tag{8.51}$$

where V_f is the volume fraction of second-phase particles. The above equation predicts that ε_g becomes nil when V_f attains 0.159.

This analysis can also be modified to incorporate a strain required to nucleate the initial void, ε_n. Brown-Embury analysis of void nucleation is based on the uniform distribution of particles. The strain to nucleation would decrease with increase in particle size, particle brittleness, clustering of particles

8.3 Failure mechanisms

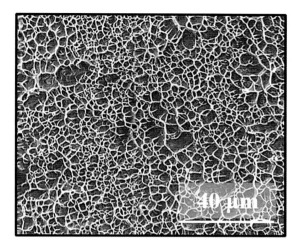

Figure 8.34

Secondary electron scanning electron microscopy image of a ductile fracture surface of a Fe-based alloy showing dimple morphology.

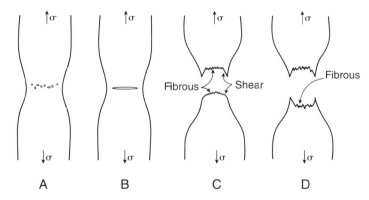

Figure 8.35

Stages in cup-and-cone-type fracture encountered in ductile fracture as a result of uniaxial tensile loading: (a) Microvoids are nucleated and grow predominantly near the externally necked region, (b) these voids link up causing void interlinkage or coalescence creating a penny-like crack, and (c) final fracture takes place when the crack has become big enough that the ligaments can no longer sustain the high local stress and shear failure occurs with appearance of "shear lips." (d) Sometimes, instead of the cup-and-cone fracture, the material ligaments can neck down gradually instead of shear rupture creating an appearance of "double cup"–type fracture.

(increases the effective particle size), and irregularity of particle shape. Approximate nucleation strains at room temperature are listed in Table 8.3. Note that most particles have low strains for nucleation.

As noted previously, microvoid nucleation takes place at the particles. Researchers have found propensity for void nucleation to happen at large second-phase particles such as inclusions or constituent particles. However, void nucleation can occur also at very fine particles (even at particles as

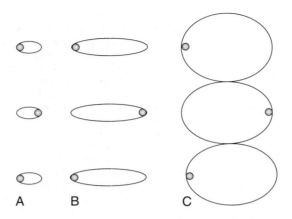

Figure 8.36
(a) The nucleation, (b) growth, and (c) coalescence of voids at second-phase particles during ductile failure.[42]

Table 8.3 A summary of void nucleation strains at room temperature under uniaxial tension for various types of second-phase particles.[43]

Material	Second-phase particle	Matrix	Void nucleation strain	Reference
Internally oxidized copper alloys	SiO_2	Cu	0.1–0.2	Atkinson et al.[44] (1974)
	Al_2O_3	Cu	0.1–0.2	Gould and Humphreys[45] (1974)
	BeO	Cu	0.2–0.4	,,
Spheroidized carbon steel	Fe_3C	Ferrite	0.4	Brown and Embury[46] (1973)
	Fe_3C	Ferrite	0.3–1.0	Inoue and Kinoshita[47] (1973)
Cu-0.6 wt.% Cr	Cu-Cr	Cu	≈1.0	Argon, Im, and Safoglu[48] (1975)

small as 5 nm), albeit they may not be viewed by usual techniques of observing the voids. Local stress due to stress concentration at the particles can reach some critical value and cause either the inclusion to fracture or debonding (i.e., decohesion) at the interface. Strong bonding at the particle-matrix interface may promote particle cracking, whereas weak particle-matrix interface would create void nucleation. Of course, other factors would be important. For example, an interesting case of 6061 alloy

---

[42]Adapted from L.M. Brown and J.D. Embury, "The microstructure and design of alloys: Proceedings of the 3rd International Conference on the Strength of Metals and Alloys" (Cambridge), The Institute of Metals, vol. 1 (1973) 164.
[43]Ashby MF. In: Taplin DMR, ed. *Fracture 77, ICF4*. University of Waterloo Press; 1977.
[44]Atkinson JD, Brown LM, Stobbs WM. The work-hardening of copper-silica: IV. The Bauschinger effect and plastic relaxation. *Phil Mag*. 1974;30(6):1247-1280.
[45]Gould D, Hirsch PB, Humphreys FJ. The Bauschinger effect, work-hardening and recovery in dispersion-hardened copper crystals. *Phi Mag*. 1974;30(6):1353-1377.
[46]Brown LM, Embury JD. *The Microstructure and Design of Alloys. Proceedings of the 3rd International Conference on the Strength of Metals and Alloys*. Vol. 1. Cambridge: The Institute of Metals; 1973:164.
[47]Inoue T, Kinoshita S. Observations of ductile fracture processes and criteria of void initiation in spherodized and ferrite/pearlite steels. *J Iron Steel Res Japan*. 1973;62:875-884.
[48]Argon AS, Im J, Safoglu R. Cavity formation from inclusions in ductile fracture. *Met Trans*. 1975;6:825-837.

8.3 Failure mechanisms

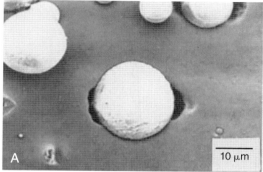 

Figure 8.37

Damage in an AA6061 Al alloy matrix reinforced with spherical Al$_2$O$_3$ particles. (a) Debonding and (b) particle cracking.

matrix reinforced with Al$_2$O$_3$ particles was originally studied by Kanetake et al.[49] in 1995. Fig. 8.37 (a) shows the debonding effect at the matrix-Al$_2$O$_3$ interface while the tensile load is applied horizontally. It shows the creation of the void shape very similar to the model of Brown-Embury. On the other hand, Al$_2$O$_3$ particle fracture is shown in Fig. 8.37 (b).

According to the Brown-Embury analysis, the total strain to failure (ε_f) is expressed as the sum of two strains—strain to void nucleation and strain to void growth:

$$\varepsilon_f = \varepsilon_n + \varepsilon_g = \ln\left(A_0/A_f\right) \tag{8.52}$$

where A_o is the original cross-sectional area and A_f is the final cross-sectional area. Following Brown-Embury analysis, for three constant strains to void nucleation (0, 0.2, and 0.4), the total strains to failure are plotted against the particle volume fraction (Fig. 8.38 [a]). Clearly, the ductility of materials suffers as the particle volume fraction increases. Experimental results shown in Fig. 8.38 (b) for various copper-based oxide dispersion–strengthened alloys follow the overall trend predicted by the Brown-Embury analysis. However, generally the analysis overestimates fracture strain. In practice, it has been usually noted that the ductility decreases as the particle volume fraction increases, and vice versa. The presence of higher particle volume fraction provides for more void nucleation sites and enhanced propensity for fracture. Le Roy and coworkers[50] subsequently rationalized the Brown-Embury analysis to include the triaxial state stress.

Void growth occurs as shown in Fig. 8.39 by making the void elongated to take an ellipsoidal form. The localization of plasticity appears to occur after achieving a critical strain, and at that stage the voids interlink quickly and lead to fracture. Mathematically, the localization of plasticity takes place when two adjacent voids reach a critical distance of approach. Refer this situation to Fig. 8.39 (c), where void length is given by $2h$, width of $2r_v$ and the distance of void separation of $2l$ is assumed to

---

[49]Kanetake N, Nomura M, Choh T. Continuous observation of microstructural degradation during tensile loading of particle reinforced aluminium matrix composites. *Mater Sci Technol.* 1995;11(12):1246-1252.
[50]LeRoy Y, Embury J, Edwards G, Ashby MF. A model of ductile fracture based on the nucleation and growth of voids. *Acta Metall.* 1981;29:1509-1522.

196 Chapter 8 Yielding and work hardening: strength limiting design

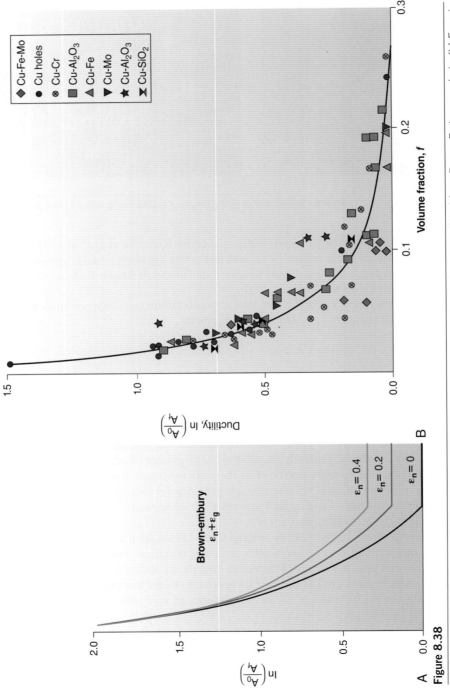

Figure 8.38

(a) Ductility as a function of particle volume fraction at three assumed nucleation strains as predicted from Brown-Embury analysis. (b) Experimental data of ductility as a function of particle volume fraction for a number of copper-based alloy systems.[51]

[51]Edelson B, Baldwin W. The effect of second phases on the mechanical properties of alloys. *Trans ASM.* 1962;55:230.

8.3 Failure mechanisms

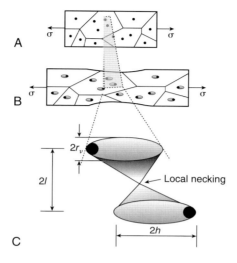

Figure 8.39

A model proposed by Ashby et al. (1979)[52] for ductile fracture initiation by voids due to decohesion at the particle-matrix interface.

be approximately close to the void length. The true strain (ε_{cs}) for coalescence to occur is given by the following relationship:

$$\varepsilon_{cs} = \frac{1}{2}\ln\left\{\Psi\left(\frac{2l-2r_v}{2r_v}\right)\right\} \approx \frac{1}{2}\ln\left\{\Psi\left(\frac{1}{V_f^{\frac{1}{2}}}-1\right)\right\} \tag{8.53}$$

where Ψ is an empirical constant usually close to 1.

One alternative theory suggests that shear bands across the microvoids form and deform by shear and lead to coalescence as depicted in Fig. 8.40. In this case, however, plastic work needed is less as the intervoid volume engaged in shear is smaller than in the earlier theory proposed by Ashby et al. (1979).

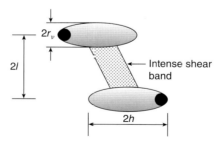

Figure 8.40

An alternative theory of void interlinkage for ductile fracture.

---

[52]Ashby MF, Gandhi C, Taplin DMR. Fracture-mechanism maps and their construction for FCC metals and alloys. *Acta Metall*. 1979;27:699.

Table 8.4 A list of critical normal stresses in certain monocrystalline metals.

Metal	Crystal structure	Cleavage plane	Temperature (°C)	Critical normal stress (MPa)
Zn-0.03% Cd	HCP	(0002)	−185	1.9
Zn-0.13% Cd	HCP	(0002)	−185	3.0
Zn-0.53% Cd	HCP	(0002)	−185	12.0
Fe	BCC	(100)	−185	275
			−100	260

8.3.1.2 Cleavage fracture

This brittle mode of failure involves separation along specific crystallographic planes (cleavage planes). The brittle fracture of single crystals can be understood by estimating the resolved normal stress on the fracture plane. Here, Sohncke's law (1869) is the equivalent theory, which states that fracture will occur when the resolved normal stress has reached a critical value. The resolved normal stress (σ_n) is expressed as

$$\sigma_n = \frac{F \cos\theta}{A/\cos\theta} = \frac{F}{A}\cos^2\theta \tag{8.54}$$

Here F is the applied force, A is the cross-section area, and θ is the angle between the stress axis and the slip plane. While the simplicity of Sohncke's law is attractive, not all experimental evidence unequivocally agrees with the law. The studies on Zn crystals by Deruyttere and Greenough[53] did show greater discrepancy in which to a great extent crystal normal stress changes as a function of orientation. The effect of microplasticity even in brittle fracture could be a factor that could have skewed the prediction. Table 8.4 summarizes critical normal stresses for cleavage in certain single crystals.

The fracture in BCC and HCP crystals associated with high strain rates, low temperatures, and/or presence of triaxial stress states (i.e., the presence of notches) changes to cleavage mechanism. A cleavage crack can be nucleated by dislocation motion and reactions. Zener[54] was the first to propose that microcrack can be created as a result of dislocation pileup. The theory of such fracture was advanced by Stroh[55] and Cottrell.[56] The likely scenario is shown by the schematic model of dislocation pileup-induced microcrack formation at a grain boundary. This will be discussed in greater detail in **Chapter 9**, which also includes ductile-brittle transition behavior.

8.4 Microstructural distribution and consequent effects

In **Chapter 4**, many examples of microstructure in engineering alloys were discussed. Preceding sections have presented various mechanisms that govern yielding and certain aspects of failure mechanisms. However, the microstructure is almost never the ideal microstructure that might be preferred. Most often, the distribution of microstructural elements such as distribution of grain size and particle

---

[53] Deruyttere AE, Greenough GB. Cleavage fracture of zinc single crystals. *Nature*. 1953;172:170-171.
[54] Zener C. In: Fracturing of metals, Amer. Soc. Metals, Cleveland, Ohio. 1948:3-31.
[55] Stroh AN. The formation of cracks as a result of plastic flow. *Proc R Soc London A Math Phys Sci*. 1954;223:404-414.
[56] Cottrell AH. Theory of brittle fracture in steel and similar metals. *Trans Met Soc AIME*. 1958;212.

8.4 Microstructural distribution and consequent effects

size must be addressed. Here the microstructural framework is introduced briefly to take into account the effects of grain size and particle size distribution.

Grain size: At a given stress, only a fraction of grains start to undergo plastic deformation. As deformation proceeds, this leads to percolation of deformation throughout the volume of the material. The hypothesis here is that there will be a stress dependence on fraction of microstructural distribution participating in deformation.

Particle size: Larger particles lead to stress concentration. Local yielding can then take place at those particles because of the stress concentration at those particles. So, the hypothesis here is that stress dependence exists at a fraction of particles that can initiate deformation.

Example: Consider A206-T4 cast aluminum alloy. Fig. 8.41 (a) shows an optical micrograph of the as-cast alloy. Grain size distribution in the form of a histogram in Fig. 8.41 (b) shows the average grain

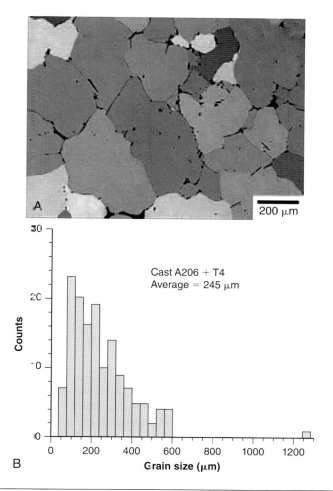

Figure 8.41

(a) An optical micrograph of A206-T4 cast aluminum alloy. (b) The grain size distribution in the as cast alloy.

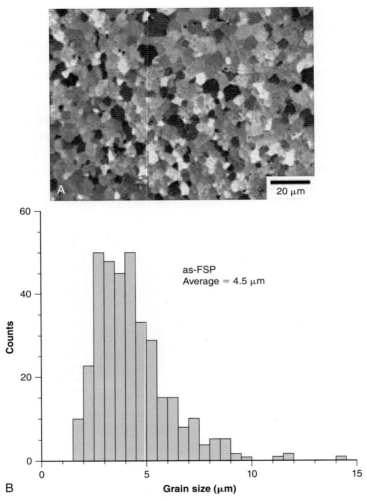

Figure 8.42

(a) An optical micrograph of friction stir-processed A206 aluminum alloy. (b) Grain size distribution in the friction stir processed as-cast alloy.

size of 245 μm and the associated distribution. After friction stir processing (a thermomechanical grain-refinement technique) grain size of the alloy was drastically reduced to a mean grain size of 4.5 μm (Fig. 8.42 [a]). Fig. 8.42 (b) shows the corresponding grain size distribution histogram. The spread of grain size in the as-cast and FSP material is quite remarkable.

In Fig. 8.43 (a), large second-phase particles are distributed in the A208-T6 alloy; the corresponding particle size distribution is shown in Fig. 8.43 (b). After friction stir processing, particle microstructure is obviously different (Fig. 8.44 [a]), and particle size distribution is much different, with mean particle length going to significantly lower value (Fig. 8.44 [b]).

8.4 Microstructural distribution and consequent effects

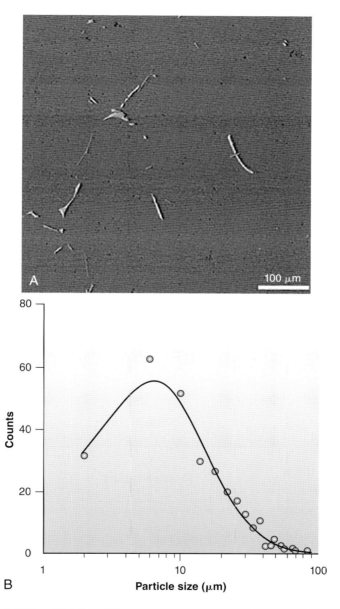

Figure 8.43
(a) SEM micrograph of A208-T4 showing particles; (b) the corresponding particle size distribution histogram.

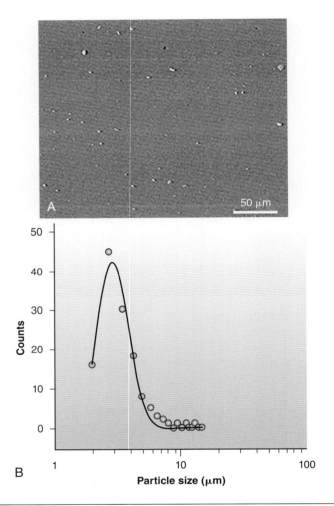

Figure 8.44
(a) SEM micrograph of the A208 Al alloy after friction stir processing, and (b) the corresponding particle size distribution.

8.5 Effect of multiaxial loading on yielding

The principal aim in the study of multiaxial stresses is to describe the stresses necessary to produce plastic deformation. We have used uniaxial tensile tests to obtain yield stress/strength values to determine how much stress could lead to yielding of the material. If low load is applied, material will behave elastically and return to its original shape upon removal of load. On the other hand, if load is increased, it will eventually reach the state of plastic deformation, and from there removal of load would not result in its returning to the original shape. However, real situations in most cases are quite different from laboratory-scale uniaxial tests. Multiaxial loading is encountered in most service conditions and during deformation processing. A material or an engineering structure may yield locally (or

8.5 Effect of multiaxial loading on yielding

globally), depending on the stress state. Thus we can use the calculated principal stresses to define the criteria for yielding or failure.

A quantitative description of the elastic-plastic boundary point is known as yield criterion. If the stress or combination of stresses reaches a certain value, the material will enter the plastic range. That is the essential of any yield criterion. The two well-known yield criteria applicable to ductile materials are discussed below.

8.5.1 von Mises criterion

The von Mises criterion (1913) states that yielding would occur when the second invariant of deviatoric stress tensor (J_2) reaches a critical value (say k^2). Earlier analyses in **Chapter 6** (Eq. 6.6c) enable expressing $J_2 = D_{ij}D_{ij}$, in the following form:

$$J_2 = \frac{1}{6}\left[(\sigma_{11} - \sigma_{22})^2 + (\sigma_{22} - \sigma_{33})^2 + (\sigma_{33} - \sigma_{11})^2\right] + \sigma_{12}^2 + \sigma_{23}^2 + \sigma_{31}^2 = k^2 \quad (8.55)$$

The value of k can be evaluated by considering the uniaxial tensile test. When the uniaxial tensile test condition relates to $\sigma_{11} = \sigma_{ys}$, $\sigma_{22} = 0$, $\sigma_{33} = 0$, and $\sigma_{12} = \sigma_{23} = \sigma_{31} = 0$, we can write

$$\frac{1}{6}\left\{(\sigma_{ys})^2 + (-\sigma_{ys})^2\right\} = k^2,$$

$$\text{or, } 2\sigma_{ys}^2 = 6k^2,$$

$$\text{or, } \sigma_{ys} = \sqrt{3}k,$$

$$\text{or, } k = \sigma_{ys}/\sqrt{3} \quad (8.56)$$

By replacing Eq. (8.56) in Eq. (8.55), we obtain the general mathematical expression for the von Mises criterion as

$$\sigma_{ys} = \frac{1}{\sqrt{2}}\left[(\sigma_{11} - \sigma_{22})^2 + (\sigma_{22} - \sigma_{33})^2 - (\sigma_{33} - \sigma_{11})^2 + 6(\sigma_{12}^2 + \sigma_{23}^2 + \sigma_{13}^2)\right]^{1/2} \quad (8.57a)$$

In terms of principal stresses, the von Mises yield criterion can be expressed as

$$\sigma_{ys} = \frac{1}{\sqrt{2}}\left[(\sigma_1 - \sigma_2)^2 + (\sigma_2 - \sigma_3)^2 + (\sigma_3 - \sigma_1)^2\right]^{1/2} \quad (8.57b)$$

where σ_1, σ_2, and σ_3 are the principal stresses in three orthogonal directions.

Yielding will occur when the expression on the right side of Eq. (8.57b), also called effective stress (σ_e) or von Mises stress, becomes equal to or greater than the left side, that is, uniaxial yield stress.

One convenient aspect of the von Mises criterion is that the principal stresses do not need to be known *a priori* to arrive at the conclusion of whether a given state of stress would cause yielding. Knowing that other interpretations for arriving at the von Mises relationship are valid is instructive, too. Hencky[57] in 1924 analyzed the yield criterion-based attainment of a critical distortion energy

---

[57]Hencky H. Zur Theorie plastischer Deformationen und der hierdurch im Material hervorgerufenen Nachspannungen. *J App Maths Mech*. 1924;4(4):323-334.

value, which is the distortion energy value per unit volume in effecting shape change in a body. As noted in **Chapter 6**, distortion energy can be expressed as follows:

$$U_d = \frac{1}{12G}\left[(\sigma_{11}-\sigma_{22})^2 + (\sigma_{22}-\sigma_{33})^2 + (\sigma_{33}-\sigma_{11})^2 + 6(\sigma_{12}^2 + \sigma_{23}^2 + \sigma_{13}^2)\right] \quad (8.58)$$

The above equation can also be written as

$$U_d = \frac{2\sigma_e^2}{12G} \quad (8.59)$$

The above equation demonstrates clearly that when σ_e value reaches uniaxial yield stress (σ_{ys}), U_d would essentially assume the critical value.

8.5.2 Tresca's yield criterion

According to Tresca criterion, yielding occurs when the difference between maximum and minimum principal stresses reaches a specific value, the uniaxial yield strength, or exceeds it. This is known as Tresca criterion. Mathematically, it can be written as

$$\sigma_{max} - \sigma_{min} = \sigma_1 - \sigma_3 \geq \sigma_{ys} \quad (8.60a)$$

In these section, single subscript notation is used to designate the principal normal stresses. So, σ_{11} is same as σ_1. Alternatively, Tresca yield criterion can also be described as follows. A body will yield if the maximum shear stress (τ_{max}) reaches or exceeds the shear yield stress (τ_o), which is half the yield stress. That is why this yield criterion is also known as the maximum shear stress criterion.

$$\tau_{max} = \frac{\sigma_1 - \sigma_3}{2} \geq \tau_{ys} = \frac{\sigma_{ys}}{2} \quad (8.60b)$$

This yield criterion neglects the effect of the intermediate principal normal stress (i.e., σ_2) value. It is not used extensively even though its form is much simpler than alternatives. However, two principal stresses should be known. The yield stress cannot be calculated from any given state of stress until the maximum and minimum principal stresses are known.

8.5.3 Yield locus

Usually yield criterion is visualized by a surface in principal stress space. Yield locus is a graphical representation of the limit of elastic behavior and start of plastic behavior based on the yield criteria discussed above under the biaxial state of stress. For simplicity, consider a biaxial state of stress with only two principal normal stresses of σ_1 and σ_3 (assuming $\sigma_2 = 0$) are in action. The von Mises criterion can be derived from the following equation:

$$\sigma_1^2 - \sigma_1\sigma_3 + \sigma_3^2 = \sigma_{ys}^2 \quad (8.61)$$

The above equation corresponds to an ellipse with a semi-major axis of $\sqrt{2}\sigma_{ys}$ and the semi-minor axis of $\sqrt{\frac{2}{3}}\sigma_{ys}$. The yield locus is shown in Fig. 8.45 in σ_1-σ_3 plane space in the form of an ellipse. Some important points on the yield locus actually correspond to particular stress ratios, α (defined as the ratio of σ_3 to σ_1). For 3D state of stress (i.e., three non-zero principal stresses), the yield locus becomes yield surface (ellipsoid), which we will not discuss further.

8.5 Effect of multiaxial loading on yielding

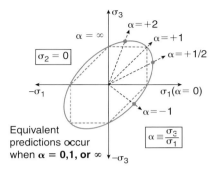

Figure 8.45

The yield locus based on Tresca (hexagonal shape) and von Mises criteria (elliptical shape). Note that the term "α" is different from the same-looking term used in dislocation theory–related topics; here the term "α" represents stress ratio.[58]

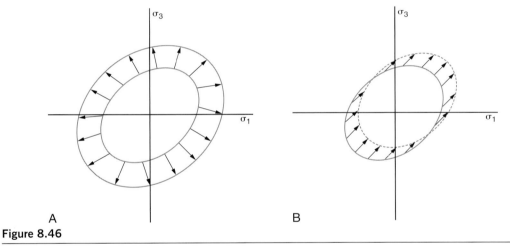

Figure 8.46

(a) Isotropic hardening and (b) kinematic hardening.

The same figure above allows drawing the yield locus as predicted by the Tresca criterion. Assuming plane state of stress as before (i.e., $\sigma_2 = 0$), the Tresca criterion can be represented as a hexagon outlined by straight lines, $\sigma_1 = \pm\sigma_{ys}$, $\sigma_2 = \pm\sigma_{ys}$, and $\sigma_1 - \sigma_2 = \pm\sigma_{ys}$. At particular stress ratios, Tresca and von Mises criteria coincide. Note, however, that regardless of stress state, the von Mises and Tresca yield criteria do not differ by more than 15% and are usually closer. Remember that the above discussion applies to isotropic materials where σ_{ys} is the same regardless of the direction along which the particular property is measured. For anisotropic materials, yield anisotropy is guided by Hill's treatment. Anisotropy in polycrystalline materials is often created by the presence of strong crystallographic texture.

The effect of strain hardening on yield locus is shown in Fig. 8.46. In continuum mechanics, two types of hardening are recognized: isotropic hardening and kinematic hardening. In isotropic hardening, the

---

[58]Bower AF. *Applied Mechanics of Solids*. 1st ed. CRC Press, Taylor and Francis Group; 2009.

yield locus expands by the same factor irrespective of the loading path; that is, the yield locus maintains its shape but dilates as the material strain hardens (Fig. 8.46 [a]). It is important to note that isotropic hardening model can be applied to anisotropic materials also; nobody should get confused by its name. The isotropic hardening has been found to be effective in describing large strains. On the other hand, in kinematic hardening the yield locus translates in the direction of loading path while keeping its size and shape (Fig. 8.46 [b]). This kind of hardening is generally used sparingly when there is small strain involved following a change in the loading path.

8.6 Principles and examples of strength limiting design

The stress level at which a particular material can be used under certain service conditions is called working stress (σ_w). For static component applications, σ_w is based on yield strength (σ_y) or ultimate tensile strength (σ_u). First, a statistical approach is taken: average strength minus three times the standard deviation. Working stress is generally specified by governmental agencies or by technical society standards such as the American Society of Mechanical Engineers (ASME). For ductile materials, the design is based on yield strength, and for brittle materials it is based on σ_u. Working stress can be expressed by the following relations.

$$\text{For ductile materials: } \sigma_w = \frac{\sigma_{ys}}{(SF)_d} \tag{8.62a}$$

$$\text{For brittle materials: } \sigma_w = \frac{\sigma_u}{(SF)_b} \tag{8.62b}$$

where $(SF)_d$ and $(SF)_b$ are the factor of safety for ductile and brittle materials, respectively. The safety factor depends on many aspects, including the type of application. Generally, the factor of safety values ranges from 1.2 to 1.8. A lower number is used by aerospace and high-performance industries. It is indeed counterintuitive, but it is based on higher control on specifications of materials and processes in the aerospace industry. Overall, this approach uses significant extra material. Overdesign is a risk-averse approach. The balance is based on cost vs. performance. For static applications such as in a building, the factor of safety will be lower than what is needed for a machine that is exposed to further severe conditions including dynamic loading.

In selecting materials for particular applications, a number of factors may need to be considered. The modulus-strength chart in Fig. 8.47 shows elastic modulus, E, plotted against yield strength, σ_y, or elastic limit, σ_{el}, for various classes of materials (MOR is modulus of rupture). As seen before, yield strength is an appropriate design parameter for ductile materials and for brittle materials σ_{el}. Elastic modulus may be an important criterion where, along with strength, stiffness is important. Note that most engineering metallic alloys (nickel-based alloys, steels) are at the top of the chart. Some ceramics (WC, SiC) can also be seen at higher location. This chart also shows us to another useful material characteristic, the yield strain, σ_y/E, meaning the strain at which the material ceases to be linearly elastic. Contours of fixed yield strain can be represented by a group of parallel straight lines. Engineering polymers tend to have large yield strains (0.01–0.1); however, engineering metallic alloys appear to have yield strains lower than the engineering polymers. Composites show very similar behavior to metals. Elastomers with quite low elastic moduli have yield strain between 1 and 10, much greater value than any other class of materials shown on the chart.

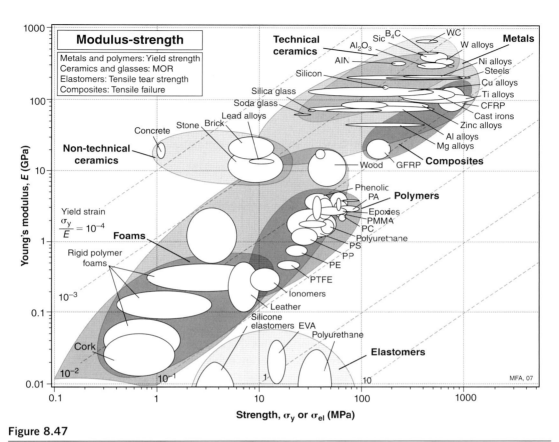

Figure 8.47

The Young's modulus-strength chart. The contours show the strain at the elastic limit, σ_y/E.[59]

8.7 Key chapter takeaways

This chapter described the important aspects of yielding and relevant mechanical characteristics. Movement of dislocations when impeded by some obstacle/barrier in the material leads to strengthening of the crystalline materials. The discussion began with emphasis on understanding the strain hardening behavior of single crystal's stress-strain curve, and then the discussion was extended to include more complex polycrystalline materials and their mechanical behavior. But the majority of the chapter is devoted to the discussion of various dislocation-dominated strengthening mechanisms (work hardening, grain boundary strengthening, solid solution strengthening, and fine particle strengthening) and their mathematical formulations. The difficulties of predicting yielding via superposition rules are discussed. Some fracture concepts related to low-temperature tensile fracture are highlighted. More will be presented in later chapters. Then some basic ideas of engineering design based on strength measures are introduced. Two yield criteria—Tresca and von Mises—are discussed.

---

[59] Ashby M, Shercliff H, Cebon D. *Materials: Engineering, Science, Processing and Design*. 3rd ed. Butterworth-Heinemann (Elsevier); 2014.

8.8 Exercises

1. (a) Why would decrease in grain size improve both strength and ductility of a polycrystalline material in the conventional grain size range?

   (b) Alloying can lead to decrease in stacking fault energy, leading to solid solution strengthening, known as Suzuki effect—explain why.

   (c) Name the two main factors influencing solid solution strengthening.

2. (a) Describe the difference between the particle shearing and dislocation by-passing (or Orowan looping) mechanisms with proper schematics.

   (b) Explain why the aging curve in precipitation hardenable aluminum systems is generally bell shaped.

3. (a) Perfect dislocations are gliding on a slip plane in aluminum (FCC). What is the Burgers vector magnitude of a perfect dislocation in Al if the lattice constant of the metal is 4.05 Å?

   (b) We have talked about sintered aluminum powder (SAP), a dispersion-strengthened aluminum alloy. The material has an aluminum metal matrix that is strengthened by a uniform dispersion of fine alumina particles. Assume these particles are spherical and have a mean diameter of 100 nm. The weight percentage of alumina phase in aluminum is 10%. Estimate the dispersion strengthening effect (consider that Orowan particle bypassing or dislocation bowing mechanism is active). Given: G_{Al} (shear modulus of aluminum) = 28 GPa, density (Al) = 2.70 g/cm^3, and density (Al_2O_3) = 3.96 g/cm^3. Take the Burgers vector (b) magnitude in Al matrix as 2.86 Å.

4. State what strengthening mechanism is mainly operating in the following materials: Thoria-dispersed nickel (TD-Ni); Al-4Mg alloy; Al-Cu-Mg (2024 Al); nanocrystalline aluminum; and highly cold-worked (90% cold reduction) commercially pure copper.

5. (a) Describe how a typical Frank-Read (F-R) source works.

   (b) Assume that you have two FCC crystals. Crystal-1 has twice the shear modulus of Crystal-2 but the lattice constant of Crystal-1 is half that of Crystal-2. Which crystal would activate F-R source at a lower shear stress if a preexisting perfect dislocation of the same length is the F-R source? If not lower, then what? State any assumptions involved.

   (c) If the grain diameter of a material is very small (in the extreme nanocrystalline range), the F-R sources may not be able to operate at all. Why? Think of a scenario of how that can be possible.

6. Why is ductile iron more ductile than gray iron? Discuss in terms of stress concentration.

7. (a) State Schmid's law clearly (please refer to your introductory Materials Science book).

   (b) Let's consider a cylindrical nickel single crystal (FCC) with diameter of 0.12 mm pulled in tension with a stress of 1000 MPa. The loading direction is along $[\bar{1}01]$ while the slip system is $(1\bar{1}1)[\bar{1}01]$. (i) What is the resolved shear stress along the slip direction in the slip plane? (ii) If τ_{CRSS} for nickel is 550 MPa, would the crystal deform plastically?

8.8 Exercises

8. A single crystal of copper is deformed in tension. The loading axis is [112].
   (a) Calculate the Schmid factor for the different slip systems.
   (b) If the critical resolved shear stress is 50 MPa, what is the tensile stress at which the material will start to deform plastically?

9. (a) Calculate the Schmid factor for a single crystal FCC metal loaded along its [010] direction on a slip system (111)[$\bar{1}$10]. Refer to the figure below.

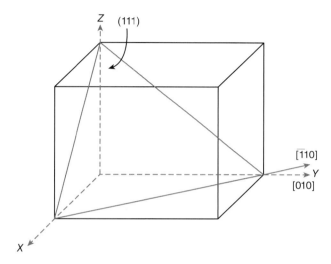

   (b) If a tensile stress of 80 MPa is applied along [100] direction of the above FCC metal (lattice constant of 0.286 nm), causing yielding in the crystal, what is the critical resolved shear stress (CRSS)?

10. A single crystal of aluminum (lattice constant: 0.405 nm) is oriented for a tensile test such that its slip plane normal makes an angle of 28.1° with the tensile loading axis. Three possible slip directions make angles of 62°, 71°, and 84° with the same tensile axis. (a) Which of these three slip directions is the most favored for initiating slip? (b) If the plastic deformation ensues at a tensile stress of 3.2 MPa, calculate the critical resolved shear stress (CRSS) for aluminum. (c) Consider that there is a perfect dislocation on the relevant slip plane along the most favored slip direction. Under the action of the shear stress of 3.2 MPa, what would be the normal force per unit length to the dislocation?

11. The critical resolved shear stress for iron is 27 MPa. Determine the maximum possible yield strength for a single crystal of iron pulled in tension.

12. (a) Define Peierls-Nabarro (P-N) stress.
    (b) Why do BCC crystals generally have higher values of P-N stress than FCC crystals?

13. (a) Why does HCP single crystal (like Zn) have a more easy-glide regime compared to FCC single crystals?

(b) Draw schematic stress-strain curves of a Zn single crystal and Al single crystal on the same graph to compare their stress-strain behavior.

14. Draw a typical stress-strain curve for an FCC single crystal. You have now studied quite a bit of dislocation theory. Describe what is expected to happen intrinsically in the specimen at each stage of deformation. Use illustrations to bolster your explanation.

15. Discuss the Hall-Petch strengthening effect.

16. Define solid solution strengthening. Why does carbon strengthen iron lattice more than Cr when present in the same amount?

17. Compare and contrast precipitation strengthening and dispersion strengthening.

18. A material has an original dislocation density of 10^8 mm^{-2}. Due to fresh plastic deformation, the dislocation density increased to 10^{10} mm^{-2}. By how many times will the yield strength increase with increasing dislocation density? State any assumptions involved.

19. (a) Describe why polycrystalline materials exhibit higher strain hardening rate compared to their single crystal counterparts. Explain your answer with the support of mathematical expressions of strain hardening rates.

    (b) Why do carbon interstitial solutes strengthen alpha-iron matrix much more than nickel substitutional solutes?

    (c) Give examples of two substitutional solutes in iron, which produce substitutional solid solution strengthening effect.

20. Identify the main strengthening mechanism operating for each scenario. Edge dislocations moving in their slip plane find themselves in the following situations: (a) They meet a sessile dislocation; (b) dislocations enter coherent precipitates; (c) they interact with interstitial atoms in a BCC metal; (d) they encounter a high angle grain boundary; and (e) they encounter subgrain boundary (say, tilt boundary).

21. The stress state $\sigma_{ij} = \begin{vmatrix} 21 & 0 & 0 \\ 10 & 0 & 0 \\ 0 & 0 & 50 \end{vmatrix}$ MPa acts on an isotropic solid with shear modulus $(G) =$ 25 GPa and Poisson's ratio $(\nu) = 0.33$, and the yield strength, σ_o, is 18 MPa. Show whether the material would yield under the given state of stress based on both von Mises and Tresca criteria.

22. A steel with yield strength of 170 MPa in uniaxial tension is tested under a principal state of stress state where $\sigma_2 = 0.5\sigma_1$ and $\sigma_3 = 0$. What are the principal stresses at which yielding will occur if (a) the maximum shear stress theory (Tresca) holds, or (b) the distortion energy (Von Mises) theory holds?

23. (a) What is a yield locus?

    (b) Draw a schematic yield locus for an isotropic material using both von Mises and Tresca criteria.

24. It is experimentally found that a certain material does not change in volume when subjected to an elastic state of stress. What is the Poisson's ratio?

25. Determine the engineering strain, true strain, and reduction in area for each of the following situations: (a) Extension from L to $1.2L$; (b) compression from L to $0.8L$.

Qualify the following statements as "True" or "False." Give adequate reasons for supporting your choice.

26. Decrease in grain size decreases yield strength of polycrystalline material.   True   False

27. The equicohesive temperature for a metal is usually at $0.1T_m$, where T_m is the melting point of the metal.   True   False

28. Homologous temperature is the ratio of a temperature of interest to the melting temperature of a metal (both temperatures given in °F).   True   False

29. Higher strain hardening exponent means lower uniform ductility.   True   False

30. Increase in volume fraction of second-phase particles would reduce the total ductility of a particle-strengthened alloy.   True   False

31. Guinier-Preston (GP) zones can form during early stage of age hardening of the precipitation-hardenable aluminum alloys.   True   False

Further readings

Ashby M, Shercliff H, Cebon D. *Materials: Engineering, Science, Processing and Design*. 3rd ed. Butterworth-Heinemann (Elsevier); 2014.
Bower AF. *Applied Mechanics of Solids*. 1st ed. CRC Press, Taylor and Francis Group; 2009.
Courtney TH. *Mechanical Behavior of Materials*. 2nd ed. CRC Press, Waveland Press; 2000.
Dieter GE, Bacon D. *Mechanical Metallurgy*. SI Metric ed. McGraw-Hill; 2001.
Frenkel J. The Theory of Elasticity and Strength of Crystalline Corps. *Z Phys*. 1926;37:572.
Hosford WS. *Materials Science: An Intermediate Text*. 1st ed. Cambridge University Press; 2006.
Seeger A. *Dislocations and Mechanical Properties of Crystals*. New York: John Wiley & Sons; 1957.

CHAPTER 9

Toughness of materials: toughness limiting design

Chapter outline

9.1 Various definitions of toughness..213
   9.1.1 Tensile toughness...214
   9.1.2 Impact toughness..215
      9.1.2.1 Transition temperature approach to fracture control *215*
      9.1.2.2 Various factors affecting transition temperature *219*
   9.1.3 Fracture toughness 224
9.2 Stress intensity and role of mechanics 227
9.3 Details of fracture toughness testing 231
   9.3.1 For less ductile materials: Plane strain fracture toughness testing procedure............231
   9.3.2 Ductile materials – plastic zone..233
9.4 Elastic-plastic fracture mechanics..235
   9.4.1 Crack tip opening displacement...236
   9.4.2 Experimental determination of CTOD 237
   9.4.3 J-integral..238
   9.4.4 Experimental determination of J-integral 239
9.5 Damage-tolerant design approach based on assumption of flaws.............................240
   9.5.1 Design considerations (Irwin's leak-before-break approach) 240
9.6 Unintended consequence of constituent particles/inclusions 241
9.7 Toughening of ceramics 244
9.8 Exercises 249

Learning objectives

- Understand the basics of three different toughness measures in materials.
- Understand how toughness measurement is done including K_{Ic}.
- Develop an understanding of the relationship between various microstructural characteristics and toughness.

9.1 Various definitions of toughness

Toughness is an important mechanical property, but the term is often confusing. Thus a reference to the definition of toughness from the Oxford Dictionary should be helpful. Of the four definitions, the most appropriate in the context of our discussion is: "The state of being strong enough to withstand adverse conditions or rough handling." However, an engineer should not see toughness as merely an alternative term for strength. From a materials engineering perspective, the most appropriate definition of toughness is: "The ability of a material to absorb energy in the plastic deformation regime until fracture." This means that a tough material should be able to plastically deform without catastrophic failure. Many machine components, such as gears, chains, and crane hooks, as well as various kinds of structures, require adequate toughness to sustain the service conditions they are subjected to. Toughness can be thought of as a composite parameter dependent on both strength and ductility. For example, a very strong material with negligible ductility cannot be tough, and, by the same token, a material with very low strength but substantial ductility would not be particularly attractive as a tough material. Neither ceramics nor lead is used to make a car body!

While the qualitative definition of toughness appears to be simple, it is not easy to pin it down, and quantitative descriptions would vary depending on the type of conditions under which the particular property (in this case toughness) is evaluated. Another key aspect to remember is that toughness is a microstructure-sensitive property, given that both strength and ductility depend on microstructural characteristics. So, as study of the topic becomes deeper, the reader is reminded to ponder the microstructural features that can be controlled in order to tailor toughness for meeting the need of a particular application. The three main categories of toughness are tensile toughness, impact toughness, and fracture toughness, the last one being the most important as a flaw-tolerant design criterion. Each will be considered one by one in the subsequent sections.

9.1.1 Tensile toughness

Tensile toughness is the ability of a material to absorb energy and plastically deform until fracture in its tensile testing. The area under the stress-strain curve is a direct measure of toughness, which in turn is given by the energy per unit area. Mathematically, it can be written as

$$U_T = \int_0^{\varepsilon_f} \sigma d\varepsilon \tag{9.1}$$

where σ is the true stress, ε is the true strain, and ε_f is the final fracture strain (Fig. 9.1).

The unit for tensile toughness from the above definition is J/m³ or Pa. The overall deformation includes uniform plastic deformation (without localized necking) and non-uniform deformation (with necking). If only the pre-necking regime is considered, replacing σ by $k\varepsilon^n$ is governed by Hollomon's equation as shown in **Chapter 5**.

$$U_T = \int_0^{\varepsilon_f} \sigma d\varepsilon = \int_0^{\varepsilon_f} k\varepsilon^n d\varepsilon = \left.\frac{k\varepsilon^{n+1}}{n+1}\right|_{\varepsilon=\varepsilon_f} = \frac{k\varepsilon_f^{n+1}}{n+1}. \tag{9.2}$$

Because of the difficulty in describing the whole true stress-strain curve with a single definitive equation, many mathematical approximate equations have used engineering stress-strain parameters (yield strength or YS, ultimate tensile strength or UTS, and engineering strain to fracture or e_f).

For ductile materials, both equations are used:

$$U_T \approx \text{UTS}(e_f) \tag{9.3a}$$

and

$$U_T \approx \left(\frac{\text{UTS} + \text{YS}}{2}\right)(e_f) \tag{9.3b}$$

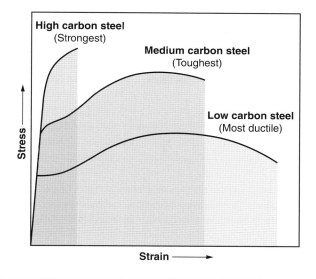

Figure 9.1

Comparison of the tensile toughness of high carbon steel, medium carbon steel, and low carbon steel, schematically shown as areas under the stress-strain curves.

For brittle materials, approximate relation is given by

$$U_T \approx \frac{2}{3}(\text{UTS})(e_f). \tag{9.4}$$

9.1.2 Impact toughness
9.1.2.1 Transition temperature approach to fracture control
An earlier approach to measuring toughness has been (notched-bar) impact testing. The philosophy behind developing the test was to devise an easy-to-use, inexpensive way to compare fracture resistance among low-strength metals. Also, in this type of testing the test conditions are made as demanding as possible and include high loading rates, thick specimens, and sharp starter flaws (notch). The most useful data is the transition temperature across which the nature of fracture changes from ductile to brittle mode. From a design perspective, the lower the ductile-to-brittle-transition temperature (DBTT), the larger the temperature window over which the material is useful. The concept of a transition temperature is valuable for relative comparisons as well as defining critical temperature for use. Historically, two main tests can be used to measure impact properties: Charpy and Izod. The most common test is the Charpy V-notch (CVN) test that measures toughness of metals and alloys. In these tests, an impact pendulum is used as shown in Fig. 9.2. The knife-edge bearing pendulum is released from height h_1 and hits the test specimen near the base of the machine (C). After fracturing the specimen, the pendulum swings to height h_2. The dial on the machine actually shows the energy absorbed during fracture as a function of the height differential ($h_1 - h_2$). The underlying assumption here is that the difference between static energies of the pendulum at those two heights represents the absorbed energy during fracture or impact energy, which is a measure of impact toughness.

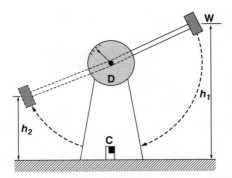

Figure 9.2

A schematic diagram shows impact hammer of mass W is released from height, h_1, impacting the specimen at location C and fracturing the specimen, following which the hammer rises to height h_2. The potential energy difference, i.e., the energy absorbed or impact energy, corresponding to the height difference, $h_1 - h_2$, is recorded in dial, D.[1]

Fig. 9.3 shows the test configurations of the Charpy and Izod tests. Charpy impact specimens are kept horizontal, while Izod impact test specimens are kept vertical (cantilever beam loaded). Charpy impact testing is widely used for metals/alloys, whereas Izod impact testing is used for polymeric materials.

ASTM standard E23 states that the standard impact test specimen has a square cross-section with dimensions of 10 mm × 10 mm and a length of 55 mm for Charpy impact testing. A 45° V-notch with a root radius of 0.25 mm at 2 mm depth is machined midway along the length-side edge (Fig. 9.3 [a]). The specimen experiences a high strain rate in the order of 10^3 s^{-1} at fracture. The specimens can also be subsize. Of course, impact toughness depends on the dimensions of the specimen. Sometimes, subsize specimen use is a necessity when only small amount of material is available. Furthermore, materials that are irradiated in nuclear reactors may need impact testing with subsize specimens to decrease the prolonged duration of irradiation-induced activation. Smaller-sized specimens can be handled safely within a much shorter time duration.

Impact testing has lost its applicability in modern mechanical design methodologies because of high variability in datasets and the relative nature of the impact energy. This scatter may come from sources such as local variations of properties in materials, inability to generate machine-reproducible notches (in terms of notch depth and shape), and difficulty in aligning the notch and pendulum motion (i.e., improper placement location of specimen). But impact testing can be a good quality indicator when used as a comparative tool. Particularly, impact testing provides information about transition temperature. The importance of transition temperature stems from the concept that the material should be used above this transition temperature to avoid brittle fracture. In other words, the lower transition temperature will ensure that the component will work safely under service conditions and the material would not be susceptible to catastrophic failure. Basically, impact energy is plotted over a temperature range, and the transition temperature is determined as shown in the diagram in Fig. 9.4. While some materials (such as FCC-based metals) do not show a sharp transition, some bcc-based materials (like

---

[1]Tetelman AS, McEvily AJ Jr. *Fracture of Structural Materials*. New York: Wiley; 1967.

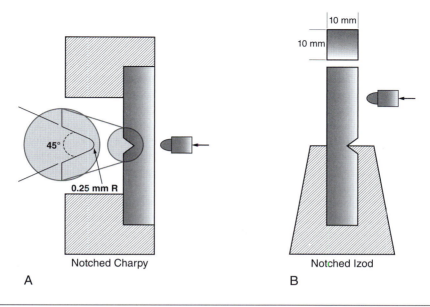

Figure 9.3

Flexed beam impact specimens: (a) Charpy impact testing (under three-point loading) and (b) Izod impact testing (cantilever-beam loading).[2]

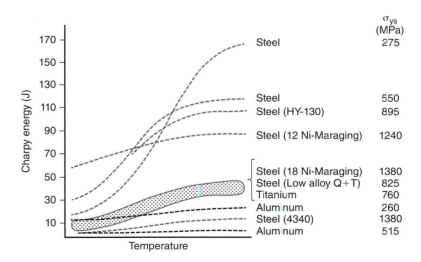

Figure 9.4

Charpy impact energy as a function of temperature for different materials.[3]

---

[2]*E23-18, Standard Test methods for Notched Bar Impact testing of Metallic Materials*. ASTM International.
[3]Davis JR, ed. *ASM Metals Handbook Desk Edition*. 2nd ed. Ohio, USA: ASM International; 1998.

carbon steels) generally exhibit a sharp transition from notch-brittle to notch-tough regime as the temperature is increased. This is widely known as ductile-brittle transition behavior, and the corresponding shift in impact energy occurs below the DBTT. Fig. 9.4 shows the impact (Charpy) energy data as a function of temperature for different materials. Note that although FCC based materials like aluminum alloys do not show DBTT, the absolute values of impact toughness of these alloys are much lower!

Transition temperature can be classified in different ways (Fig. 9.5). One criterion defines the transition temperature at which the fracture surface is fully (i.e., 100%) fibrous or ductile. This type of transition temperature is known as *fracture transition point* (FTP) as shown by T_1 in Fig. 9.5. Clearly the FTP criterion is quite conservative. So, a less conservative criterion is based on the transition temperature at which the fracture surface of the Charpy impact test specimen consists of 50% cleavage and 50% ductile fracture. It is designated by T_2 in Fig. 9.5. This transition temperature is called *fracture appearance transition temperature* (FATT) and is very close to temperature T_3, which is basically the mean of the transition temperatures corresponding to the upper shelf and lower shelf energies. Another way of obtaining transition temperature is to fix an arbitrary impact energy or absorbed energy value (say 20 J or 15 ft-lb) and determine the corresponding temperature (see T_4). This criterion may not work for all materials. Another criterion is based on the temperature where the fracture surface becomes fully brittle (cleavage). This transition temperature (T_5) is known as *nil ductility temperature* (NDT).

The fracture surface that is produced during impact testing generally consists of brittle (shiny, faceted, bright, flat overall, no or little deformation) and ductile (fibrous, grey, dull, possibly ridged) features. Sides may be pulled in, which can be an important visual indication of higher toughness. Note Fig. 9.6 (a), which features a series of fractured impact specimens (tested at different temperatures). Below that picture is a schematic standard comparative chart (Fig. 9.6 [b]). The shear lip region is the outside region that expands in area as test temperature increases. This represents the extent of the ductile nature of the behavior. The method of quantifying percentage shear lip is shown in Fig. 9.6 [c]. The variation of lateral expansion in the fracture surface represents some aspect of transition behavior (Figs. 9.7 [a] and [b]).

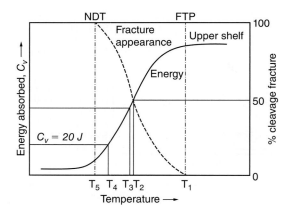

Figure 9.5

Various definitions of impact transition temperatures shown on the Charpy plots.[4]

---

[4]Tetelman AS, McEvily AJ Jr. *Fracture of Structural Materials*. Wiley; 1967.

9.1 Various definitions of toughness 219

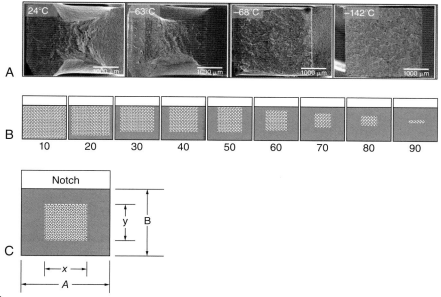

Figure 9.6
(a) Transition of the fracture surface of the tested impact specimens from the ductile to cleavage fracture as test temperature decreases for API 5L X80 steel.[5] (b) Standard comparison chart of the shear lip (dark areas)[6] and (c) the method of calculation of percentage shear lip.

9.1.2.2 Various factors affecting transition temperature

Since most steels exhibit sharp ductile-brittle transition temperatures, various factors affecting the transition temperatures of steels are discussed here. However, the fundamental principles would also be applicable to non-steel materials that show transition behavior. Quick examples of these are BCC refractory metals that routinely exhibit high transition temperatures.

Metallurgical factors such as composition and microstructure can change transition temperatures. On a composition front, increasing carbon content in steels can decrease upper shelf energy and increase transition temperature. Fig. 9.8 shows the effect of carbon concentration on Charpy impact energy for steels. On the other hand, transition temperature decreases by about 5.5°C for each 0.1% increase in manganese content. Of note is that the ratio of Mn:C should be kept at 3:1, to get adequate impact toughness in steels. Other elements in steels that influence transition temperature include phosphorus and silicon (>0.25 wt.%), while molybdenum raises transition temperature. However, nickel improves impact toughness up to about 2% concentration. Oxygen impurities (>0.003 wt.%) in steels promote intergranular fracture and cause lower toughness. That is why deoxidation of steels is important during the melt processing stage. Adding aluminum/silicon to control oxygen content has shown good improvement in transition temperature.

---

[5] Avila JA, Lucon E, Sowards J, et al. Assessment of ductile-to-brittle transition behavior of localized microstructural regions in a friction-stir welded X80 pipeline steel with miniaturized charpy v-notch testing. *Metall Mater Trans A*. 2016;47:2855-2865.
[6] *E23-18, Standard Test methods for Notched Bar Impact testing of Metallic Materials*. ASTM International.

Chapter 9 Toughness of materials: toughness limiting design

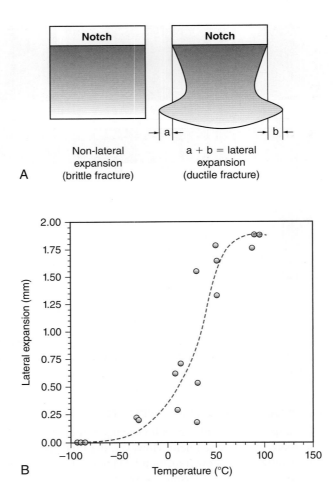

Figure 9.7
(a) Measurement of lateral expansion at the compression side of a tested Charpy notched-bar specimen,[7] and (b) the lateral expansion of the specimen increases as temperature increases.[8]

Apart from composition, microstructural features, such as grain size, influence transition temperature. A fine-grained material generally gives rise to a decrease in transition temperature, however contradictory data can also be observed. For example, an increase in ASTM grain size number (meaning decreasing grain size) in mild steel has been reported to lead to an increase of transition temperature by about 16°C. The presence of stable alloy carbides/nitrides, such as NbC, increases yield strength but also restricts grain growth and maintains impact toughness. Fig. 9.9 shows the comparison among the martensitic, pearlitic, and bainitic steels as well as the effect of carbon content in the steels.

---

[7]*E23-18, Standard Test methods for Notched Bar Impact testing of Metallic Materials*. ASTM International.
[8]Nishikawa S, Hasegawa T, Takahashi M. Effect of PWHT conditions on toughness and creep rupture strength in modified 9Cr-1Mo steel welds. *High Temp Mate Proc*. 2019;38:739-749.

9.1 Various definitions of toughness   221

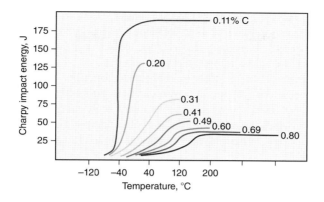

Figure 9.8
Effect of carbon content on the impact (Charpy V-notch) energy curves of plain carbon steels.

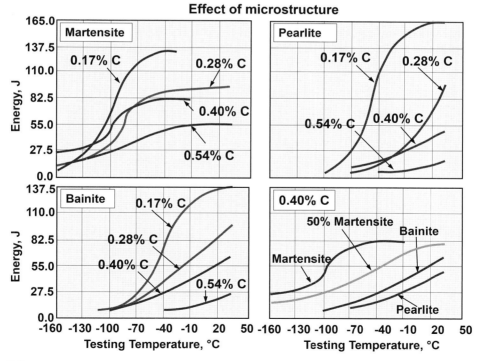

Figure 9.9
Effect of different microstructures in Cr-Mo steel with different carbon content on the impact toughness curves.[9]

---

[9]Davis JR, ed. *ASM Metals Handbook Desk Edition*. 2nd ed. ASM International; 1998.

Interestingly, for a given carbon content, martensitic phase owing to its very fine micro-constituent distribution shows lower transition temperature. Pearlitic and bainitic steels show slightly higher transition temperature compared to martensitic steel. As the carbon content in steels increases, the impact toughness appears to decrease across all the steels.

Tempering of martensitic phase in steels helps improve ductility and toughness while reducing strength/hardness. Tempering changes the microstructure of the martensite. Such treatment generally leads to an increase in upper shelf energy (as well as lower shelf energy) and a decrease in transition temperature. Fig. 9.10 illustrates this aspect in 4340 steel tempered for 1.5 hours at different temperatures. However, in certain cases, tempering heat treatment can lead to a decrease in impact toughness, especially under strain aging.

Impact toughness can also depend on the orientation of the Charpy specimens machined out of rolled/forged plates. Fig. 9.11 shows impact energy transition temperature curves of Charpy specimens with notches in different orientations in an as-rolled low carbon (0.12 wt.% C) steel. The right side of the figure shows schematics of the Charpy specimen configurations, whereas the left side reveals transition temperature curves for the specimens (A, B, and C). While transition temperature does not appear to be influenced, upper shelf energy did vary. This can be used as a strategy if the loading of a component can be tailored to take advantage of grain orientation effect. At the lower end of the curves, they are pretty close together, and that is where the ductility transition temperature criterion is used.

Transition temperature can be thickness dependent as well. Fig. 9.12 shows the impact-energy curves for different thickness specimens. With increasing specimen thickness, increase in the upper

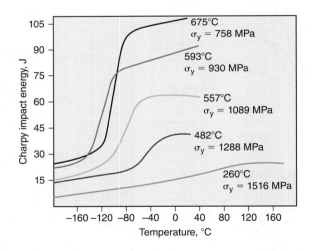

Figure 9.10

The effect of tempering temperature on transition temperature curves (the upper and lower shelf energies increase and transition temperature decreases). Yield strength and tempering temperature are shown inside the figure. Note that the tempering temperature is given in degrees F in many literature: 1250°F = 677°C, 1100°F = 593°C, 1035°F = 557°C, 900°F = 482°C, and 500°F = 260°C.[10]

---

[10]Adapted from Larson FR, Nunes J. *Watertown National Laboratories, Technical report No. WAL TR 834.2/3*. December, 1961.

9.1 Various definitions of toughness   223

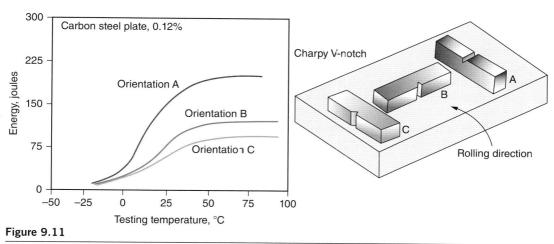

Figure 9.11

Effect of orientation on the Charpy impact energy (toughness) for an as-rolled low-C steel plate.[11]

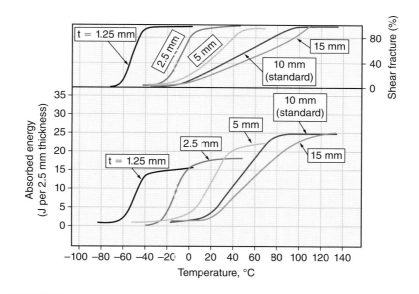

Figure 9.12

Effect of thickness of impact specimen on impact toughness (Charpy V-notch) vs. temperature curves for an A283 steel.[12]

---

[11] Adapted from Puzak PP, Eschbacher EW, Pellini WS. *Welding Research Supplement*. December, 1952.
[12] Adapted from McNicol R., 1965. Correlation of Charpy test results for standard and nonstandard size specimens. *Weld Res Suppl.*, 385-393.

shelf energy takes place; it also pushes the transition temperature to higher values. That is why using a standard specimen size is important to obtain consistent results, and particularly to compare different alloys and alloys processed under different conditions. Other test methods, such as drop weight test, can accommodate much thicker specimens.

Another important aspect is that some types of environments have direct consequences on the impact behavior of material. Radiation exposure (such as neutrons) inside a nuclear reactor can embrittle reactor pressure vessel steels by decreasing upper shelf energy and increasing transition temperature, which is also referred to as radiation embrittlement. Similarly, there are also examples of chemical environments that can cause loss in material toughness, e.g., hydrogen embrittlement and liquid metal embrittlement.

9.1.3 Fracture toughness

Fracture toughness is an important quantitative property for design, just like any other mechanical property, such as yield stress. Understanding this property and related test methods is based on fundamental fracture mechanics concepts. Fracture toughness has become an essential property metric for damage-tolerant design in a wide range of load-bearing applications including aerospace. It also has some intrinsic advantages as compared to tensile toughness and impact toughness as well as some complexities.

The elastic stress field around the crack tip can be described by a single parameter known as "stress intensity factor (K)." As the concept is based on linear elastic fracture mechanics (LEFM), the fracture toughness obtained is applicable only to materials that are brittle or have limited ductility. Thus most ceramics are good candidates for such an approach. But fracture toughness of high-strength steels, titanium alloys, and high-strength aluminum alloys is routinely measured using such a concept. At this time, although understanding of the stress intensity factor may be limited, the concept has been discussed further later in the chapter. The stress intensity factor is dependent on many factors, including the geometry of crack-containing solids, crack size/location, and the magnitude and distribution of the load used. However, what is clear-cut is that an unstable rapid failure would occur if a critical value of K is reached. That critical value is the measure of fracture toughness.

We start with a focus on understanding the three independent modes of crack deformation as first observed by Irwin. Thus our ability to evaluate more clearly and reliably which mode of loading is appropriately used for fracture toughness measurement will be extended. Three modes of crack propogation or deformation are illustrated in Fig. 9.13. Other types of modes must arise from various combinations of these three basic modes. In mode-I (tensile opening mode), the crack propagates perpendicular to the applied tensile stress. For mode-II (sliding or in-plane shear mode), the crack lips displace perpendicular to the leading edge but in the same plane. This is caused by shear stress. In mode-III (tearing or antiplane shear mode), the displacement remains parallel to the crack faces and the leading edge. This displacement is also caused by shear stress.

Although the critical stress intensity factor (K_C) varies for each mode of loading ($K_{IC} \neq K_{IIC} \neq K_{IIIC}$), in practice, the opening mode (i.e., mode-I) is the most important. Most metallic as well as nonmetallic materials are more prone to fracture under mode-I loading as compared to other two shear modes. Hence, K_{IC} is observed to be lower than K_{IIC} and K_{IIIC}. That is why most tests for evaluation of fracture toughness are performed under mode-I loading. The critical value of K, K_{Ic}, is known as plane strain fracture toughness. For a given type of loading and geometry, the relation is

$$K_{Ic} = Y\sigma\sqrt{\pi a_c} \tag{9.5}$$

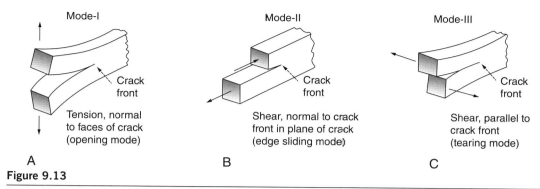

Figure 9.13

Three standard modes of loading: (a) mode-I (opening mode), (b) mode-II (sliding mode), and (c) mode-III (tearing mode).[13]

where Y is a dimensionless parameter dependent on loading geometry and crack configuration, a_c is the critical crack length (i.e., the crack length at which crack propagation becomes unstable), and σ is the applied stress. Note that the unit of fracture toughness MPa$\sqrt{m}$ is a bit unusual in its form. The significance of this equation is clear in that the maximum tolerable crack length size can be computed if applied stress and K_{Ic} are known. In other words, maximum allowable stress can be evaluated for a given crack size provided the K_{Ic} value of the material is known. K_{Ic} generally decreases with decreasing temperature and increasing strain rate, and vice versa. It is also strongly dependent on microstructural and compositional variables such as crystallographic texture, grain size, impurities, inclusions, and other parameters.

Let us take a bit more detailed look at the effect of thickness on the K_{Ic} parameter. A notch in a thick plate is generally known to be more detrimental than that in a thin plate because the notch in a thick plate results in a plane strain condition with a high degree of stress triaxiality. The material away from the crack tip imposes geometrical constraint that restricts lateral contraction near the crack tip. The extent of the constraint depends on the ratio of material thickness (B) and a measure of plastic zone size (r_y). Plastic zone is the region of plastic deformation created due to stress concentration around the crack tip in moderately ductile material. Fracture toughness evaluated under plane strain conditions is obtained under the maximum constraint or material brittleness. ***This is the lowest value of fracture toughness that can be observed, and therefore the designers rely on this value for fracture toughness limiting design.***

As illustrative examples, Fig. 9.14 shows variation of fracture toughness as a function of material thickness for two precipitation-hardenable aluminum alloys. Similar trends are also observed for other materials. Three regions can be observed in this figure. On the extreme left side, region-1 represents plane stress fracture toughness where B is less than or equal to r_y. The crack tip is under plane stress loading condition, and the relief of any constraint is complete. In this region, the plastic zones from each crack face merge, and fracture takes place through a macroscopic shearing mechanism that gives rise to a slant fracture (combined mode-I and mode-III) (Fig. 9.15 [a]). Note the plane stress condition

---

[13]Megson THG. *Introduction to Aircraft Structural Analysis*. Butterworth-Heinemann; 2010.

226 Chapter 9 Toughness of materials: toughness limiting design

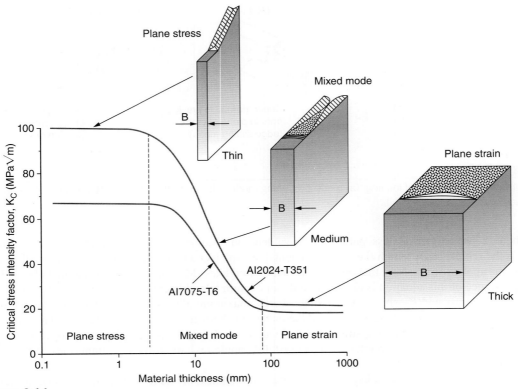

Figure 9.14

The critical stress intensity factor as a function of material thickness. Thickness must be large enough to achieve plane strain condition and to obtain plane strain fracture toughness. Trend lines from two well-known aerospace aluminum alloys, 2024Al and 7075Al, are plotted as examples.[14]

denoted by $\sigma_z = 0$. Maximum fracture toughness is obtained when B essentially becomes equal to r_y. In region-1, plastic deformation emanating from the crack tip would be limited by the thickness of the specimen. That is why with decreasing thickness, fracture toughness decreases in region-1. In region-2, where B is $>r_y$, fracture toughness decreases with associated increase in constraint at the crack tip with increasing thickness. The fracture here will be flat in the mid-region and slanted (known as shear lip) at the near-surface regions. The width of each shear lip region is generally half of the plastic zone size ($0.5r_y$). In essence, region-2 is a transition region between region-1 (plane stress) and region-3 (plane strain) conditions. Region-3 is obtained when B is much larger than r_y. Here the crack tip will be under plane strain condition and will be the minimum, i.e., independent of specimen thickness. When the specimen has critical thickness, the fracture surface becomes largely flat. (See

---

[14]Mouritz AP. *Introduction to Aerospace Materials*. Woodhead Publishing; 2012.

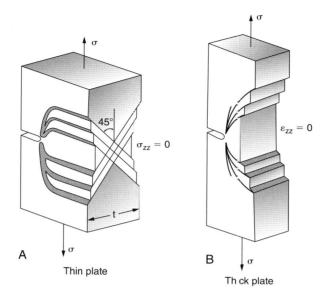

Figure 9.15

Crack deformation configuration under (a) plane stress (thin plate) and (b) plane strain (thick plate).[15]

Fig. 9.15 [b]; note condition $\varepsilon_z = 0$, denoting plane strain condition.) The mid-thickness of the specimen is subjected to plane strain condition due to lateral constraints which results in increase in stress levels ahead of crack tip. Fracture toughness reaches a minimum value that remains constant with increasing specimen thickness (Fig. 9.14). Thus as mentioned earlier, the plane-strain fracture toughness is called K_{Ic}, which is considered an inherent material property. The minimum thickness to achieve plane strain condition is given by

$$B \geq 2.5(K_{Ic}/\sigma_{ys})^2 \tag{9.6}$$

where σ_{ys} is yield stress. Note that K_{Ic}/σ_{ys} is usually very small for ceramic materials because of its limited or negligible ductility, and thus the condition in Eq. (9.6) is almost always satisfied. However, a thick enough material still must be used so that the microstructure contained in the specimen section truly represents bulk material behavior.

9.2 Stress intensity and role of mechanics

While fracture toughness has already been introduced previously, stress intensity factor was not discussed. George Irwin[16,17] first proposed the concept of stress intensity factor and developed solutions

---

[15]Hahn GT, Rosenfield AR. Local yielding and extension of a crack under plane stress. *Acta Metallurgica*. 1965;13(3):293-306.
[16]Irwin GR. *Fracture Dynamics, Fracturing of Metals*. American Society of Metals; 1948:147-166.
[17]Irwin GR. Analysis of stresses and strains near the end of a crack traversing a plate. *J App Mech*. 1957;24:361-364.

for stress distribution around the crack tips using Westergaard stress function.[18] Discussion on stress functions is outside the scope of this book; some references are included in the "Further reading" section at the end of the chapter. Stress distribution near the crack tip under mode-I loading condition is given by

$$\sigma_{xx} = \sigma \left(\frac{a}{2r}\right)^{\frac{1}{2}} \left[\cos\frac{\theta}{2}\left(1 - \sin\frac{\theta}{2}\sin\frac{3\theta}{2}\right)\right] \qquad (9.7a)$$

$$\sigma_{yy} = \sigma \left(\frac{a}{2r}\right)^{\frac{1}{2}} \left[\cos\frac{\theta}{2}\left(1 + \sin\frac{\theta}{2}\sin\frac{3\theta}{2}\right)\right] \qquad (9.7b)$$

$$\tau_{xy} = \sigma \left(\frac{a}{2r}\right)^{\frac{1}{2}} \left[\cos\frac{\theta}{2}\sin\frac{\theta}{2}\cos\frac{3\theta}{2}\right] \qquad (9.7c)$$

$$\sigma_{zz} = v(\sigma_{xx} + \sigma_{yy}) \text{ and } \tau_{xz} = \tau_{yz} = 0 \qquad (9.7d)$$

where σ is far-field stress and the point of interest in polar coordinates (r, θ) around the crack tip (Fig. 9.16). The equations above are exact solutions only when r approaches zero but still are adequate

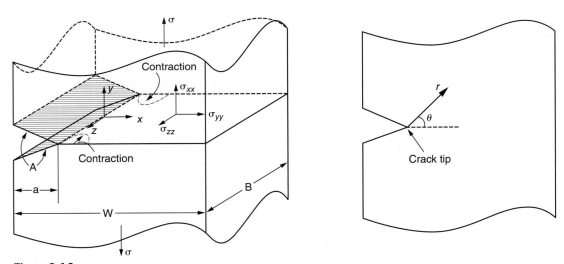

Figure 9.16

The geometrical configuration of a crack containing body in the relevant coordinate systems.

---

[18]Westergaard HM. Bearing pressures and cracks. *J App Mech.* 1939;6:49-53.

9.2 Stress intensity and role of mechanics

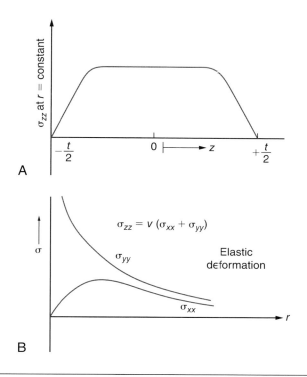

Figure 9.17
The distribution of stresses for obtaining expressions for stress intensity factors: (a) σ_{zz} and (b) σ_{xx} and σ_{yy}.

approximations when $r \ll a$ and $\ll$ the smallest planar dimension of the specimen. The distribution of different stresses is plotted as a function of position in Fig. 9.17.

Eq. (9.7a–d) makes clear that when angle θ becomes zero (just ahead of the crack along X-axis), normal stresses at any point become $\sigma \left(\dfrac{a}{2r}\right)^{1/2}$ and shear stress is zero. Irwin also noted that local stresses near the crack tip involve the product of the applied (global) stress and the square root of the crack size (half crack size for the center crack) and came up with an expression of a new term called stress intensity factor (K) given by the following expression:

$$K = \sigma \sqrt{(\pi a)}. \tag{9.8}$$

Replacing the above equation into Eq. (9.7a–c) enables writing the following for mode-I loading conditions:

$$\sigma_{xx} = \dfrac{K_I}{\sqrt{2\pi r}} \left[\cos\dfrac{\theta}{2} \left(1 - \sin\dfrac{\theta}{2}\sin\dfrac{3\theta}{2}\right) \right] \tag{9.9a}$$

$$\sigma_{yy} = \dfrac{K_I}{\sqrt{2\pi r}} \left[\cos\dfrac{\theta}{2} \left(1 + \sin\dfrac{\theta}{2}\sin\dfrac{3\theta}{2}\right) \right] \tag{9.9b}$$

Chapter 9 Toughness of materials: toughness limiting design

$$\tau_{xy} = \frac{K_I}{\sqrt{2\pi r}}\left[\cos\frac{\theta}{2}\sin\frac{\theta}{2}\cos\frac{3\theta}{2}\right] \quad (9.9c)$$

where $\sigma_{zz} = \nu(\sigma_{xx} + \sigma_{yy})$ for plane strain and $\sigma_{zz} = 0$ for plane stress, and $\tau_{xz} = \tau_{yz} = 0$.
For mode-II loading, similar expressions are obtained:

$$\sigma_{xx} = -\frac{K_{II}}{\sqrt{2\pi r}}\left[\sin\frac{\theta}{2}\left(2 + \cos\frac{\theta}{2}\cos\frac{3\theta}{2}\right)\right] \quad (9.10a)$$

$$\sigma_{yy} = \frac{K_{II}}{\sqrt{2\pi r}}\left[\sin\frac{\theta}{2}\cos\frac{\theta}{2}\cos\frac{3\theta}{2}\right] \quad (9.10b)$$

$$\tau_{xy} = \frac{K_{II}}{\sqrt{2\pi r}}\left[\cos\frac{\theta}{2}\left(1 - \sin\frac{\theta}{2}\sin\frac{3\theta}{2}\right)\right] \quad (9.10c)$$

where $\sigma_{zz} = \nu(\sigma_x + \sigma_y)$ and $\tau_{xz} = \tau_{yz} = 0$.

Concept Builder – Visualization and Magnitude of Stress Concentration: A key concept in this chapter is stress intensity at cracks. The concept also applies to voids and second-phase particles. Essentially, whenever you have elastic field discontinuity, you can expect some form of stress concentration. The form and details of stress concentration will change with the type of discontinuity. For an easy visualization of stress concentration, consider an analog of the stress flow lines shown in (a) of this box figure as fluid flow lines in a pipe. And think of the hole shown in (b) as a piece of marble blocking the flow of fluid! For the amount of fluid flowing through the pipe to stay constant through different parts of the cross-section, the fluid around the marble blockage will get squeezed through a narrower opening at higher velocity. Similarly, the stress flow lines converge in the regions next to the void. Remember that the void cannot support any stresses. So, the applied stress in that cross-section has to be carried by reduced cross-section of the material. As shown in (c) of the box figure, that leads to **stress concentration** in that region. The magnitude of the stress concentration depends on the shape of the void. As shown in the (c) of the box figure, the general expression for stress concentration for an elliptical void is given by $K_{conc} = 1 + 2(a/b)$, where a and b are the major and minor axes of the ellipsoid (students are encouraged to see Peterson RE. *Stress Concentration Factors*. John Wiley & Sons; 1974, for impact of various geometrical shapes and design features on stress concentration values). For a circular hole, a = b, which gives the maximum stress concentration of 3 at the edge of the circular void. The stress concentration reduces to the nominal stress level when the distance from the void reaches around its size. The stress concentration, K_t, is defined as the ratio of the local stress to the applied nominal stress, i.e., $K_t = \sigma_{local}/\sigma_{nominal}$. For our discussion purposes in this chapter on toughness and the next one on fatigue, this concept of K_t is very important. For simplicity, just try to remember the maximum value of K_t as 3 for conceptual discussion that influences plasticity and fracture behavior.

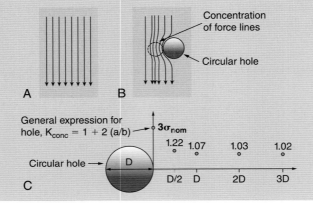

9.3 Details of fracture toughness testing

9.3.1 For less ductile materials: Plane strain fracture toughness testing procedure

Three types of specimens shown in Fig. 9.18 are commonly used to measure K_{Ic}, even though there are more specimen types such as center-cracked, double-edge-cracked, and single-edge-cracked specimens; double

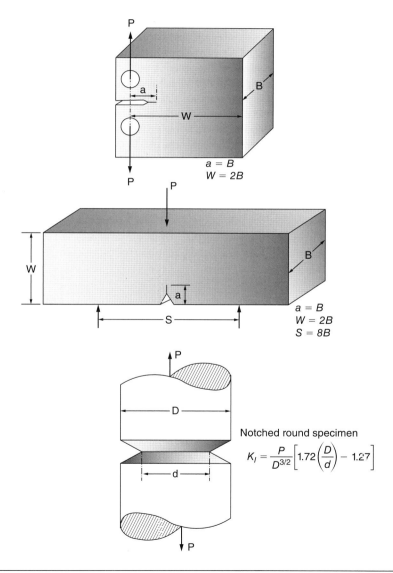

Figure 9.18

Three specimen designs (compact tension, three-bend, and notched round specimen) for K_{Ic} measurement.

cantilever beam specimens; and double torsion specimens. ASTM standard E-399 describes the test procedure and specifications for fracture toughness measurement. The compact test specimen (CT) is the most popular for fracture toughness measurement of metals/alloys. The presence of notch is not enough to produce the needed stress concentration. After the notch is machined into the specimen, low cycle (about 100 cycles) fatiguing is carried out to create the sharpest possible crack at the notch root. With $a = 0.45$–$0.55W$ (stress ratio is generally kept at 0.1), fatiguing should be done with proper care to avoid plastic deformation at the crack tip that may lead to blunting even before fracture toughness can be evaluated.

Plane strain fracture toughness determination is not straightforward in that there is no advance assurance that a valid K_{Ic} can be measured by doing a single test. This situation is quite different from a tensile test where data on yield strength, ultimate tensile strength, and percentage elongation to fracture of a material can be obtained. An estimate of the expected K_{Ic} is made using Eq. (9.6) to determine specimen thickness required for plane strain loading conditions. The specimen is pulled in tension with clip-in displacement gage for *crack mouth opening displacement* (CMOD), which is proportional to increase in crack length.

Fracture toughness testing is usually performed in a testing machine that provides for continuous recording of load (P) using load cell and relative displacement across the open end of the notch using the clip-in gage. Fig. 9.19 offers three kinds of load/crack displacement curves (Type-I, Type-II, and Type-III) depending on the type of material tested. The ASTM procedure requires first drawing a secant line OP_s from the origin with a slope that is 5% less than tangent OA. This fixes the P_s. Next is to draw a horizontal line at a load equal to 80% of P_s and measure the distance along this line from tangent OA to the actual curve. If the value of x_1 exceeds one-fourth of the corresponding distance x_s at P_s, the material is too ductile to obtain a valid K_{Ic}. If the material is not too ductile, the load P_s is then designated as P_Q and is used in the calculation.

The value of P_Q determined from the load-displacement curve is used to calculate a conditional value of fracture toughness denoted by K_Q using the equation (for CT specimen) below.

$$K_Q = \frac{P_Q}{B\sqrt{W}} f\left(\frac{a}{W}\right) =$$

$$\frac{P_Q}{B\sqrt{W}}\left[29.6\left(\frac{a}{W}\right)^{\frac{1}{2}} - 185.5\left(\frac{a}{W}\right)^{\frac{3}{2}} + 655.7\left(\frac{a}{W}\right)^{\frac{5}{2}} - 1017.0\left(\frac{a}{W}\right)^{\frac{7}{2}} + 638.99\left(\frac{a}{W}\right)^{\frac{9}{2}}\right] \quad (9.11)$$

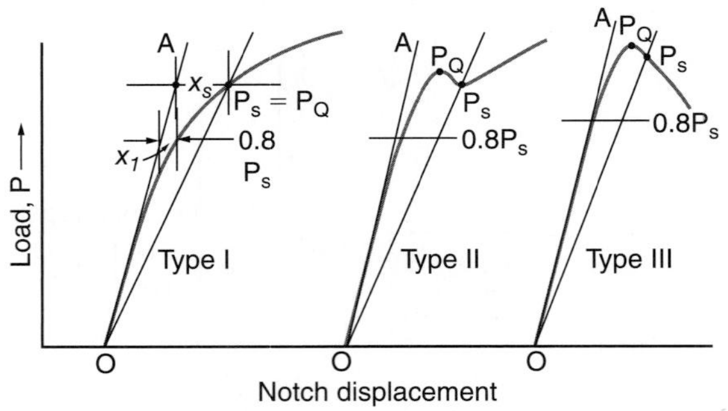

Figure 9.19

Load-displacement curves (type-II and type-III show "pop-in" behavior).[19]

---

[19]*E399-20a, Standard test method for linear-elastic plane strain fracture toughness if metallic materials*. ASTM International.

Crack length (*a*) used in this equation is measured after fracture. Then the factor $2.5(K_Q/\sigma_y)^2$ is calculated. If this quantity is less than both thickness and crack length of the specimen, then K_Q is actually K_{Ic}. Otherwise, thicker specimens must be used until a valid K_{Ic} value can be determined. The measured value of K_Q can be used to estimate the new specimen thickness through Eq. (9.6).

Type-II and type-III load-displacement curves exhibit *pop-in* behavior (sharp decrease in load), which is represented by sudden, unstable fast crack propagation. For type-II curve, the load is recovered after a *pop-in* event. Here also is the same procedure as the method applied to type-I curve. Note that here P_Q is the maximum load. However, for type-III curve, no recovery of load leads the *pop-in* behavior to complete failure and applies to elastically very brittle materials.

Table 9.1 shows fracture toughness and yield strength values of various materials along with critical crack lengths, which can be determined using Eq. 9.5.

9.3.2 Ductile materials – plastic zone

Although the term plastic zone (r_y) was introduced in a previous section, the concept was not elaborated. The original concept was introduced by Irwin to develop a plasticity zone correction factor to take into account the inherent ductility of a material. Fig. 9.20 shows stress concentration created near the crack tip, which can go above the yield stress and can create a plastically deformed zone around

Table 9.1 A list of room temperature fracture toughness and yield strength and critical crack length.

Material	K_{IC} MPa√m	σ_{ys} MPa	a_c mm
2014-T651	24.2	455	3.6
2024-T3	~44.0	345	~21.0
2024-T851	26.4	455	4.3
7075-T651	24.2	495	3.0
7178-T651	23.1	570	2.1
7178-T7651	33.0	490	5.8
Ti-6Al-4V	115.4	910	20.5
Ti-6Al-4V	55.0	1035	3.6
4340	98..9	860	16.8
4340	60.4	1515	2.0
4335+V	72.5	1340	3.7
17-7PH	76.9	1435	3.6
15-7Mo	49.5	1415	1.5
H-11	38.5	1790	<0.6
H-11	27.5	2070	0.23
350 Maraging	55.0	1550	1.6
350 Maraging	38.5	2240	<0.4
52100	~14.3	2070	~0.06

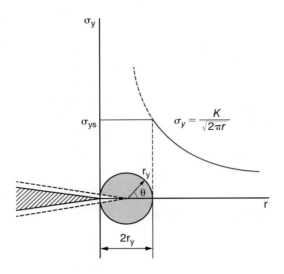

Figure 9.20
Effective crack length concept including plastic zone size.

the crack tip. Fig. 9.20 shows the situation with the plastic zone ahead of the crack tip. Irwin considered an elastic-plastic solution in analyzing materials with moderate ductility and assumed an effective crack length (a_e) instead of a physical crack size (a).

$$a_e = a + r_y \tag{9.12}$$

Plastic zone size is estimated by equating normal stress $\sigma_{yy}(r,0)$ equal to uniaxial yield stress. Mathematically then:

$$\sigma_{yy}(r_y,0) = \frac{K}{\sqrt{2\pi r_y}} = \sigma_y.$$

Thus plastic zone size is given by

$$r_y = \frac{1}{2\pi}\left(\frac{K}{\sigma_{ys}}\right)^2 \quad \text{(for plane stress condition)} \tag{9.13a}$$

and

$$r_y = \frac{1}{6\pi}\left(\frac{K}{\sigma_{ys}}\right)^2 \quad \text{(for plane strain condition).} \tag{9.13b}$$

The expression of plastic zone size for plane strain is an approximate expression of various analyses given the complex nature of the constraint imposed by the plane strain condition. However, when plastic zone size becomes an appreciable fraction of crack length size, the above correction ceases to be valid. Under those circumstances, other concepts like *crack tip opening displacement* (CTOD) and *J*-integral should be used.

9.4 Elastic-plastic fracture mechanics

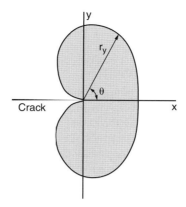

Figure 9.21
The shape of the plastic zone.

Plastic zone size was derived at $\theta = 0$. However, to understand the shape of the plastic zone, its dimension at other angular values must be known. So, to find the locus of yielding (i.e., the boundary between the plastically and elastically deformed regions) as a function of θ for mode-I loading, the following equation can be used from an analysis that assumes von Mises criterion:

$$r_y = \frac{K_I^2}{2\pi\sigma_{ys}^2}\left[\frac{1}{2}\left(1+\cos\theta + \frac{3}{2}\sin^2\theta\right)\right], \text{ for plane stress} \tag{9.14a}$$

and

$$r_y = \frac{K_I^2}{2\pi\sigma_{ys}^2}\left[\frac{1}{2}\left\{(1-2\nu)^2(1+\cos\theta) + \frac{3}{2}\sin^2\theta\right\}\right], \text{ for plane strain.} \tag{9.14b}$$

The equation for the plane strain condition predicts a smaller plastic zone size as the geometrical constraint is larger. Fig. 9.21 shows the shape of the plastic zone ahead of crack tip for plane strain loading. Although these equations can predict plastic zone size and shape, they are governed by other factors including work hardening behavior of the material. However, actual size and shape can be measured only by experiments.

9.4 Elastic-plastic fracture mechanics

The analysis and discussion given in the previous sections are applicable only when the non-linear material deformation is confined to a very small region near the crack tip as given by Eqs. (9.13a) and (9.13b) for plane stress and plane strain conditions, respectively. The above analysis is called *linear elastic fracture mechanics* (LEFM) approach. However, for materials that exhibit severe nonlinear material behavior, LEFM approach is no longer applicable as the plastic zone size ahead of the crack tip is larger than those proposed by LEFM. For such materials *elastic-plastic fracture mechanics* approach is utilized. This section introduces two fracture-related elastic-plastic parameters called CTOD and J-integral, both of which can be used as measures of fracture toughness when there is severe crack tip plasticity.

9.4.1 Crack tip opening displacement

Wells[26] was the first to notice the limitation of applicability of LEFM in materials with high toughness in which crack tip blunting of the sharp crack was observed due to opening of crack under load, as shown in Fig. 9.22 (a). He also observed that the extent of crack tip blunting was proportional to the toughness of material and established that the crack tip opening is a measure of fracture toughness, which is called the CTOD. An approximate analysis for determining the relationship of CTOD (δ) with stress intensity was performed by Wells. He assumed the effective crack length as $(a + r_y)$ as shown in Fig. 9.22 (b) according to Irwin[20] and calculated the displacement u_y at a distance r_y from the effective crack tip.

Similar to Eqs. (9.9 a–c), which describe the stress field around the crack tip in mode-I loading condition, Eqs. (9.15 a, b) describe the displacement field around the crack tip.

$$u_x = \frac{K_I}{2\mu}\sqrt{\frac{r}{2\pi}}\cos\left(\frac{\theta}{2}\right)\left[\kappa - 1 + 2\sin^2\left(\frac{\theta}{2}\right)\right] \tag{9.15a}$$

$$u_y = \frac{K_I}{2\mu}\sqrt{\frac{r}{2\pi}}\sin\left(\frac{\theta}{2}\right)\left[\kappa + 1 - 2\cos^2\left(\frac{\theta}{2}\right)\right] \tag{9.15b}$$

where μ is the shear modulus, $\kappa = 3 - 4\nu$ for plane strain, and $\kappa = (3 - \nu)/(1 + \nu)$ for plane stress. Value of u_y at the original crack tip is obtained from Eq. (9.15b) as follows:

$$u_y = \frac{(k+1)K_I}{2\mu}\sqrt{\frac{r}{2\pi}} \tag{9.16}$$

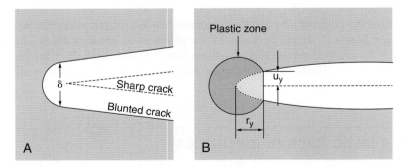

Figure 9.22

(a) Blunting of an original sharp crack and associated crack tip opening displacement (CTOD), and (b) determination of CTOD using Irwin's effective crack length correction.[21]

---

[20]Wells AA. Unstable crack propagation in metals: cleavage and fast fracture. In: Crack propagation symposium, eds. *Proceedings of the Crack propagation symposium*. College of Aeronautics and the Royal Aeronautical Society Cranfield; 1961.
[21]Irwin GR. Plastic zone near a crack and fracture toughess. 7th Sagamore Ardance Materials Research Conference. Syracuse, University Press; 1960.

Under plane stress condition, the radius of the plastic zone, r_y, is given in Eq. (9.13a), which when substituted into Eq. (9.16), gives the value of CTOD as below:

$$\delta = 2u_y = \frac{4}{\pi E} \frac{K_I^2}{\sigma_{ys}}. \tag{9.17}$$

9.4.2 Experimental determination of CTOD

CTOD measurements need to be performed as per ASTM E1290 standard that specifies single-end notched bend (SENB) specimen and compact tension (CT) specimen as preferred geometries. Due to difficulty in accurate measurement of displacement at the crack tip, CTOD is determined by measuring the opening at the crack mouth (Δ) utilizing a hinge model where the two halves are assumed to rotate about the hinge at point O as shown in Fig. 9.23. CTOD is determined using a similar triangle construction as:

$$\frac{\delta}{\omega(W-a)} = \frac{\Delta}{\omega(W-a)+a}. \tag{9.18}$$

Hence,

$$\delta = \frac{\omega(W-a)\Delta}{\omega(W-a)+a}, \tag{9.19}$$

where ω is the rotational factor, value of which lies between 0 and 1.

ASTM standard specifies a modified hinge model in which the total displacements are separated into elastic and plastic components since the hinge model showed inaccuracies when the displacements are primarily elastic. The hinge analogy is applied only to the plastic part of the displacement. The general form of CTOD using the modified hinge model for a work hardening material is represented as

$$\delta = \delta_e + \delta_p = \frac{K_I^2}{m\sigma_{ys}E'} + \frac{\omega_p(B-a)\Delta}{\omega_p(B-a)+a}, \tag{9.20}$$

where $m = 1$, $E' = E$ and $m = 2$, $E' = E/(1-\nu^2)$ for plane stress and plane strain, respectively.

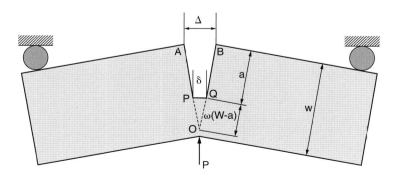

Figure 9.23

Illustration of CTOD estimation using hinge model.

9.4.3 J-integral

Introduction of a path-independent integral called J-integral by Rice[22] has been a huge leap in the fracture mechanics approach for analyzing cracks in nonlinear materials. Knowledge of J-integral enabled fracture property measurement in subsized samples which otherwise cannot be qualified for LEFM-based analysis. Additionally, computational implementation of fracture mechanics and determination of fracture properties has been eased with the advent of J-integral. J-integral is deliberated as both an energy-based parameter as well as a stress-related parameter. The energy-based approach has been proposed by Rice in which he showed that the value of a path-independent contour J-integral is equal to the energy release rate in a nonlinear material with a crack.

Consider a crack with the origin of the coordinate axis fixed at the crack tip as shown in Fig. 9.24. Ω is an arbitrary contour around the crack in counterclockwise direction, enclosing the crack tip. Rice has defined J-integral as

$$J = \int_\Omega \left(w\, dy - T_i \frac{\partial u_i}{\partial x} ds \right), \qquad (9.21)$$

where w is the strain energy density, T_i is the traction vector components along the contour Ω, u_i are the displacements, and ds is the length of the small segment along the contour. Strain energy density function w is defined as

$$w = \int_0^{\varepsilon_{ij}} \sigma_{ij} d\varepsilon_{ij}, \qquad (9.22)$$

where σ_{ij} is the Cauchy stress tensor and ε_{ij} is the strain tensor. Traction vector T_i is defined as $T_i = \sigma_{ij} n_j$, where n_j are the components of the normal vector along the contour Ω.

Following are some of the characteristics of J-integral:

1. J-integral is path independent, which means that the value of J-integral is independent of the path considered around the crack tip.
2. J-integral along a closed contour around the crack is zero.

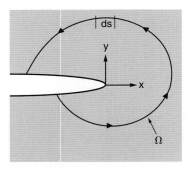

Figure 9.24

Contour around a crack tip along which J-integral is evaluated.

---

[22]Rice JR. A path independent integral and the approximate analysis of strain concentration by Notches and cracks. *J Appl Mech*. 1968;35:376-386.

3. J-integral is equal to the energy release rate in an elastic-plastic (approximated as nonlinear elastic) material which is given by

$$J = -\frac{d\Pi}{dA} \tag{9.23}$$

where Π is the potential energy of the cracked body and A is the crack area.

4. For a linear elastic material, J reduces to G, which is the energy release rate in linear elastic material, and $J = G = K_I^2/E'$.

9.4.4 Experimental determination of J-integral

Experimental determination of J-integral is easier when the material behavior is linear elastic. In such case, $J = G$, and G can be determined through evaluation of K_I. However, for nonlinear materials calculation of J is comparatively difficult. ASTM E1820[23] specifies the standard guidelines for determination of J. Similar to CTOD experiments, SENB and CT specimens are recommended for J-integral determination too. Although multiple ways are available to determine J, ASTM standard specifies a much simpler method from the load vs. displacement curve, as shown in Fig. 9.25, of a cracked SENB or CT specimen. The J can be expressed as

$$J = J_e + J_p \tag{9.24}$$

where subscripts e and p stand for elastic and plastic contributions, respectively. The elastic and plastic components are

$$J_e = \frac{K_I^2}{E'}, \text{ and} \tag{9.25a}$$

$$J_p = \frac{\eta_p A_p}{B(W-a)}, \tag{9.25b}$$

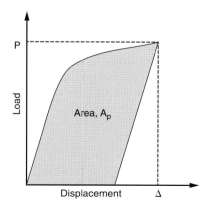

Figure 9.25

Calculation of area under the load-displacement curve for experimental determination of J-integral.[24]

---

[23]*E1820-18, Standard test method for measurement of fracture toughness.* ASTM International.
[24]*E1820-18, Standard test method for measurement of fracture toughness.* ASTM International.

where η_p is a dimensionless constant, A_p is the area under the load-displacement curve for a specific P and Δ, B is the nominal specimen thickness (after reducing the depth of side grooves if any), and W and a are as shown in Fig. 9.18. $\eta_p = 1.9$ for SENB specimen and $(2 + 0.522\,(W - a)/W)$ for CT specimen. ASTM specifies method to determine J-integral from load vs. CMOD as well.

9.5 Damage-tolerant design approach based on assumption of flaws

9.5.1 Design considerations (Irwin's leak-before-break approach)

The traditional design approach as discussed in **Chapter 8** is based on strength, which generally makes sure that plastic deformation is averted by keeping service stresses below yield stress. The other approach in the context of this chapter deals with averting fracture. Irwin proposed the leak-before-break criterion.

Let's first discuss a conceptual situation. Consider a wide plate with a center crack length of $2a$ subjected to an applied tensile stress of σ. If yielding and fracture are assumed to occur simultaneously, the minimum fracture toughness is given by

$$(K_{IC})_{min} = \sigma_{ys}\sqrt{\pi a}. \tag{9.26}$$

Thus the minimum fracture toughness required would depend on yield stress and allowable crack size that will depend on the effectiveness and resolution of the measurement of the nondestructive evaluation techniques used.

Irwin suggested an alternative approach geared toward a practical approach with the possibility of design, material selection, and operation of pressure vessels in power plants. Such an approach assumes the safety consideration that would preclude periodic inspection of the pressure vessel. The idea is that if a puddle of water forms at the bottom of the pressure vessel, a crack has grown to such a size that it has penetrated the wall. So, the safe design would be to have the material's fracture toughness sufficient to avoid fracture at a particular crack size and stress. Irwin assumed a central crack to be along the axis of the vessel with the crack faces being at 90° to the circumferential direction (Fig. 9.26). The crack length $(2a)$ will be equal to twice the wall thickness $(2B)$, and the crack becomes a through-thickness crack.

For optimum design, yielding and fracture would have to happen simultaneously. Then hoop stress and crack tip stress intensity factors have to reach yield stress and minimum fracture toughness, respectively. Irwin's approach takes into account plastic zone size (r_y) and crack size. So, a_{eff} needs to be considered for minimum fracture toughness.

$$(K_c)_{min} = \sigma_{ys}\sqrt{\pi a_{eff}},\ \text{or}\ a_{eff} = \frac{(K_c)^2_{min}}{\pi \sigma_{ys}^2} \tag{9.27}$$

$$\text{Or,}\ a_{eff} = a + r_y = a + \frac{(K_c)^2_{min}}{2\pi \sigma_{ys}^2} \tag{9.28}$$

From Eqs. (9.27) and (9.28), physical crack length (a) can be written in terms of minimum fracture toughness and yield stress at the onset of leakage:

$$a = \frac{(K_c)^2_{min}}{2\pi \sigma_{ys}^2},\ \text{or}\ (K_c)_{min} = \sigma_{ys}\sqrt{2\pi a}$$

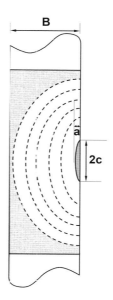

Figure 9.26
A schematic diagram showing Irwin's "leak-before-break" approach with a semielliptical surface crack – at leak condition ($a = B$) and unbroken ligaments (shown by the shaded area) break open to form a through-thickness crack.[25]

As stated before, Irwin assumed that if half-crack length (a) becomes equal to wall thickness (B) of the vessel, it would leak. So, the relation becomes

$$(K_c)_{min} = \sigma_{ys}\sqrt{2\pi B}. \tag{9.29}$$

From Eq. (9.29), some practical understanding can be gained. Greater minimum fracture toughness is required for higher yield strength and higher section thickness. However, sometimes fracture toughness is sacrificed to permit use of high-strength material. So, during design and materials selection, this should be kept in mind.

9.6 Unintended consequence of constituent particles/inclusions

Fracture toughness is a microstructure-sensitive property and is affected by microstructural anisotropy, alloy cleanliness, and grain size, to name a few. Here the effect of inclusions on fracture toughness is discussed. Inclusions are very much prevalent in many commercial alloys. Inclusions are coarse, second-phase particles present in the microstructure. If the material is strong and needs to have good toughness as well, inclusions must be eliminated or drastically minimized. If the inclusions

---

[25]Raju IS, Newman Jr JC. Stress-intensity factors for internal and external surface cracks in cylindrical vessels. *J Press Vessel Technol*. 1982;104:293-298.

do not have good bonding with the matrix, large inclusions with wide interparticle spacing may be less damaging than fine inclusions with close interparticle spacing. Below, we will consider some examples from commercial aluminum alloys to describe the effect of the inclusions on toughness.

Coarse particles (*aka* constituent particles) form interdentrically by eutectic decomposition during ingot solidification. One group is virtually insoluble (even after homogenization treatments) and appears to contain iron and/or silicon. Examples include $Al_6(Fe, Mn)$, Al_3Fe, $\alpha Al(Fe, Mn, Si)$, and Al_7Cu_2Fe. The second group, which is known as soluble constituents, consists of equilibrium intermetallic compounds of the major alloying elements. Examples include Al_2Cu, Al_2CuMg, and Mg_2Si. During subsequent fabrication of cast ingot, the size of these particles broken down to 0.5–10 μm causes them to become aligned as stringers in the direction of working or metal flow.

Fracture toughness depends on the distribution of particles, resistance to decohesion, and stress concentration. Fracture toughness decreases with increasing Fe and Si content in aluminum alloys because of the formation of coarse Fe-rich and Si-rich inclusions. These inclusions serve as sites for cracking and thus reduce fracture toughness. Fig. 9.27 shows crack path touching the coarse Si particles in an aluminum casting alloy.

Impurity content (such as Fe and Si) significantly affects the fracture toughness of aluminum alloys. As shown in Fig. 9.28, fracture toughness values of two Al-Cu-Mg sheets with Fe + Si concentration <0.5 wt.% and another one with Fe-Si content with >1 wt.% are plotted as a function of yield strength. Clearly the alloy with lower Fe and Si content has consistently higher fracture toughness even at the same yield strength level. This occurs because Fe and Si create coarse and brittle constituent particles that happen to decrease fracture toughness. The same figure also makes clear that toughness and strength in these alloys are inversely related. The conflict between strength and toughness is highlighted in the next concept box.

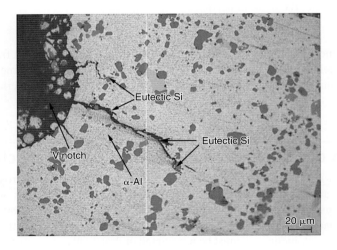

Figure 9.27

Microstructure of an aluminum casting alloy with a crack passing through the coarse Si particles.[26]

---

[26]Chao Wei, Guang-lei Liu, Hao Wan, Yu-shan Li, Nai-chao Si. Effect of heat treatment on microstructure and thermal fatigue properties of Al-Si-Cu-Mg alloys. *High Temperature Materials and Processes*, 2018;37(4):289-298.

9.6 Unintended consequence of constituent particles/inclusions

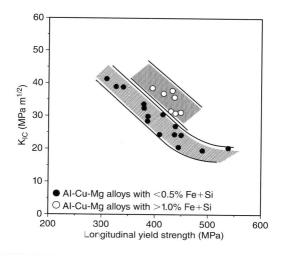

Figure 9.28

Plane strain fracture toughness vs. yield strength in two Al-Cu-Mg alloy sheets – one with low Fe and Si content and another with higher impurity level.[27]

Concept Reinforcement – The Conflict Between Strength and Toughness: Ritchie (Ritchie RO. The conflicts between strength and toughness. *Nat Mater*. 2011;10:817) has highlighted the tradeoff between strength and toughness. We use his paper for concept reinforcement boxes in this chapter. As highlighted in Chapter 2, Ashby charts are a nice way to capture the overall trend. The Box Figure below captures the toughness-strength variation in many materials. Irwin's "leak-before-break" criterion gives us $a = \dfrac{(K_c)^2_{min}}{2\pi\sigma_{ys}^2}$, as highlighted before. This relationship can be plotted to show that as we increase the strength of a material, to keep the same level of damage tolerance, we must increase the toughness proportionately! However, as you see from the Ashby bubble charts for individual alloy group, the toughness decreases with increase in strength for an alloy. This provides opportunities to design alloys that beat this traditional paradigm!

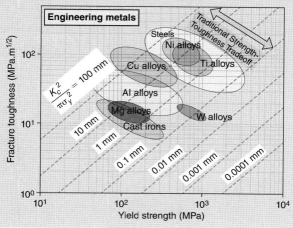

[27]After: Speidel MO. *Proceedings of Sixth International Conference on Light Metals*. Austria: 1975. Taken from Polmear IJ. *Light Alloys*. Elsevier.

Generally, toughness for a given microstructure increases with increasing temperature and decreases with increasing strain rate. However, in situ metallurgical phenomena can impact this behavior.

In a ductile metallic alloy, if the fracture is assumed to be always due to microvoid nucleation and coalescence related, we can write a simple mathematical expression for the fracture toughness parameter, J_{Ic}, as follows:

$$J_{Ic} = \sigma_{ys}\, \varepsilon_f\, l_{mv} \tag{9.30}$$

where σ_{ys} is the yield stress, ε_f is the fracture strain, and l_{mv} is the mean distance between microvoids before the crack tip that establishes the plastic process zone.

9.7 Toughening of ceramics

Unlike metals, ceramics are generally considered brittle and toughness is not something that we associate with them. In reality, however, the scenario changed quite a bit starting in the 1980s or so when scientific/technological breakthroughs have rendered toughening ceramics a reality. Some of the earliest breakthroughs were achieved with the development of partially stabilized zirconia and fiber-reinforced ceramic composites based on fracture mechanics approach. It is not the tensile toughness or impact toughness; it is the fracture toughness that is of importance for understanding ceramic toughness. Microstructural control has been found to be key in toughening of ceramics, which enhances the energy absorption during crack growth. Fracture toughness of a few ceramics and related materials is shown in Table 9.2. In this list, materials with easy cleavage planes, such as silicate glasses and single crystals, exhibit low toughness. However, materials that are polycrystalline and/or do not cleave easily appear to show bit higher fracture toughness. Transformation-toughened ceramics like PSZ, TZP, Al_2O_3-ZrO_2 exhibit higher fracture toughness due to microstructural control. Nevertheless, metallic alloys, including aluminum alloys, steels, and titanium alloys, have still much higher fracture toughness as shown in Table 9.1.

Table 9.2 Room temperature fracture toughness of some ceramics and glasses.

Material	K_{IC} (MPa$\sqrt{m^{1/2}}$)
Silicate glasses	0.7–0.9
Glass ceramics	~2.5
Single crystal MgO	~1
Polycrystalline Al_2O_3	3.5–4
Sintered, hot-pressed SiC	4–6
Reaction-bonded Si_3N_4	2.5–3.5
Cubic stabilized ZrO_2	~2.8
MgO-partially stabilized ZrO_2 (PSZ)	9–12
Tetragonal zirconia polycrystals (Y-TZP)	6–12
Al_2O_3-ZrO_2 composites	6.5–13
Single crystal WC	~2
Metal (Ni, Co) coated WC	5–18

9.7 Toughening of ceramics

Ceramic toughening has mainly been accomplished using three approaches: crack deflection, crack bridging, and phase transformation. Here, we will not discuss all possible ceramic toughening approaches. For more details, readers are referred to the resources listed at the end of this chapter.

Crack deflection refers to the situations where cracks in polycrystalline ceramics take more energy absorbed when the cleavage plane changes path as the cleavage crack moves from one grain to another, as shown in Fig. 9.29 (a). However, intergranular fracture actually leads to more fracture energy, i.e., higher toughness if the crack moves in a zigzag fashion following the grain boundary architecture as shown in Fig. 9.29 (b). The fracture surface area would be higher than the nominal fracture surface area due to the tortuosity of the grain boundary path. The maximum increase in fracture toughness could be two times at most from this mechanism. Also, application of grain boundary engineering principles with a certain increased population of special grain boundaries as well as suitable grain boundary connectivity can lead to a greater resistance to crack propagation and thus better toughness can result. The effect of these factors on the fracture toughness in nanocrystalline ceramics needs to be carefully studied.

Crack bridging occurs when fibers, whiskers, or certain elongated grains can impart additional fracture energy by severing bonds between the matrix and those ligaments when the rest of the part of the ceramic already failed. Interestingly, the crack bridging is employed when straws are added to clay to build mud huts with toughened walls. While the crack bridging mechanism can be utilized in different instances, including in polycrystalline alumina for which it was demonstrated first, it is most elegantly utilized in two-phase materials. One very important example in this regard is fiber/whisker-reinforced composites. In these cases, pull-out of brittle reinforcing phases requires energy because of frictional work mainly. For ductile fibers, plastic deformation can have a greater role than the pull-out work. However, thermal residual stresses present within the composites may augment the process. Some of the factors by which the toughening effect of crack bridging can be enhanced are by increasing the volume fraction of the reinforcements (fibers/whiskers), reducing interfacial bond between fiber-matrix, and using reinforcements with lower elastic modulus. Fig. 9.30 (a) shows a schematic of composite undergoing crack bridging with fiber pull-outs being clearly evidenced. The effect of SiC reinforcing fibers on the fracture toughness of various ceramics and glass is shown in Fig. 9.30 (b).

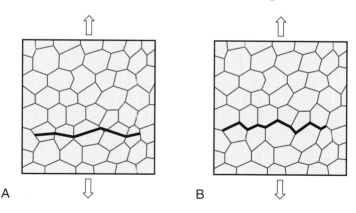

Figure 9.29

Crack propagation in a schematic polycrystalline ceramic microstructure: (a) cracks grow through grains via preferred crystallographic planes varying from one grain to another, and (b) crack propagates following grain boundaries.

Chapter 9 Toughness of materials: toughness limiting design

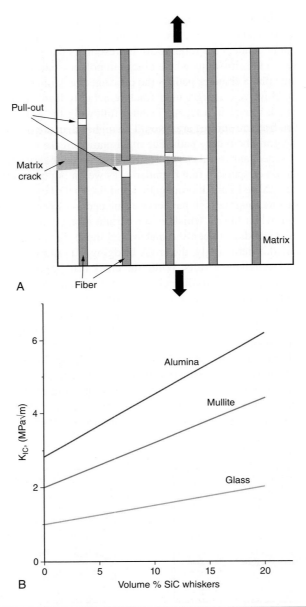

Figure 9.30
(a) A schematic depiction of toughening by crack bridging in a fiber-reinforced composite.[28] (b) Effect of SiC whiskers on the fracture toughness of ceramics and glasses via crack bridging.[29]

---

[28]Adapted from Lawn B. *Fracture of Brittle Solids*. 2nd ed. Cambridge University Press; 1993.
[29]Becher PF. Microstructural design of toughened ceramics. *J Am Ceram Soc*. 1991;74(2):255-269.

9.7 Toughening of ceramics

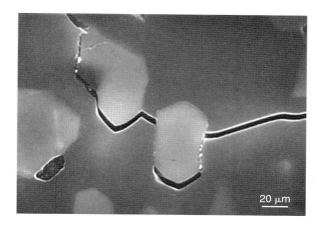

Figure 9.31

Bridged crack by TiB$_2$ particle in a SiC–TiB$_2$ composite.[30]

Fig. 9.31 shows a scanning electron microscopy image of SiC–TiB$_2$ particulate composite where the bright-looking TiB$_2$ particles provide the bridges between the crack opening, much like a peg holding a structure in a hole.

It is important to note that effects similar to the crack-bridging can be achieved by introducing ductile phases in brittle ceramics. Let us take an example from WC. Table 9.2 lists the fracture toughness value of WC fracture toughness (<2 MPa$\sqrt{m}^{1/2}$). However, the introduction of metal phase (Ni, Co, etc.) into WC creates ceramic matrix composites (cermets) with considerably improved toughness (5–18 MPa$\sqrt{m}^{1/2}$).

One very important toughening mechanism that really helps elevate fracture toughness in ceramics is transformation toughening. This toughening mechanism is based on the stress-induced martensitic transformation that occurs mainly in zirconia ceramics (ZrO$_2$). ZrO$_2$ ceramics exhibits polymorphism. Monoclinic is the low-temperature phase; at a higher temperature it transforms into tetragonal phase and then to a cubic phase. However, suitable addition of dopants (Y$_2$O$_3$) can stabilize the high-temperature tetragonal phase to room temperature. Application of sufficient stress on the metastable tetragonal ZrO$_2$ phase can cause it to undergo stress-induced martensitic transformation to monoclinic ZrO$_2$. This type of transformation can involve about 4% volume expansion and significant amount of shear strain ($\sim$14%). For example, a zirconia-toughened alumina (ZTA) or a partially stabilized ZrO$_2$ (PSZ) will exhibit enhanced toughness because of the transformation toughening. These materials need to have metastable tetragonal ZrO$_2$ phase that under the influence of the crack tip stress concentration gets converted to monoclinic ZrO$_2$ phase via the stress-assisted martensitic transformation. The situation creates a compressive state of stress that significantly reduces the driving force for crack propagation. The metastability engineering to alter toughness of a ceramic phase is a good continuation of the discussion of TRIP in metals from Chapter 8. Fig. 9.32 shows a schematic situation where the metastable tetragonal ZrO$_2$ phases near the crack-tip martensitically transform to monoclinic ZrO$_2$. It is also possible to combine a set of toughening mechanisms to create synergistic enhancement.

---

[30]Faber KT. *Ceramics: Microstructural Toughening.* In: Buschow KH, et al., eds. *Encyclopedia of Materials: Science and Technology.* Elsevier; 2001.

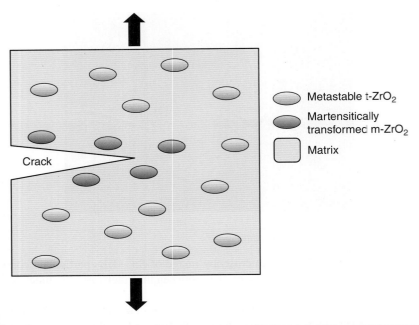

Figure 9.32
Schematic illustration of transformation toughening mechanism with metastable tetragonal zirconia (t-ZrO$_2$) phases (denoted by light gray ellipses) martensitically transforming to monoclinic zirconia (denoted by dark ellipse), creating a zone of compression around the crack tip.

Key chapter takeaways

This chapter presented three ways to measure toughness – tensile, impact, and fracture toughness. Tensile toughness can be obtained by tensile testing. However, tensile toughness is of little significance for design of components. It does not lead to a measurable value that can assist designers. Impact toughness as a quantitative material property is not perfect for design purposes, but it is still useful as a relative property comparison and provides guidance to useful temperature range for components through the transition temperature concept. The most useful toughness parameter in modern design is the fracture toughness. Actually, plane strain fracture toughness is an inherent mechanical property of material. It also provides designers with a guaranteed minimum toughness of a material. This chapter discussed some fundamental concepts and design criteria. Fracture toughness depends on microstructural characteristics such as grain size, texture, and particle microstructure. The adverse effects of coarse constituent particles on fracture toughness are highlighted. This is an important aspect of material selection of toughness-limiting design. The constituent particles form from impurities, and therefore control of these can be done with alloy specification. Cleaner compositions lead to higher toughness. Ceramics have challenges in attaining toughness given the lack of intrinsic ductility at low temperatures. Some of the toughening mechanisms for ceramics including transformation toughening are described in this chapter. The overall toughness of metals and ceramics is either based on "intrinsic" plasticity-based mechanisms or "extrinsic" approaches to enhance energy absorption during crack growth. As a recap of the entire chapter, these aspects are captured in the concept reinforcement box with the help of an overview figure.

Concept Overview – Intrinsic and Extrinsic Toughening Mechanisms: Ritchie (Ritchie RO. The conflicts between strength and toughness. *Nat Mater*. 2011;10:817) has very succinctly captured various toughening mechanisms using the terms "intrinsic" and "extrinsic." The figure below is inspired from his review and adapted to convey this distinction. The "intrinsic" mechanisms are dependent on intrinsic plasticity of the material. On the other hand, the "extrinsic" mechanisms are approaches that lead to enhanced energy absorption during the crack growth. Similar to the conceptual discussion of "hierarchical activation" of deformation mechanisms to enhance work hardening and ductility in Chapter 8, one can draw a parallel thought process to design tough alloys, ceramics, and composites by hierarchical activation of mechanism during crack growth.

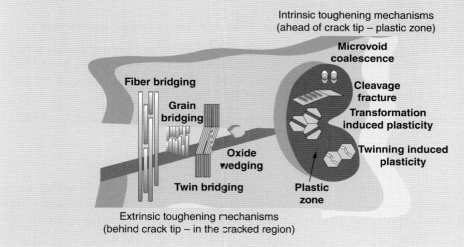

9.8 Exercises

1. Conceptual question: Describe three types of toughness definitions with the units they use. Discuss the differences on the basis of total energy absorbed during the fracture process in each method.
2. Explain the concept of stress concentration.
3. What is the basic design philosophy using stress intensity? Write relationship or explain how toughness is related to damage tolerance.
4. Draw a simple sketch of toughness vs. strength and mark relative position of lines for aluminum alloys, titanium alloys, and steel.
5. What is crack tip plasticity, and what is the shape of plastic zone ahead of a crack?
6. Explain three modes of crack opening or crack growth.

7. Why is ductile-brittle transition temperature (DBTT) important? How does the fracture mode change during DBTT?
8. What is the influence of carbon concentration and microstructure on DBTT of steels?
9. What are the steps involved in plane strain fracture toughness testing?
10. Describe plane strain vs. plane stress fracture mode transition. What is the importance of specimen thickness?
11. What microstructural features influence toughness and why?
12. Why does improving the alloy cleanliness increase the toughness? Do a Google search of 2XXX and 7XXX alloys. Can you list a set of alloys with similar matrix alloying composition but different level of impurity element specification?
13. Why ductile iron is more ductile than gray iron? Discuss in terms of stress concentration.
14. (a) Describe the essence of Griffith's law. (b) An aluminum oxide specimen is pulled in tension. If the specimen contains a flaw of size 80 µm in the center, what is the fracture stress (σ_f)? If the crack was at the edge, would that change the fracture stress? Given: The surface energy of alumina (γ_s) is 0.8 J/m² and elastic modulus (E) of 380 GPa.
15. Assume that the whole flow curve (true stress – true strain or σ vs. ε curve) can be described by simple Hollomon's equation:

$$\sigma = K\varepsilon^n$$

where K is the strength coefficient and n is the strain hardening exponent. Derive an expression for tensile toughness (defined as the area below the stress-strain curve). You can use appropriate upper limit (ε_f) and lower limit (0) of true strain. Neglect the elastic deformation region.

16. Consider a material that exhibits deformation induced transformation (TRIP effect). (a) How will crack growth be affected if there is volume change during this transformation? Consider both possibilities, volume expansion and volume contraction. (b) What will be the impact on fracture toughness value?

Further readings

Anderson TL. *Fracture Mechanics: Fundamentals and Applications*. CRC Press; 1995.
Chiang YM, Birnie D III, Kingery WD. *Physical Ceramics: Principles for Ceramic Science and Engineering*. John Wiley & Sons; 1997.
Dieter GE, Bacon D. *Mechanical Metallurgy*. McGraw-Hill; 2001.
Hannink R, Kelly PM, Muddle BC. Transformation toughening in zirconia-containing ceramics. *J Am Ceram Soc*. 2000;83(3):461-487.
Ritchie RO. The conflicts between strength and toughness. *Nat Mater*. 2011;10:817.
Tetelman AS, McEvily AJ Jr. *Fracture of Structural Materials*. Wiley; 1967.
Wei RP. *Fracture Mechanics: Integration of Mechanics, Materials Science and Chemistry*. Cambridge University Press; 2010.

CHAPTER 10

Fatigue behavior of materials: fatigue limiting design

Chapter outline

- 10.1 Constant stress and constant strain amplitude testing 252
  - 10.1.1 Stress-life fatigue (S-N curve) 255
  - 10.1.2 Strain-controlled fatigue testing and Bauschinger effect 261
  - 10.1.3 Strain life approach 265
- 10.2 Fatigue crack initiation and growth 268
  - 10.2.1 Fatigue crack initiation 269
  - 10.2.2 Crack growth models 273
- 10.3 Life prediction 276
- 10.4 Fatigue deformation and influence of microstructure 278
- 10.5 Fatigue fracture characteristics 285
- 10.6 Design aspects 289
  - 10.6.1 Fatigue design approaches 291
    - *10.6.1.1 Infinite-life design* *291*
    - *10.6.1.2 Safe-life design* *291*
    - *10.6.1.3 Fail-safe design* *291*
    - *10.6.1.4 Damage-tolerant design* *291*
  - 10.6.2 Use of Ashby charts in materials selection and design 292
  - 10.6.3 Geometrical stress concentrations 293
  - 10.6.4 Avoidance/minimization of fatigue damage 296
- 10.7 Key takeaways 296
- 10.8 Exercises 298

Learning objectives

- Know about the need for studying fatigue.
- Learn about different types of fatigue testing and approaches to fatigue data analysis.
- Introduce design approaches for fatigue-tolerant design.

Chapter 10 Fatigue behavior of materials: fatigue limiting design

In the age of hectic human civilization, the "fatigue" (tiredness) that we feel amid our busy schedules taxes our physical and mental states. Analogically, materials in cyclic loading or straining situations experience damage and subsequent failure, known as material fatigue. In general, over 90% of failures were estimated to occur due to fatigue. However, this estimate is not so valid anymore, given the complexity and variety of the materials, processes, and designs used in the modern world. Table 10.1 summarizes the frequency of failures in engineering components and specifically in aircraft components. Even though fatigue is an important mode of failure overall, fatigue as a direct failure mode is exceedingly important for the metallic materials used in aircraft structures. However, such direct failure would change if structural design and materials used become different.

Early examples of fatigue failure resulted in accidents involving Comet airplanes, one of the earliest commercial jetliners. One of the many recent examples of fatigue failures is aircraft fuselage cracking in a Boeing 737 aircraft, as shown in Fig. 10.1.

Most often, fatigue failures occur while the component serves below the yield stress of the material and sometimes catastrophically without warning. The term *fatigue* was coined in the early part of the 19th century. The advent of railways and industrialization brought to the fore the importance of fatigue as an issue to be studied with care. Various types of fatigue can take place in engineering applications. While the focus of this chapter is on mechanical fatigue, depending on the conditions, other types of fatigue-related phenomena include creep fatigue, thermo-mechanical fatigue, corrosion fatigue, and fretting fatigue, to name a few.

10.1 Constant stress and constant strain amplitude testing

Constant stress and constant strain amplitude fatigue testing are the methods of measuring total fatigue life. This information can itself provide the basis of the classical total-life approach of fatigue design.

Table 10.1 Frequency of occurrence of various engineering failures in metallic materials systems.[1,2]

Failure type	Percentage of failures	
	Engineering components	**Aircraft components**
General corrosion	29	16
Fatigue	25	55
Brittle fracture	16	-
Overload	11	14
High-temperature corrosion	7	2
Stress corrosion cracking/corrosion fatigue/hydrogen embrittlement	6	7
Creep	3	-
Wear/abrasion/erosion	3	6

---

[1] Findlay SJ, Harrison ND. Why aircraft fail? *Mater Today*. 2002;5:18-25.
[2] Gorelik M. Additive manufacturing in the context of structural integrity. *Int J Fatigue*. 2017;94(2):168-177.

10.1 Constant stress and constant strain amplitude testing

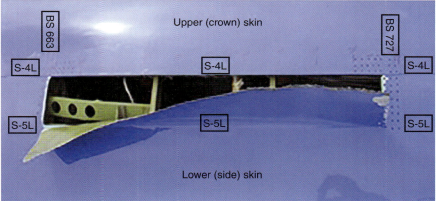

Figure 10.1

Fuselage cracking in B737-300 aircraft; occurrence of accident on April 1, 2011.[3,4]

In these tests, standardized specimens with smooth surfaces (no pre-existing cracks) are tested under controlled stress or strain amplitudes.

Stress-controlled fatigue testing has been used extensively to generate fatigue design data and is very useful when large numbers of stress cycles are involved. A quick review of the nomenclature of various terms of importance related to stress cycles used in stress-controlled fatigue tests is presented below. In laboratory fatigue tests, generally cyclic loading is accomplished by applying stress cycles that are typically in sinusoidal or some other wave forms. Note a repeated stress cycle shown in Fig. 10.2.

---

[3]Tavares SMO, De Castro PMST. An overview of fatigue in aircraft structures. *Fatigue Frac Eng Mater Struct*. 2017. doi:10.1111/ffe:12631.
[4]National Transportation Safety Board (NTSB), B737 depressurization while enroute, Accident Number: DCA11MA039, 2013.

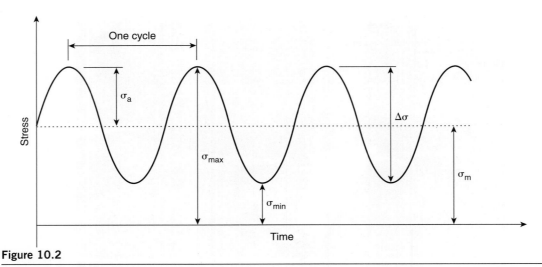

Figure 10.2
A typical fatigue stress cycle.

A cyclic stress cycle can be described as composed of two terms—*mean stress* (σ_m) and *alternating stress* or *stress amplitude* (σ_a). In Fig. 10.2, the maximum stress is symbolized as σ_{max} and the minimum stress as σ_{min}. The mean stress is the algebraic average of the maximum and minimum stresses, given by the following relation:

$$\sigma_m = \frac{\sigma_{max} + \sigma_{min}}{2} \tag{10.1a}$$

Alternating stress is defined as one-half of the range of stress ($\Delta\sigma$):

$$\sigma_a = \frac{\Delta\sigma}{2} = \frac{\sigma_{max} - \sigma_{min}}{2} \tag{10.1b}$$

From Fig. 10.2, the following relations can also be written:

$$\sigma_{max} = \sigma_m + \sigma_a \tag{10.1c}$$

$$\sigma_{min} = \sigma_m - \sigma_a \tag{10.1d}$$

Of two ratios that need to be considered in the context of stress-controlled fatigue testing, one, the *stress ratio,* is defined as the ratio of the minimum and maximum stress:

$$R = \frac{\sigma_{min}}{\sigma_{max}} \tag{10.1e}$$

Another ratio is the *amplitude ratio* (*A*) that is basically defined as the ratio of the stress amplitude and the mean stress. However, it is not so widely used as the stress ratio:

$$A = \frac{\sigma_a}{\sigma_m} = \frac{1-R}{1+R} \tag{10.1f}$$

If the maximum and minimum stresses are equal but opposite in nature (i.e., $\sigma_{max} = -\sigma_{min}$), the stress cycle is called fully reversed cycle. In that case, the mean stress becomes zero. The stress ratio becomes −1. On the other hand, for the zero-tension fatigue cycle (pulsating tension), $R = 0$. In tension, a typical stress ratio of 0.1 is used to avoid fully relaxing the specimen loading, which can become an issue for certain types of gripping. However, in real applications the stress cycles are often random.

10.1.1 Stress-life fatigue (S-N curve)

Engineering fatigue data are often presented in terms of S-N curves. The term S here stands for stress and N for number of cycles to failure. This approach, which has been used in design for over 150 years, takes into account the total fatigue life, which means the time span over damage accumulation, crack nucleation, and crack propagation leading to the final failure. Wöhler first introduced the stress-life approach during the 1860s while studying fatigue characteristics of rail axles. Generally, these types of tests use smooth specimens without pre-existing cracks. Wöhler carried out rotating beam bending experiments. Other types of fatigue tests include cantilever-type bending, uniaxial tension-compression, or tension-tension.

Different loading modes used in fatigue testing include a cantilever beam, rotating bending machine as shown in Fig. 10.3 (a). Fig. 10.3 (b) depicts the axial fatigue test configuration, which is basically a push-pull type test. The ASTM E466-E468 standard provides details of the test procedures used in axial fatigue tests.

The S in the S-N curve can be stress amplitude, maximum stress, or minimum stress. However, stress amplitude is the most widely used stress term for S-N curves. Almost always N is plotted in log-scale along the x-axis. Mild steel, titanium alloys, and some other materials, which undergo strain

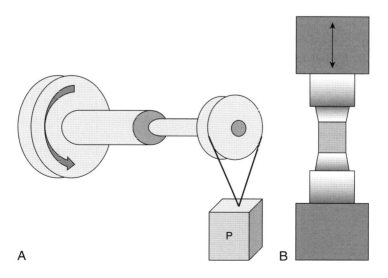

Figure 10.3

Various loading modes used in fatigue testing: (a) Single point rotating beam test; (b) uniaxial push-pull type testing (tension-compression and pulsating tension).

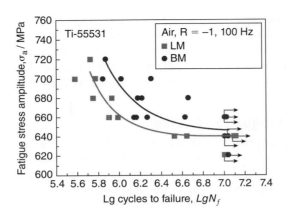

Figure 10.4

A typical *S-N* curve obtained from fatigue tests of a Ti-5Al-5Mo-5V-3Cr-1Zr (Ti-55531) alloy in two microstructural states (LM—lamellar microstructure—and BM—bimodal microstructure).[5]

aging, exhibit a plateau in the *S-N* curve at high fatigue cycles. This generally occurs in these materials above 10^5 to 10^6 fatigue cycles, as depicted for a Ti-5Al-5Mo-5V-3Cr-1Zr alloy in Fig. 10.4. The significance of the plateau region is that any testing done below this stress level would not fail even under an infinite number of fatigue cycles. That is why this plateau stress level has a special meaning and thus can be termed as *endurance limit* or *fatigue limit*. In reality, however, there is no such thing as infinite fatigue life.[6] Under very high numbers of loading cycles, significant decrease in fatigue strength has been observed. Fig. 10.4 also signifies the effect of microstructure on fatigue properties, which will be discussed in a later section.

To create *S-N* curves may require 8 to 12 fatigue specimens for scientific understanding, although the ASTM standard recommends testing 6 specimens at one stress level. The first test is done under a higher stress amplitude (generally two-thirds of the ultimate tensile strength) so that the specimen can fail in a reasonable amount of time, and a datum point for the *S-N* curve can be obtained quickly. Gradually stress amplitude is reduced till the sample survives 10^7 loading cycles. That is considered to be the last data point on the *S-N* curve and generally happens above 10^7 cycles. Scatter in fatigue data is an issue of concern, especially at the lower stresses where fatigue limits are generally determined. The origin of scatter is attributed to a multitude of factors including but not limited to test machine, specimen variability, surface preparation, environments, and metallurgical variables.[7] The origin of scatter and approaches to counter the issue will be discussed in some detail in a later section. Many non-ferrous alloys, aluminum, magnesium, and copper alloys along with some high-strength steels do not show a distinct fatigue limit. The S-N curves of those alloys do not show a stress plateau

---

[5]Huang C, Zhao Y, Xiu S, et al. Effect of microstructure on high cycle fatigue behavior of Ti-5Al-5Mo-5V-3Cr-1Zr alloy. *Int J Fatigue*. 2017;94:30-40.
[6]Bathias C. There is no infinite fatigue life in metallic materials. *Fatigue Fract Eng Mater Struct*. 1999;22:559-565.
[7]Sonsino CM. Course of SN-curves especially in the high-cycle fatigue regime with regard to component design and safety. *Int J Fatigue*. 2007;29:2246-2258.

as shown in Fig. 10.5. In those cases, an endurance or fatigue strength at a pre-defined life such as 10^7 or 10^8 is specified as an arbitrary number of cycles to failure.

Table 10.2 includes fatigue endurance limits (σ_e) of different engineering metallic alloys. This table also lists corresponding ultimate tensile strengths (σ_{UTS}) of the alloys and corresponding ratios (σ_e/σ_{UTS}). Generally, endurance limit is within 30% to 50% of ultimate tensile strength. This empirical observation also provides a quick guide for testing range in initial stages of data generation of a new alloy.

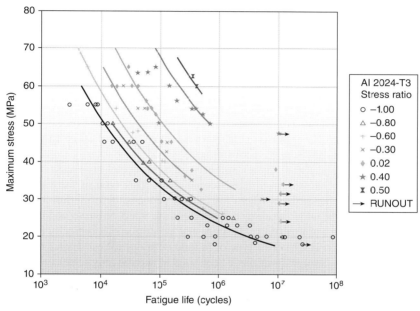

Figure 10.5

Stress-life *(S-N)* curves for 2024-T3 Al alloy sheets under various stress ratios.[8]

Table 10.2 Endurance limit of various engineering alloys.[9]

Alloy	Condition	σ_{UTS} (MPa)	σ_e (MPa)	σ_e/σ_{UTS}
AA 2024 Al	T3	483	138	0.29
AA 6061 Al	T6	310	97	0.31
AISI 1015 Steel	Annealed	455	240	0.53
AISI 1015 Steel	60% CW	710	350	0.49
AISI 1040 Steel	Annealed	670	345	0.51
AISI 4340	Annealed	745	340	0.46
AISI 4340	Quenched and tempered	1950	480	0.25

[8]Rice RC, Jackson JL, Bakuckas J, Thompson S. *"Metallic Materials Properties Development and Standardization (MMPDS),"* Scientific Report DOT/FAA/AR-MMPDS-01. Washington, DC: U.S. Department of Transportation—Federal Aviation Administration—Office of Aviation Research; 2003.

[9]Source: Suresh S. *Fatigue of Materials.* Cambridge Press; 2004.

> *Food for thought* – Are typical values of <0.33 for σ_e/σ_{UTS} in precipitation strengthened aluminum alloys a coincidence? Students are encouraged to create a table like Table 10.2 with data from various studies on precipitation strengthened aluminum alloys. Include 2XXX, 7XXX, and 6XXX, alloys in various tempers and microstructural conditions. Look at the σ_e/σ_{UTS} ratio. Are they mostly in 0.3 to 0.35 range? Why? We learned about the concept of stress concentration at the second phase particles in **Chapter 9**. The maximum stress concentration at constituent particles can be as high as three times the applied stress! So, when the fatigue specimen is loaded at 0.3 to 0.35 of σ_{UTS}, are the regions around hard particles above the σ_{YS}? Could that be the fundamental reason behind the run-off stress? Can you justify this answer by separating the data for aluminum alloys with tighter impurity element specification?

At the higher stress cycles, when the *S-N* data are plotted on a double logarithmic scale, they follow the relation known as Basquin's equation where stress amplitude is given as a function of stress reversals (i.e., twice the number of cycles):

$$\sigma_a = \sigma_f'(2N)^b \tag{10.2}$$

where σ_f' is the fatigue strength coefficient, *N* is the total number of stress cycles, and *b* is the Basquin exponent (or fatigue strength exponent) with a typical value of −0.05 to −0.12. Note that for most purposes, σ_f' is same as the true fracture strength.

The *S-N* curves are quite useful at higher numbers of stress cycles (typically $>10^5$ cycles) where deformation remains mainly elastic on a macroscopic scale. This type of fatigue is known as high cycle fatigue (HCF). At higher stresses where fatigue life reduces, deformation involves both elastic and plastic deformation, and understanding the fatigue data via *S-N* curve becomes difficult. Thus fatigue behavior at the lower number of cycles (thus shorter fatigue life, typically $<10^4$–10^5 cycles) is described as low cycle fatigue (LCF). This type of fatigue behavior is best studied by strain-controlled fatigue tests. Also, the relatively new terms have been used to ascribe fatigue behaviors to both extreme ends of the numbers of loading cycles: very low cycle fatigue (VLCF) or extremely low cycle fatigue, typically for $N < 10^2$ cycles, and very high cycle fatigue (VHCF), $>10^7$ cycles.

Specifically, the scientific community has become interested in understanding fatigue behavior of materials at much higher numbers of stress cycles (i.e., beyond 10^7) due to the need for damage assessment modes and life prediction methodologies in applications ranging from aircraft turbine disks to medical devices. In many instances, such components in service are exposed to stress cycles of 10^9 to 10^{10} or above.[10] The bottleneck surrounding standard laboratory fatigue testing methods noted earlier in this chapter is that the number of cycles possible to apply in a realistic time-frame is limited to 10^7 cycles or so. Thus novel test methods capable of accessing the VHCF test regime are needed. Of late, ultrasonic fatigue testing systems run at a much higher frequency range (20 kHz), and new servo-hydraulic devices allow for closed-loop control up to 1 kHz. Table 10.3 summarizes fatigue test types, test frequency, and duration.[11] Note the lower frequency of stress cycles used in the traditional fatigue tests.

Crack initiation and propagation are the two main stages of fatigue failure in an otherwise smooth specimen. Fig. 10.6 schematically shows separately the *S-N* curve for both processes. Note that the relative number of stress cycles that the material spends in the crack initiation stage (N_i) relative to the crack propagation phase (N_p) increases considerably at lower stress levels compared to higher stress levels.

---

[10] Christ HJ. Preface – Advances in very high cycle fatigue. *Int J Fatigue*. 2011;33(1):1.
[11] Pyttel B, Schverdt D, Berger C. Very high cycle fatigue: is there a fatigue limit? *Int J Fatigue*. 2011;33:49-58.

10.1 Constant stress and constant strain amplitude testing

Table 10.3 Various fatigue test methods and upper frequency/test duration ranges.[9]

Test facility	Frequency (Hz)
Rotating bending machine	50
Resonant frequency machine	150
Servo-hydraulic test machine	400
Ultrasonic test machine	20,000

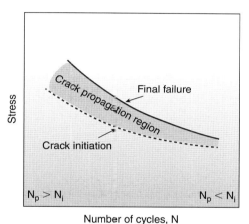

Figure 10.6
Schematic of S-N curves illustrating separately the relative contributions of crack initiation and crack propagation.[12]

As mentioned before, the mean stress is zero for the completely reversed stress cycles. However, in engineering applications, this is rarely true. Generally, mean stress is not zero in real applications. That is why knowing the effect of non-zero mean stress on the fatigue behavior of materials is important. In general, as mean stress increases, overall fatigue strength diminishes (Fig. 10.7).

From the S-N curves in Fig. 10.7, stress amplitude versus mean stress for different constant life levels could be derived. Haigh analysis shows that all constant-life curves converge at a single value. Fig. 10.8 is an example of such a diagram. However, Haigh diagrams are very expensive to create experimentally. Hence, different empirical methods are used to generate the necessary plots to predict fatigue life under non-zero mean stress.

Fig. 10.9 shows different empirical models that can predict the effect of mean stress on fatigue behavior. Some mathematical relations are listed as follows:

$$\text{Gerber (1874):} \quad \frac{\sigma_a}{\sigma_e} + \left(\frac{\sigma_m}{\sigma_u}\right)^2 = 1 \quad (10.3\text{a})$$

---

[12]Adapted from Suresh S. *Fatigue of Materials*. 2nd ed. Cambridge University Press; 1998.

Goodman (1899): $\quad \dfrac{\sigma_a}{\sigma_e} + \dfrac{\sigma_m}{\sigma_u} = 1 \qquad (10.3b)$

Soderberg (1939): $\quad \dfrac{\sigma_a}{\sigma_e} + \dfrac{\sigma_m}{\sigma_y} = 1 \qquad (10.3c)$

Morrow (1968): $\quad \dfrac{\sigma_a}{\sigma_e} + \dfrac{\sigma_m}{\sigma_f} = 1 \qquad (10.3d)$

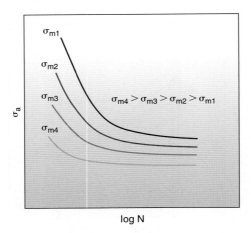

Figure 10.7

The effect of mean stress on the S-N curves.[13]

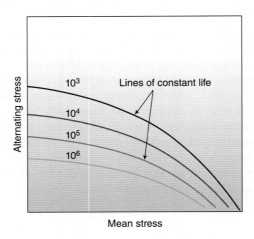

Figure 10.8

Haigh diagram.

---

[13] Adapted from Suresh S. *Fatigue of Materials*. 2nd ed. Cambridge University Press; 1998.

10.1 Constant stress and constant strain amplitude testing

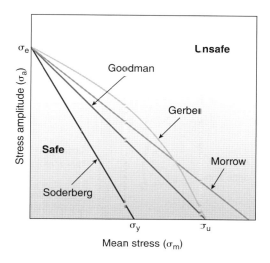

Figure 10.9

Goodman, Gerber, Soderberg, and Morrow diagrams for predicting the effect of mean stress.[14]

In Eq. (10.3a-d), the terms σ_u and σ_f are ultimate tensile strength and (true) fracture stress, respectively. Note that σ_e is the fatigue limit determined for the fully reversed loading cycle. In Fig. 10.9 Gerber developed a parabolic model, whereas Goodman demonstrated a linear model, both based on ultimate tensile strength. Most materials will fall between Goodman and Gerber lines. Ductile materials are closer to Gerber, whereas harder materials are closer to Goodman criterion. As the scatter in fatigue data and the notched data are closer to the Goodman criterion, the more conservative Goodman relationship is often used in practice. If component design is based on yield strength rather than ultimate strength, as most are, the even more conservative model is the one proposed by Soderberg. Morrow's model, however, is based on true fracture stress.

10.1.2 Strain-controlled fatigue testing and Bauschinger effect

Another important type of fatigue testing is constant strain-amplitude or strain-controlled fatigue testing. The stress-life approach, which is best suited for high cycle fatigue (HCF), involves mostly gross elastic strains and constant amplitude loading. However, surface finish effects and size effect must be included. The stress-life approach does not work effectively when residual stresses are present in the material or gross plasticity is involved. Strain-controlled fatigue testing is very pertinent during thermal cycling when the repeated expansion and contraction of materials can introduce fatigue strains. The design approach predicts initiation of fatigue crack using strain-life behavior and uses crack growth data to predict life to fracture.

Most fatigue failures are initiated at notches where the stress concentration is large (Fig. 10.10). Local strains in the notch have significant plastic strain. The magnitude of the plastic strain is

---

[14]Adapted from Suresh S. *Fatigue of Materials*. 2nd ed. Cambridge University Press; 1998.

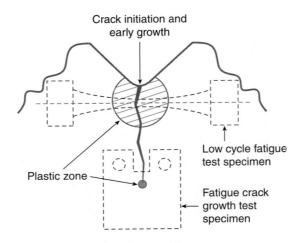

Figure 10.10
A schematic illustration of the different phases of fatigue failure in a component and approaches used to estimate fatigue life.

controlled by the elastically stressed material surrounding the notch. However, S-N data fail to predict failure. The strain-life approach brings with it a real advantage in that the experimental method involves strain analysis, for example, strain gages, photoelastic coatings, etc.

The underlying principle of strain-life behavior is that plastic deformation produces damage, and each reversal in strain application contributes to the damage. As a result, fatigue damage is measured with respect to the number of reversals, not the number of cycles (N_f). Considerable plastic strain occurs during each half-cycle. One interesting effect that is observed during the initial cycle is that yielding in the second half-cycle occurs at a stress well below the monotonic yield stress. This is known as Bauschinger effect (1886) and is generally attributed to a dislocation-assisted phenomenon. When a material is deformed in one direction, dislocation pileups are formed against barriers present in the microstructure. However, when deformation in the reverse direction takes place within the same cycle, yielding of the material starts at a lower flow stress because of already available dislocations, and they just need to move collectively in the reverse direction to achieve yielding. Another theory discusses the Bauschinger effect in terms of internal stress development and internal stress relaxation. Fig. 10.11 shows a stable cyclic stress-strain hysteresis loop.

In constant stress amplitude testing that involves fully reversed stress cycle, cycling hardening manifests in a gradual decrease in the axial strain amplitude, whereas cyclic softening leads to increase in axial strain amplitude (Fig. 10.12 [a]). However, by using the constant strain amplitude test mode, cyclic hardening manifests in terms of increasing axial stress amplitude, and the cyclic softening effect results in decreasing axial stress amplitude. Such transient effects are illustrated in Fig. 10.12 (b). However, these situations constitute only a transient state. After initial "shakedown," the hysteresis loop becomes stable after reaching saturation. The cyclic hysteresis loop will remain the same following saturation through subsequent cycles until the specimen fails. During shakedown, the dislocation

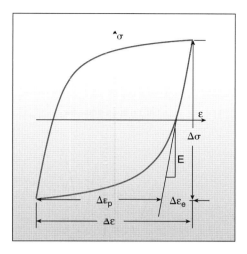

Figure 10.11

A schematic cyclic stress-strain stable hysteresis loop.[15]

substructure changes continuously until a dynamically stable configuration characteristic of the saturation phase is achieved.

If the tips of the stable hysteresis loops are joined, the locus of those points should provide the cyclic stress-strain curve (Fig. 10.13). The strain-controlled mode of testing has become increasingly important in engineering applications. The three methods for obtaining cyclic stress-strain curves are as follows. In the first method, a single test specimen is used to generate stress-strain hysteresis curve for constant strain amplitude strain cycle. Then more specimens are needed to create other relevant hysteresis curves. Subsequently, the cyclic stress-strain curve can be developed from the individual hysteresis curves. In the second method, a single specimen is subjected to constant strain limits to create stable hysteresis loop; the strain limits are then increased until another saturation loop is obtained, and the process is continued until the whole stress-strain curve is created. The third method, called the incremental step method, uses repeated stress cycle patterns consisting of increasing (zero to a specific maximum strain) and decreasing strain (from the maximum total strain to zero) amplitudes.

Before further describing strain-controlled fatigue testing and relevant data analysis premises, a quick review of the monotonic stress-strain curve is in order. Total strain (ε_{total}) is given by the summation of elastic strain ($\varepsilon_{elastic}$) and plastic strain ($\varepsilon_{plastic}$):

$$\varepsilon_{total} = \varepsilon_{elastic} + \varepsilon_{plastic} \tag{10.4a}$$

Elastic strain ($\varepsilon_{elastic}$) is

$$\varepsilon_{elastic} = \sigma/E$$

where σ is the true stress and E is the elastic modulus.

---

[15] Adapted from Suresh S. *Fatigue of Materials*. 2nd ed. Cambridge University Press; 1998.

264 Chapter 10 Fatigue behavior of materials: fatigue limiting design

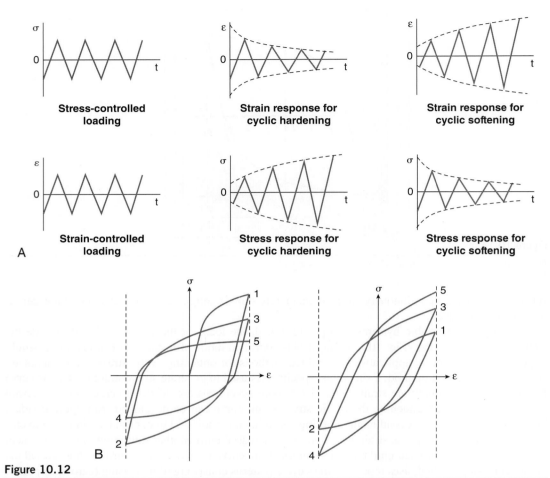

Figure 10.12
(a) Transient effects relevant to fatigue under strain-controlled loading and stress-controlled loading.[16] (b) Cyclic stress-strain response showing cyclic softening characteristics *[left]* and cyclic hardening characteristics *[right]*.

The strain hardening behavior of most metals obeys the following power law (Hollomon's equation):

$$\sigma = k(\varepsilon_{plastic})^n \implies \varepsilon_{plastic} = \left(\frac{\sigma}{k}\right)^{1/n}$$

where *n* is the strain hardening exponent and *k* is the strength coefficient. Generally, *n* value for metallic alloys ranges between 0 and 0.5.

---
[16] Adapted from Suresh S. *Fatigue of Materials*. 2nd ed. Cambridge University Press; 1998.

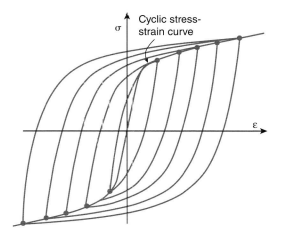

Figure 10.13

Cyclic stress-strain curve by joining the tips of stable hysteresis loops.[17]

So, now Eq. (10.4a) can be written as

$$\varepsilon_{total} = \frac{\sigma}{E} + \left(\frac{\sigma}{k}\right)^{1/n} \tag{10.4b}$$

Generally, cyclic softening occurs when $\sigma_{UTS}/\sigma_{yield} < 1.2$ and $n < 0.1$, whereas cyclic hardening takes place when $\sigma_{UTS}/\sigma_{yield} > 1.4$ and $n > 0.2$. The strain amplitude, ε_{amp}, is expressed as

$$\varepsilon_{amp} = \frac{\Delta\varepsilon_{tot}}{2} = \frac{\Delta\varepsilon_e}{2} + \frac{\Delta\varepsilon_p}{2} \tag{10.5a}$$

where $\Delta\varepsilon_e$ is elastic strain range and $\Delta\varepsilon_p$ is plastic strain range.

Similarly, as in Eq. (10.4b), the following relation for cyclic stress-strain behavior can be written:

$$\frac{\Delta\varepsilon_{tot}}{2} = \frac{\Delta\sigma}{2E} + \left(\frac{\Delta\sigma}{2k'}\right)^{1/n_f} \tag{10.5b}$$

where n_f is the cyclic strain hardening exponent (generally varies in a narrow range of 0.1–0.2) and k' is the cyclic strength coefficient.

10.1.3 Strain life approach

Coffin[18] and Manson[19] were the first to describe a strain-based approach for fatigue life determination while analyzing thermal stresses. Note that they developed the approach independently of each other. The necessity of understanding the strain-life approach given the importance of LCF in engineering

---

[17]Adapted from Suresh S. *Fatigue of Materials*. 2nd ed. Cambridge University Press; 1998.

applications has been described previously. The *Coffin-Manson relationship*[18,19] between plastic strain amplitude ($\Delta\varepsilon_p/2$) and the number of stress cycle reversals up to failure (N_f) is given as

$$\frac{\Delta\varepsilon_p}{2} = \varepsilon_f'\,(2N_f)^c \tag{10.6}$$

where ε_f' is the fatigue ductility coefficient, which is essentially equal to the true fracture strain in the uniaxial tension test. The term c represents the fatigue ductility exponent, typically of the value between -0.7 and -0.5 for the majority of metallic materials. Generally, a longer fatigue life results in a smaller value of c.

In the HCF range (elastic), Basquin's equation is the most appropriate. From Eq. (10.2), then ensues

$$\sigma_a = \frac{\Delta\sigma}{2} = \frac{\Delta\varepsilon_e E}{2} = \sigma_f'\,(2N_f)^b$$

or

$$\frac{\Delta\varepsilon_e}{2} = \frac{\sigma_f'}{E}(2N_f)^b \tag{10.7}$$

Because of the validity of Eq. (10.5a), Eqs. (10.6) and (10.7) can be added via linear superposition, and the following equation for the full range of fatigue life can be obtained:

$$\frac{\Delta\varepsilon_{tot}}{2} = \frac{\sigma_f'}{E}(2N_f)^b + \varepsilon_f'\,(2N_f)^c \tag{10.8}$$

A schematic plot of Eq. (10.8) is shown in Fig. 10.14. See Table 10.4 for typical b and c values for different engineering metallic materials. Note that the fatigue life curve aligns itself with the plastic

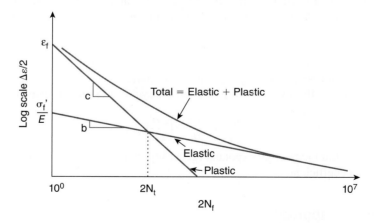

Figure 10.14

Schematic plot of Eq. (10.8) showing strain amplitude as a function of the total number of stress reversals.[20]

---

[18] Coffin Jr. LF. A study of the effects of cyclic thermal stresses in a ductile metal. *Trans ASME*. 1954;76:931-950.
[19] Manson SS. Behavior of materials under conditions of thermal stress, Technical Report NACA-TR-1170, National Advisory Committee for Aeronautics.
[20] Adapted from Suresh S. *Fatigue of Materials*. 2nd ed. Cambridge University Press; 1998.

Table 10.4 Summary of Basquin coefficient (*b*) and Coffin-Manson coefficient (*c*) values for different engineering alloys.[20]

Alloy	Condition	b	c
AA 2024 Al	T351	−0.124	−0.59
AA 7075 Al	T6	−0.126	−0.52
Ti-6Al-4V	Solution treated and aged	−0.104	0.69
Inconel-X (Nickel base superalloy)	Annealed	−0.117	−0.75
AISI 1015 Steel	Air cooled	−0.110	−0.64
AISI 4340 Steel	Tempered	−0.076	0.62

curve at larger strain amplitudes, whereas the curve aligns with the elastic curve under low strain amplitudes. Ductile materials conform the best for high cyclic strains (LCF), whereas strong materials appear to follow the best performance for low cyclic strain (HCF) conditions.

Fig. 10.15 showing the total strain amplitudes from 2.5×10^{-3} to 2×10^{-2} as a function of total number of stress reversals illustrates the fatigue resistance of Fe-15Mn-10Cr-8Ni-4Si seismic damping alloy. As discussed earlier, the intersection of the elastic and plastic regimes is depicted as transitional fatigue life (N_t), which is 2×10^5 reversals. When $2N_f$ is lower than N_t, plastic strain regime is dominant, and ductility influences fatigue resistance for such LCF deformation. On the other hand, at fatigue life greater than N_t, the elastic deformation region is dominant, and this HCF region is

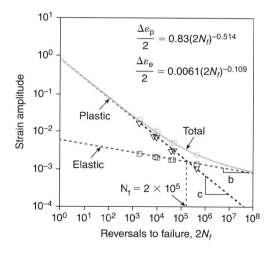

Figure 10.15

Variation of strain amplitude versus number of reversals to failure in a Fe-15Mn-10Cr-8Ni-4Si seismic damping alloy.[21]

---

[21]Nikulin I, Sawaguchi T, Kushibe A, Inoue Y, Otsuka H, Tsuzaki K. Effect of strain amplitude on the low cycle fatigue behavior of a new Fe-15Mn-10Cr-8Ni-4Si seismic damping alloy. *Int J Fatigue*. 2016;88:132-141.

controlled by strength. The following equation shows total strain amplitude in terms of the number of stress reversals to failure:

$$\frac{\Delta \varepsilon_t}{2} = \frac{1130}{184000}(2N_f)^{-0.109} + 0.85(2N_f)^{-0.514} \tag{10.9}$$

As shown in Fig. 10.15, the transition occurs at N_t. An expression for N_t can be obtained by equating Eqs. (10.6) and (10.7):

$$2N_t = \left(\frac{\varepsilon_{f'}E}{\sigma_{f'}}\right)^{1/(b-c)} \tag{10.10}$$

The above approach requires careful attention regarding these observations:

1. Not all materials, for example, high-strength Al and Ti alloys, may be represented with a four-parameter plot.
2. Power law relations that have been discussed so far do not have much physical meaning; most often they are used for empirical convenience.
3. The more real data that is available, the more reliable will be estimates of life.

Morrow[22] incorporated a correction when the mean stress is not zero. The mean stress (σ_m) in his strain-life analysis assumed that tensile mean stress reduces fatigue strength:

$$\frac{\Delta \varepsilon_{tol}}{2} = \frac{(\sigma_f' - \sigma_m)}{E}(2N_f)^b + \varepsilon_f'(2N_f)^c \tag{10.11}$$

However, mean stress is generally relaxed during fatigue tests performed under strain-controlled mode. It is observed predominantly at longer lives. When $\varepsilon_p > 0.5\%$, mean stress relaxation occurs, and the mean stress tends to zero. This is not cyclic softening.

> *Concept reinforcement – Engineering versus Science of fatigue.* You read through various approaches for fatigue life estimation. As you can note, many of the relationships are well established for a really long time! As an engineer, your job is primarily to make sure that components do not fail or perform as per the designed life. So, empirical approaches work well for this task. However, when we are developing next generation materials to push performance, the "science" becomes important. If we do not have understanding of the micromechanisms, we cannot design microstructural features to defeat it! Hopefully, you are developing the appreciation for both the science and engineering aspects.

10.2 Fatigue crack initiation and growth

As previously mentioned, crack initiation and crack propagation (or growth) are the two most important stages of fatigue life. Although a prevalent assumption is that the crack initiation stage consumes almost 90% of fatigue life, such is not always the case! For example, in cast aluminum alloys, initiation may occur in less than 10% of the total life, followed by slow crack propagation. More of these aspects will be discussed in later sections when the interlinkage between fatigue behavior and microstructure is analyzed.

---

[22]Morrow JD. *Fatigue Design Handbook – Advances in Engineering.* Vol. 4, Sec. 3.2. Society of Automotive Engineers; 1968:21-29.

10.2.1 Fatigue crack initiation

Fatigue cracks occur exclusively on surfaces or near-surface regions at micro- or macrostress concentration sites in defect-free specimens. Forsyth (1953)[23] was the first to report surface roughening created by slip activity during fatigue while investigating thin (just 0.1 μm thick) ribbons of a solution heat-treated Al-4Cu (wt.%) alloy. Extrusions and intrusions formed as a result of slip in fatigued copper were also reported by Cottrell and Hull (1957).[24] These unique slip steps formed on the material surface serve as the precursors of fatigue crack initiation. Wood (1958)[25] put forth the mechanistic hypothesis of the formation of these intrusions and extrusions, which is described in the next paragraphs.

Even when cyclic stress is less than yield strength, localized microscopic plastic deformation can occur. The cyclic nature of the stress causes slip to appear as extrusions and intrusions on the surface. The mechanism is perhaps better understood by referring to Fig. 10.16, which shows a simple situation where slip steps are created under static loading (unidirectional deformation). Nonetheless, these surface steps do not lead to crack formation. However, the situation changes when dynamic loading is involved. Fig. 10.16 (b) presents the case where an intrusion is created due to dynamic loading (slip occurring along two opposite directions), whereas an extrusion such as shown in Fig. 10.16 (c) can also result. That is, under fatigue loading, a set of grooves and ridges are formed on the fatigued specimen surface. If many active slip planes are involved in the reverse slip, these grooves and ridges would appear shallow and undulating. But if more closely spaced planes are operative in the reverse slip process in the local region, these grooves and ridges would appear as walls and crevices. These intrusions and extrusions act as micro stress-concentration sites (large strain localization), and fatigue cracks can initiate from these regions. These intrusions and extrusions also form at very low temperatures. Hence, the possibility of thermal activation working behind the creation of these features can be ruled out.

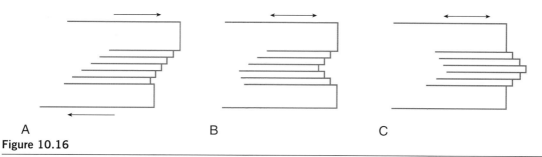

A   B   C

Figure 10.16

(a) Slip steps appear during forward loading. During reverse loading, they become either (b) intrusion or (c) extrusion.[26]

---

[23]Forsyth PJE. Exudation of material from slip bands at the surface of fatigued crystals of an aluminum-copper alloy. *Nature*. 1953;171:172-173.
[24]Cottrell AH, Hull D. Extrusion and intrusion by cyclic slip in copper. *Proc Roy Soc A*. 1957;A242:211-213.
[25]Wood WA. Formation of fatigue cracks. *Philos Mag*. 1958;3:692-699.
[26]Adapted from Suresh S. *Fatigue of Materials*. 2nd ed. Cambridge University Press; 1998.

Researchers have shown that the slip steps formed on specimen surfaces are related to the applied plastic strain. Mathematically, the following expression is valid:

$$\gamma_p = (N_d b_d)^{m_p} \qquad (10.12)$$

where γ_p is the applied plastic strain, N_d is the number of dislocations contained in the slip offset, b_d is the magnitude of the dislocation Burgers vector, and m_p is a constant. Cheng and Laird (1981) revealed that copper single crystals under an applied plastic strain amplitude of 0.001 to 0.0001 can lead to the $N_d b_d$ value in the range of 0.3 to 3 μm and m_p of about 0.78.[27]

Many models have been proposed to explain the origin of intrusions and extrusions on the surface of fatigued specimens. One early model proposed by Cottrell and Hull[28] is presented here. Figs. 10.17 (a) to (e) illustrate a model showing the slip movement involved in the formation of an intrusion and

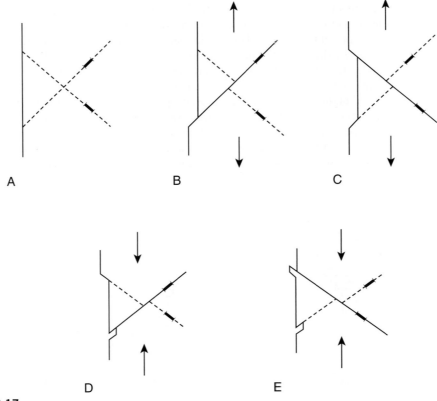

Figure 10.17
Sequence of slip movements involved in the formation of an intrusion and extrusion: (a) undeformed; (b) and (c) extension; and (d) and (e) compression.[30]

---

[27]Cheng AS, Laird C. Fatigue life behavior of copper single crystals. Part I: Observations of crack nucleation. *Fatigue Fract Eng Mater Struct.* 1981;4:331-341.
[28]Cottrell AH, Hull D. Extrusion and intrusion by cyclic slip in copper. *Proc R Soc A.* 1957;242:211-213.

extrusion. In this case, fatigue cycle is composed of two phases of loading (positive and negative phases). Two intersecting slip bands activate in sequence during the positive (extension) and negative (compression) phases of the loading cycle (Fig. 10.17 [a]). If the sequence is maintained, an intrusion is created on the band that first becomes active, and an extrusion is formed on the other one. The reason that one band becomes active first over the other is simply that band is favorably oriented to the applied stress; recall Schmid's law from **Chapter 8**. The stress increases during a half-cycle when this favorably oriented slip band undergoes further work hardening than the other. However, during the reverse half-cycle, the very slip band now becomes operative first because of the Bauschinger effect that was discussed briefly earlier. A reasonable deduction from this model would be that the intrusion and extrusion features developed in one full cycle can be repeated, thereby resulting in doubling of the depth and length of the extrusion and intrusion created, albeit without any change in their thickness. The mechanism operates at the free surface, which also occurs in practice. The back stresses arising out of dislocation pileups formed against grain boundaries within the grain interiors in the bulk of materials generally relieve themselves during the return half-cycle. So, intrusions and extrusions inside the material interior are not easy to form and thus generally do not serve as crack initiation sites during fatigue loading. In fact, fatigue life shows drastic improvement if the surface roughness is removed.

A natural progression from the emphasis on intrusions and extrusions of the surface is a follow-on discussion of slip activity. During fatigue, only a few grains exhibit heavy slip activity. Within initial numbers of cycles (a few thousand), the slip bands appear. Although new slip bands appear with increasing numbers of cycles, their numbers are not proportionate. Eventually, in most metals the number of these slip bands saturates, they become the areas of high slip activity, and fatigue cracks nucleate in these regions. Careful experiments have confirmed that some slip bands are more persistent than others. These slip bands are observed within the first few percentages of total fatigue life of the specimen. That is why these slip bands are called persistent slip bands (PSBs). Examples of intrusion and extrusion slip bands formed on the surface of a Sanicro 25 steel are presented in Fig. 10.18. Experiments have revealed that if fatigue test is interrupted and the surface is removed via electropolishing, some slip bands will be removed, while some continue to be present. These latter slip bands are the PSBs. A fatigue crack initiating at the PSB-matrix interface is also indicated in Fig. 10.18.

Extensive investigations have been performed to understand crack behavior during fatigue. Fatigue testing of a smooth specimen shows three distinct stages: Stage I (crack initiation and initial propagation), Stage II (stable crack growth), and Stage III (unstable crack propagation). Fig. 10.19 is a schematic of crack growth through the various stages. Initially, the crack follows the crystallographic shear path, followed by crack propagation on the plane perpendicular to the direction of the applied stress. Generally, fatigue cracks propagate through the grain interiors. Thus the fracture is transgranular in nature.

Stage I

In this stage, the cracks propagate along the PSBs via crystallographic shear. The rate is very small, typically on the order of nanometers per cycle. Here crack growth rates are controlled by microstructure, mean stress, and environment. Transition to a finite crack growth rate takes place from no growth below a threshold value of ΔK_{th}. Cracks propagate through only a few grains after initiating at the surface. The Stage I fracture surface appears quite featureless.

Stage II

During this Paris law stage, the crack propagation rate has a linear relationship with ΔK on a log-log scale. Extensive research has been performed to understand this stage. This part of the

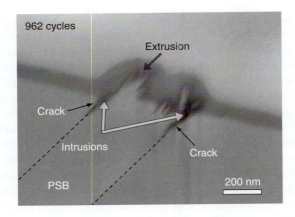

Figure 10.18

Intrusion and extrusion persistent slip band formed in a Sanicro 25 steel subjected to room temperature fatigue for 962 cycles at a constant strain amplitude of 0.0035 and the crack formed at the PSB-matrix interface.[29]

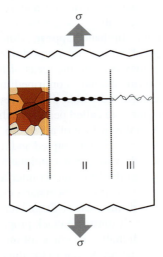

Figure 10.19

Schematic crack propagation path during the various stages of crack propagation.

fatigue fracture surface consists of a set of striations (ripples) on the fracture surface. These features develop in Stage II because of the plastic blunting process at the tip of the propagating cracks under fatigue cycles. This is a characteristic feature of fatigue failure, and such feature is generally not observed in other failure processes. This aspect is helpful in failure analysis of components.

---

[29]Polák J, Mazánová V, Heczko M, et al. The role of extrusions and intrusions in fatigue crack initiation. *Engineering Fracture Mechanics*. 2017;185:46-60.

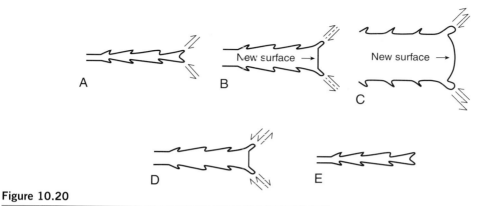

Figure 10.20

Sequential steps (a–e) of plastic blunting process during Stage II fatigue crack growth.[30]

In an effort to understand the process of striation formation during Stage II, we can now direct our attention to Fig. 10.20. Fig. 10.20 (a) illustrates an advancing crack under the tensile part of the fatigue cycle that creates shear stresses at the crack tip. As the crack opens up under tensile loading and a new crack surface is created, ear-like features are generated at the crack tip (Fig. 10.20 [b]). These ears become both extended and blunted (Fig. 10.20 [c]). After the tensile part of the loading is over, the compressive part of the loading cycle ensues; see Figs. 10.20 (d) and (e). Under the action of compressive stress, the shear stresses are reversed as the crack closes. In this process, the crack faces are "crushed" together (i.e., almost closes), and the crack face partially folds back, leaving a resharpened crack tip; this corresponds to the formation of a new striation. This process repeats itself until Stage III of the crack growth begins.

Stage II crack propagation is generally described in mathematical terms, as this stage is important from the viewpoint of a rate-controlling process and thus life prediction of components under fatigue.

Stage III

In this stage, crack growth rate picks up with increasing stress intensity factor and leads ultimately to fracture, as the net reduced cross-section can no longer sustain the applied stress. Brittle or ductile fracture may take place, depending on material characteristics. For example, a particle-containing material may involve cracking of particles or void creation and subsequent evolution as the crack propagates to failure.

10.2.2 Crack growth models

A fracture mechanics approach to understanding fatigue crack growth model is important. This enables development of the fatigue design based on loading parameters and geometry of the crack and specimen.

---

[30]Smallman RE, Ngan AHW. *Modern Physical Metallurgy*. 8th ed. Butterworth-Heinemann; 2014.

Paris and coworkers[31,32] suggested that linear fracture mechanics can be used to describe fatigue crack growth behavior and can be based on stress intensity factor range (ΔK). **Chapter 9** presented stress intensity factor (K) and fracture toughness (K_c) in some detail. What exactly ΔK means is defined mathematically as,

$$\Delta K = K_{max} - K_{min}, \tag{10.13a}$$

where K_{max} is maximum stress intensity factor corresponding to (algebraically) maximum stress (σ_{max}) in the fatigue cycle, whereas K_{min} is stress intensity factor corresponding to minimum stress. Note that K_{min} is taken as zero under the compressive part of the cycle, as compression does not grow fatigue cracks. Therefore for an edge-cracked specimen,

$$\Delta K = Y\sigma_{max}\sqrt{\pi a} - Y\sigma_{min}\sqrt{\pi a} \tag{10.13b}$$

where Y is a geometrical factor that is a function of the ratio of crack length (a) to specimen width (W).

Fig. 10.21 (a) shows the variation of fatigue crack length as a function of the number of cycles to failure for different stress ranges for the same initial crack length, a. The derivative of the curve gives the crack growth rate. When crack growth rate, $\frac{da}{dN}$, is plotted against ΔK on a log-log scale as shown in Fig. 10.21 (b), a sigmoidal curve with three distinct stages of fatigue crack propagation emerges. These three regions of fatigue crack propagation correspond directly to the three stages just discussed above (refer to Fig. 10.19).

Region I corresponds to crack propagation, which is mostly undetectable or too slow. This region is limited by the threshold stress intensity factor range (ΔK_{th}) above which fatigue crack propagation follows a power law relation with ΔK. Region II fatigue crack growth is governed by a useful empirical power law relationship known as Paris law.

$$\frac{da}{dN} = C(\Delta K)^m \tag{10.14}$$

where C and m are scaling constants that are subject to a variety of factors, depending on test environment, material microstructure, temperature, and load ratio $\left(R = \frac{\sigma_{min}}{\sigma_{max}} = \frac{K_{min}}{K_{max}}\right)$. The value of C can be obtained by setting ΔK equals 1 MPa$\sqrt{m}$ and finding the intercept. The m value varies between 2 and 4 for most cases. As environmental conditions including temperature influence Region II, Paris law equation can also be affected by loading frequency and the nature of the waveform. Paris law has also been applicable to low-strength, highly ductile materials. The reason is that under stress intensity factor values, fatigue crack growth occurs and leads to small plastic zone sizes ahead of the crack tip. Limitations of Paris law include the noneffect of stress ratio or mean stress and its failure to describe Regime I and Regime III. Nonetheless, it is simple to use and can be easily integrated into a program.

As illustrated in Fig. 10.21 (b), Regime III is the region where crack growth rate increases significantly. Basically, fracture toughness (K_c) is reached by K_{max}. If load ratio (R) is increased, it lifts the whole curve (Fig. 10.21 [b]). The effect of R is more in Region III than in Region II. Thus correction

---

[31] Paris PC, Gomez MP, Anderson WP. A rational analytic theory of fatigue. *The Trend in Engineering*. 1961;3:9-14.
[32] Paris PC, Erdogan F. A critical analysis of crack propagation laws. *J Basic Eng*. 1963;85:528-534.

10.2 Fatigue crack initiation and growth

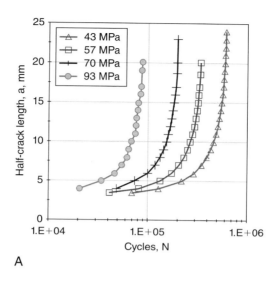

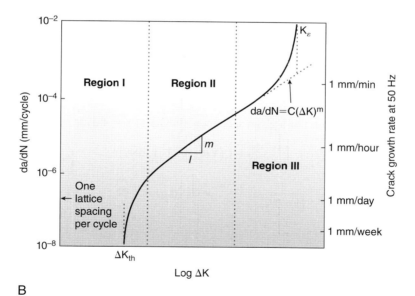

Figure 10.21

(a) Crack growth as a function of stress range ($\Delta\sigma$) for aluminum alloy 6063-T6.[33] (b) Schematic of the fatigue crack propagation rate as a function of the stress intensity factor range on a log-log scale.[34]

---

[33]Ravi Chandran KS. A universal functional for the physical description of fatigue crack growth in high-cycle and low-cycle fatigue conditions and in various specimen geometries. *Int J Fatigue*. 2017;102:261-269.
[34]Adapted from Macdonald KA, Ed. *Fracture and Fatigue of Welded Joints and Structures*. Woodhead Publishing; 2011.

needs to be made to incorporate the effect of R on the Paris law. Forman et al. (1967)[35] developed an empirical relation to address the effect:

$$\frac{da}{dN} = C_f \left\{ \frac{\Delta K^{m_f}}{(1-R)K_c - \Delta K} \right\} \qquad (10.15)$$

where C_f and m_f are the material constants.

Weertman[36] proposed another form of equation that is applicable to Regime II and Regime III:

$$\frac{da}{dN} = \frac{C_w (\Delta K)^4}{K_c^2 - K_{max}^2} \qquad (10.16)$$

where C_w is the proportionality constant.

10.3 Life prediction

While total fatigue is close to the sum of crack initiation and crack propagation time, crack initiation time cannot be determined precisely. So, fatigue life prediction based on Paris law–based methodology leads to shorter life; that is, a more conservative life estimate is obtained.

Assume a simple problem of an edge cracked plate where the stress range is $\Delta\sigma = \sigma_{max} - 0$. The critical crack length, $a_{critical}$, is the crack length that produces fracture, while a_{safe} is a reasonably sized crack. So

$$\Delta K = Y \Delta\sigma \sqrt{\pi a}$$

$$\frac{da}{dN} = C\left(Y\sigma_{max}\sqrt{\pi a}\right)^m = C\left(Y\sigma_{max}\sqrt{\pi}\right)^m a^{m/2}$$

Note that Y is not a constant. For a given specimen thickness, $Y = f(a)$:

$$\int_{a_{initial}}^{a_{safe}} \frac{da}{a^{m/2}} = C\left(Y\sigma_{max}\sqrt{\pi}\right)^m \int_0^N dN$$

or

$$\frac{1}{(1-m/2)} a^{1-m/2} \Big|_{a_i}^{a_{safe}} = C\left(Y\sigma_{max}\sqrt{\pi}\right)^m N$$

If $m = 2$, then

$$N_f = \frac{\ln(a_f/a_0)}{CY^2(\Delta\sigma^2)\pi} \qquad (10.17)$$

In Paris law regime, the only effects that significantly change crack propagation rates are stress state or ΔK. For the most part, microstructure has little to do with crack growth rates. However, important to note is that final crack length does not significantly affect total life, that is, N_f at failure. Fig. 10.22 illustrates this point. However, initial crack size has a significant effect.

---

[35] Forman RG, Kearny VE, Engle RM. Numerical analysis of crack propagation in cyclic loaded structures. *J Basic Eng.* 1967;89:459-464.
[36] Weertman J. Theory of fatigue crack growth based on a BCS crack theory with work hardening. *Int J Fract.* 1973;9:125-131.

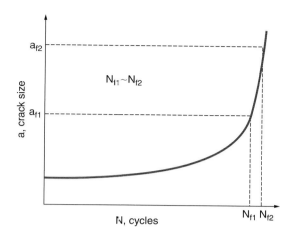

Figure 10.22

Crack size as a function of number of cycles. Note that the steep increase in crack size actually makes the N_{f1} and N_{f2} approximately the same.[37]

The term Y (calibration constant) in ΔK usually changes throughout fatigue life. That is why a numerical method is sometimes applied to take care of these variations pursuing the following steps as shown in Fig. 10.23:

1. Divide crack growth into several finite intervals.
2. Determine Y at a_n and a_{n+1} for each.
3. Calculate ΔK_n and ΔK_{n+1}.
4. For each ΔK determine da/dN from the crack growth data.
5. Average the growth rates using the equation below:

$$\left(\frac{da}{dN}\right)_{Avg} = \frac{\frac{da_n}{dN} + \frac{da_{n+1}}{dN}}{2}$$

6. Determine ΔN for crack growth a_n to a_{n+1} as

$$\Delta N = \frac{a_{n+1} - a_n}{\left(\frac{da}{dN}\right)_{Avg}}$$

7. Sum all the intervals to estimate total life.

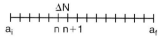

Figure 10.23

The numerical integration scheme of fatigue life estimation.

---

[37]Adapted from Suresh S. *Fatigue of Materials*. 2nd ed. Cambridge University Press; 1998.

10.4 Fatigue deformation and influence of microstructure

Fatigue is a highly microstructure-dependent property. Understanding the linkage between fatigue properties and microstructure could unleash the development of better fatigue-resistant materials. However, since microstructure-property correlations for fatigue are not well developed, mostly qualitative or semi-quantitative relations exist. Therefore fatigue resistance in materials is enhanced mainly by keeping the mechanical design of the component free from stress concentration sites (as discussed in the section dealing with geometrical effects) as much as possible and introducing residual compression stresses at surface by certain methods. Some aspects are discussed here qualitatively.

The term *fatigue ratio* is defined as the ratio of fatigue limit to tensile strength. For steels, the fatigue ratio stays close to 0.5, whereas for non-ferrous metals, it tends to be around 0.35. So, a natural temptation is to focus on the microstructural factors that will improve material tensile strength, as these will also increase fatigue limit and thus keep the fatigue ratio in the same ballpark. With notches present, the fatigue ratio decreases. Although the fatigue limit increases with increasing tensile strength in the higher strength (hardness) range, this assertion does not hold true. Rather, fatigue resistance can decrease in materials with very high strength. In the high strength range, the fatigue properties become very sensitive to surface conditions, stress-state at the surface, and inclusions. Fig. 10.24 shows such a trend in the endurance limits against Rockwell hardness of different alloy steels.

While understanding the effect of microstructure on fatigue properties is important, a focus on how microstructural factors affect crack initiation and/or propagation characteristics is critical to get an idea of how microstructure affects fatigue life. The primary influence of microstructure (including defects, surface treated condition) in Region I of the da/dN has been presented in Fig. 10.19. Microstructural

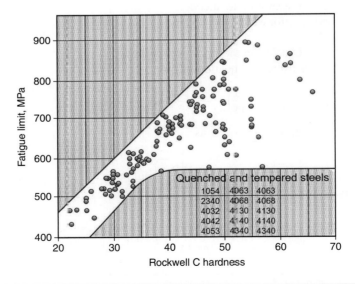

Figure 10.24

Variation of endurance limit as a function of Rockwell C hardness of various alloy steels.[38]

---

[38] Lyman T, ed. *Properties and Selection of Metals*. Vol. 1. In: *Metals Handbook*. 8th ed. American Society for Metals; 1961.

factors may lead to easier crack nucleation, which in turn results in a lower ΔK_{th}, as fracture toughness (K_c) is affected by microstructure and can influence Region III.

Microstress concentration sites inside materials can reduce the stress required to nucleate a crack and thus reduce the time needed for crack nucleation. One example is the difference between fatigue strength of coarse pearlite and spheroidized pearlite of eutectoid steel (0.77 wt.% C). Even though hardness of the two steels may be the same, the propensity for flaw nucleation at stress concentration sites of the cementite lamellae within the pearlite is larger than spheroidized pearlite that contains cementite spheroids. Furthermore, steels having more mobile solutes (such as carbon) can exhibit improved fatigue limits through dislocation interaction with solutes.

Some researchers have emphasized the beneficial aspect of slip homogenization on fatigue performance. Hence, factors promoting coarse slip band formation would be detrimental. This can also be related to stacking fault energy of materials involved. An interesting case of fatigue properties of an age-hardened aluminum alloy is discussed here. The presence of finer dispersoids results in better fatigue strength (Fig. 10.25 [a]). The S-N curves generated from fatigue testing of smooth specimens of a commercial 7075 Al and a high-purity version of X-7075 Al alloy treated under identical aging conditions are compared. The dispersoids, $Al_{12}Mg_2Cr$, present in commercial 7075 Al, help in slip homogenization and help in preventing easy fatigue crack initiation. Slip homogenization is opposite of slip localization and these two terms can be linked with particles that are sheared vs particles that are

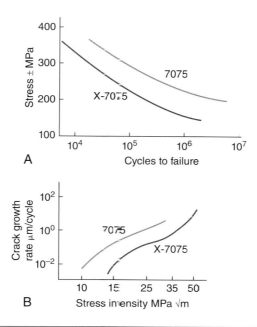

Figure 10.25

(a) Fatigue (S-N) curves for commercial 7075 Al and a high-purity version of 7075 Al (designated as X-7075); (b) crack growth rate versus stress intensity factor curves for 7075 Al and X7075 alloys.[39]

---

[39]Polmear IJ. *Light Alloys*. 4th ed. Butterworth-Heinemann; 2005.

non-shearable. The non-shearable particles lead to local dislocation loop storage and hardening. That will promote activation of other slip planes, thereby leading to slip homogenization. However, the effect reverses when precracked specimens are used to compare fatigue crack growth rates (Fig. 10.25 [b]). The high-purity X-7075 is more resistant to crack propagation. So, minimization of constituent particles is an important microstrutural engineering principle for enhanced fatigue life. It not only removes easy nucleation site for cracks, but also influences crack propogation by removing easy linkage sites.

Generally, the poor fatigue properties of these alloys stem from relative metastability of the precipitate structure under cyclic loading. Some coherent precipitates below a critical size may redissolve into the matrix during fatigue deformation. This situation can lead to coarse slip bands and result in slip localization because dislocations will have less resistance against their motion through these regions, which in turn can lead to fatigue crack initiation. Precipitate-free zones are also found in certain precipitation-hardened alloys. which can serve as sites of strain localization and may serve as precursors for fatigue crack initiation. Boyapati and Polmear[40] worked on an Al-5Mg-0.5Ag alloy and performed fatigue testing of the alloy under three different heat treatment conditions. In this case, solution heat-treated/quenched alloy shows the best fatigue limit, followed by the overaged alloy (aged for 70 days for 175°C) with the underaged alloy (aged for 1 day at 175°C) having the lowest fatigue limit (Fig. 10.26). The underaged alloy contains coherent precipitates, which promote coarser slip bands and thus easier fatigue crack initiation.

Although we discussed in the previous sections how fatigue cracks appear on the surface for HCF regime, specifically VHCF regime, internal inclusions and other defects can also contribute to fatigue

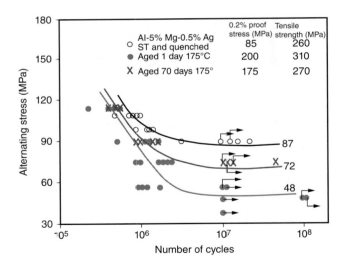

Figure 10.26

Fatigue (S-N) curves for Al-5Mg-0.5Ag alloy under various aging conditions.[41]

---

[40]Boyapati K, Polmear IJ. Fatigue microstructure relationships in some aged aluminum alloys. *Fatigue Fract Eng Mater Struct*. 1979;2(1):23-33.

[41]Pineau A, Antolovich SD. Probabilistic approaches to fatigue with special emphasis on initiation from inclusions. *Int J Fatigue*. 2016;93:422-434.

10.4 Fatigue deformation and influence of microstructure

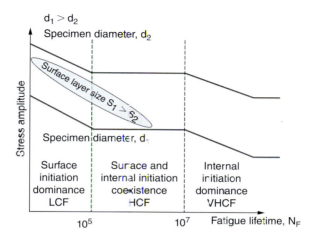

Figure 10.27
A schematic plot illustrating the importance of fatigue crack initiation sites across the different fatigue life regimes.[43]

crack initiation. Fig. 10.27 schematically shows stress amplitude as a function of the full spectrum of fatigue life of steels. Also marked are the regions where fatigue cracks nucleated internally rather than on the surface of the specimens. Also, a bigger specimen size generally exhibits poor fatigue strength because of the possibility of more stress concentration sites and stress gradients.

Pores created by gas evolution or shrinkage can also act as crack nucleation sites. Thus, powder metallurgy materials with residual porosity can also be susceptible to crack nucleation. Hence, castings generally have poorer fatigue resistance than wrought materials as shown by the S-N curves (Fig. 10.28). However, improved casting techniques, such as squeeze casting, with reduced porosity levels, can help improve fatigue strength significantly. Castings can also be subjected to hot isostatic pressing (HIP) to minimize and/or eliminate internal porosity. HIP of aluminum-based castings is generally performed at a pressure of 105 MPa using argon gas and temperature range of 480°C to 525°C. Improvement in fatigue life for C354-T6 alloy due to HIP is illustrated in Fig. 10.29 in terms of respective maximum stress versus number of cycles to failure curves before and after HIP. While HIP improves the fatigue and other properties of castings by eliminating porosity, it also increases time and cost.

Now let us discuss the effect of another important microstructural factor, *grain size*. The resistance of metallic materials to both fatigue crack initiation and propagation is generally affected by grain size. Many studies in microcrystalline materials have validated that increasing grain size generally leads to a decrease in the fatigue endurance limit. Further, a microstructure with coarser grains can result in an enhanced threshold stress intensity factor range (ΔK_{th}) and a reduction in the rate of fatigue crack propagation. Fig. 10.30 compares S-N fatigue performance in pure nickel specimens of three different grain sizes (nanocrystalline [nc], ultrafine crystalline [ufc], and conventional microcrystalline materials [mc]). The nc nickel (30 nm grain size) has a slightly better fatigue resistance than the ufc nickel (300 nm grain size). The endurance limit range of mc (grain size >1 μm) nickel is clearly less than the nanocrystalline or ultrafine grained nickel. Fig. 10.31 illustrates the effect of grain size change on the fatigue crack growth behavior of ufc and mc

282 Chapter 10 Fatigue behavior of materials: fatigue limiting design

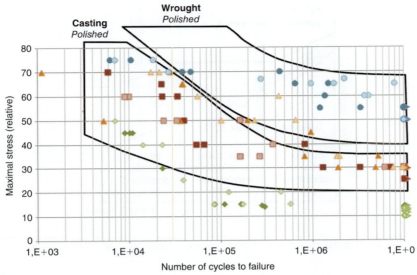

Figure 10.28

The *S-N* curves of AA 7010 Al alloy processed via gravity die casting and squeeze casting as compared to wrought condition.[42]

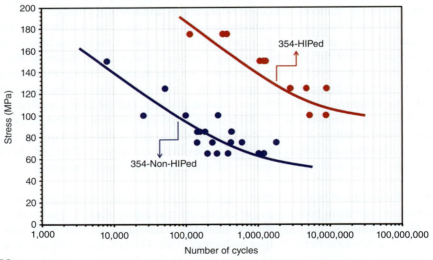

Figure 10.29

The effect of hot isostatic pressing (HIP) on the fatigue life of C354-T6 alloy.[43]

---

[42]Charkaluk, E.; Chastand, V. Fatigue of Additive Manufacturing Specimens: A Comparison with Casting Processes. Proceedings 2018, 2, 474.
[43]Ammar HR, Samuel AM, Doty HW, et al. The influence of hot isostatic pressing on the fatigue life of Al–Si–Cu–Mg 354-T6 casting alloy. *Inter Metalcast.* 2022;16:1315-1326.

10.4 Fatigue deformation and influence of microstructure

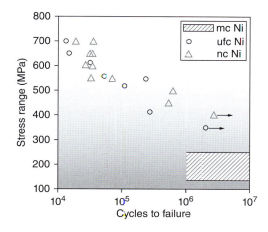

Figure 10.30
Variation of stress range as a function of the number of cycles to failure of pure Ni with grain sizes in the range of nanocrystalline (nc), ultrafine-crystalline (ufc), and microcrystalline (mc).[44]

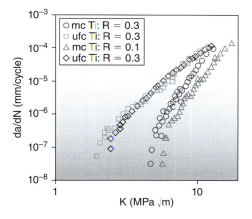

Figure 10.31
Crack growth rate versus stress intensity factor range for equal channel angular pressed (ECAP) ultrafine-grained titanium and its microcrystalline counterpart.[45]

titanium. The reason for the greater resistance to fatigue crack growth in the mc material is explained by the greater crack deflections compared to their ufc counterparts. This point can be illustrated in the difference in the crack paths in two materials as shown in Fig. 10.32. Clearly, crack path tortuosity in the mc titanium (Fig. 10.32 [a]) is significantly greater than that in the ufc one (Fig. 10.32 [b]).

---

[44]Hanlon T, Kwon YN, Suresh S. Grain size effects on the fatigue response of nanocrystalline materials. *Scr Mater*. 2003;49:675-680.
[45]Hanlon T, Tabchnikova ED, Suresh S. Fatigue behavior of nanocrystalline metals and alloys. *Int J Fatigue*. 2005;27:1147-1158.

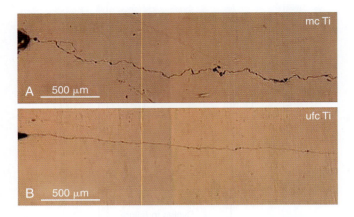

Figure 10.32

Evidence of difference in crack path tortuosity: (a) microcrystalline (mc) titanium and (b) ultrafine crystalline (ufc) titanium. Fatigue crack growth tests were run using initial stress intensity factor of 6.5 MPa$\sqrt{m}$ for mc-Ti and 5 MPa$\sqrt{m}$ for ufc-Ti. Other parameters were the same: loading frequency of 10 Hz, stress ratio of 0.3, and test temperature of room temperature.[46]

Another important microstructural feature is the nature of grain boundaries. Both fatigue crack initiation and propagation can be greatly affected by the nature of grain boundaries, that is, the local crystallography of the grain boundaries. The advent of the electron backscatter diffraction (EBSD) technique has simplified obtaining a quantitative description of grain boundary character distribution over the previous transmission electron microscopy–based method. Many EBSD studies have been dedicated to understanding the function of grain boundaries on cracking.

Initiation of fatigue cracks can take place at the grain boundaries if the material's ambient environment promotes attack on the grain boundaries or the grain boundary particles, as well as at high temperatures where grain boundary cavitation or sliding can occur. Brittle solids such as ceramics with glassy phases on the grain boundaries can lead to crack initiation at the grain boundary, or differential thermal contraction/expansion behavior can lead to strain incompatibilities in neighboring grains. Note that a majority of fatigue failure cases do not involve intergranular cracking. Also, PSBs observed near the twin-matrix interface give credence to the formation of fatigue cracks at the twin boundaries.

Concept build-up—Effect of fatigue testing method on progression of deformation during fatigue. In the previous sections, we covered stress-amplitude fatigue testing and strain-amplitude fatigue testing. Then we went through the micromechanisms in the preceding section, including the impact of microstructure on fatigue life. At this stage, try to first reflect on the mechanics of deformation percolation in a volume. Stress-amplitude fatigue testing is always done below the nominal yield strength of the material. ASTM standard specifies that the maximum initial stress during the cycling should not exceed 0.8 of the yield strength (0.8YS). So, when we think of the volume of material made of grains of different sizes,

---

[46]Sachs NW. Understanding the surface features of fatigue fractures: how they describe the failure cause and the failure history. *J Fail Anal Prev.* 2005;5(2):11-15.

precipitate and particles of different sizes, we need to visualize how the deformation starts and percolates through the volume. Let us take the stress-amplitude testing first. When a stress of 0.5YS is applied, multiple ways we can visualize the deformation to start. The coarser grains with larger slip spacing can start deforming if the local CRSS is achieved at that location. At this stage only a fraction of the overall specimen is locally deforming. A similar argument can be built toward particles and stress concentration enabled deformation around them. So, the starting stress amplitude during stress-amplitude fatigue testing would dictate the initial volume where the fatigue deformation starts. Conceptually then, the fatigue damage during a stress-amplitude testing is nucleation and percolation of deformation during the stress-amplitude testing. So, this type of microstructural evolution has concurrent influence of stress and time.

Now consider strain-amplitude fatigue testing. In this the entire volume of the specimen is driven to a predefined strain. Conceptually all the grains will now be deformed in a way more analogous to tensile testing. This type of testing is driving the entire microstructure to deformation and the role of time is diminished.

Students are encouraged to run thought experiments on this and write down your arguments. Consider the figure below, which has two grain size distributions (Yuan W, Mishra RS. Grain size and texture effects on deformation behavior of AZ31 magnesium alloy. *Mater Sci Eng A*. 2021;558:718-724). What would you expect the fatigue behavior to look like? Can you draw the distinction between the crack initiation life and crack propagation life for these two microstructural distributions?

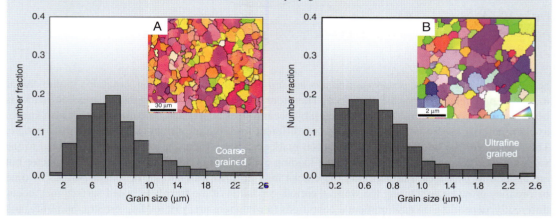

10.5 Fatigue fracture characteristics

To supplement what has been learned about fatigue testing and relevant fatigue mode, interest in further failure analysis of the specimen can be augmented by examining the fracture surface. With visual or low-magnification observation of the fracture surface, the macroscopic features of the fatigue fracture surface can be studied. As shown in Fig. 10.33, the fatigue fracture surface consists mainly of two distinct areas: (a) *fatigue zone,* which in most cases starts from the point of origin (or the site for the origin of the fatigue crack) that happens to be on the surface; it represents the region of slow crack growth; and (b) *instantaneous zone,* which is the second zone that spreads over the rest of the fracture surface. This overload zone occurs because the crack reaches a stage where the rest of the cross-section can no longer sustain the stress when catastrophic crack propagation takes place and thus leads to fast fracture. This portion of the fracture surface may have either brittle or ductile characteristics, features quite similar to tensile tested specimens.

The fatigue zone appears relatively smooth, whereas the instantaneous zone may appear rough. The fatigue zone actually contains progression marks also known as beach marks. The point of origin could

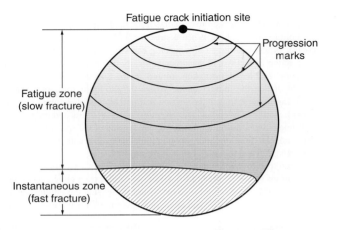

Figure 10.33

A schematic of fatigue fracture surface showing the main regions (macroscopic).[48]

be single or multiple. The progression marks are representative of substantial variations in the load as the crack propagates across the specimen. The more conventional name for the progression marks is beach marks since they look like the lines left by ocean waves on a beach. The markings are also known as clamshell marks. Note that beach marks are different from fatigue striations, which are microscopic. The reasons for the development of the beach marks can vary. First, these marks are attributed to the variation of crack growth rates because the loading cycles are not always continuous in character but are applied in packages. Furthermore, the cracks may need to wait shorter or longer periods of time before they grow again. During the waiting period, oxidation of the freshly exposed surface may occur differently, depending on environmental conditions (air temperature, pH, loading spectra, humidity, and so forth). Even the direction of load cycling can have an effect on beach mark formation. Indeed, people failed to observe any beach marks in instances of fatigue fracture where uninterrupted crack propagation without load variations took place. When there is more than one origin of fatigue cracks, the ratchet lines or ratchet marks separate into two adjacent crack fronts. More than one origin of fatigue initiation points is possible at high stress amplitudes and/or the presence of severe stress concentration sites. Other macroscopic features could be on the fatigue fracture surface depending on specific fatigue conditions.

Next, we describe a fatigue fracture surface and its macroscopic characteristics using a real-world example. Fig. 10.34 shows the fatigue fracture surface of a rail as seen by the naked eye. The rail is subsequently fatigued to fracture at a test facility. The dark region at the upper left hand corner is the initiation region of the crack, and then the circumferential rings emanate from the point of origin. Because of the two different types of load cycling, the beach mark characteristics in the first set look different from the subsequent set.

Each of the beach marks viewed at higher magnification with adequate resolution confirms that each beach mark contains more or less concentric microscopic ripple marks known as fatigue striations, which can also be described as remnants of the microplasticity-assisted process during fatigue loading. Here the crack front progresses perpendicular to the applied tensile stress. Each striation width represents the distance traveled by the crack front in each cycle. The origin of striations has been discussed previously in **Chapter 5**. Striations are usually curved out in the direction of crack propagation

Figure 10.34

The macroscopic fatigue fracture surface of a rail.[47]

and would denote stress gradient distribution upon crack initiation. The presence of striations on the fracture surface implies that the failure was a result of fatigue. However, striations not seen on the fracture surface do not automatically lead to the conclusion that the reason for the failure was not due to fatigue. Fatigue striations may not be visible due to lower resolving power of the characterization tool, lower intrinsic ductility of the material, and some unexpected damage to the fracture surface. While some idea about fatigue life can be obtained from quantitative examination of fatigue striations, precise depiction of total fatigue life is still lacking.

While light microscopy can reveal fatigue striations, transmission electron microscopy and especially scanning electron microscopy provide the best method of characterizing fatigue striations. Scanning electron microscopy has the right depth of focus and resolution to perform superior fatigue fractographic examination. A schematic array of striations is presented in Fig. 10.35. Actual scanning electron microscopy images of striations in the fatigue fracture surface of a 321 grade stainless steel show striations from fatigue testing under both strain-control and stress-control modes (Fig. 10.36). Striations have been categorized into two main types: ductile striation and brittle striation. For low-ductility materials, striations could become indistinct. Also, the effect of environment can modify the nature of the striations. For example, hydrogen embrittlement under fatigue conditions may exhibit intergranular striations, even though in the absence of the hydrogen environment, the fracture surface may show ductile striations.

A real-world example of a fatigue fractured sample from a rail end bolt hole is shown in Fig. 10.37 (b). The corresponding fatigue striations are shown in Fig. 10.37 (d). The rail was loaded to some significant extent during each load cycle when a train passed by. The crack propagated during each of such loading cycle and the characteristic striation pattern results on the fatigue fracture surface.

---

[47]Rice RC, Rungta R. Fatigue analysis of a rail subjected to controlled service conditions. *Fatigue Fract Eng Mater Struct*. 1987;10(3):213-221.

288 Chapter 10 Fatigue behavior of materials: fatigue limiting design

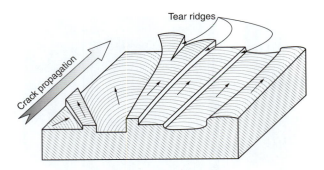

Figure 10.35
A schematic illustration of fatigue striations.[48]

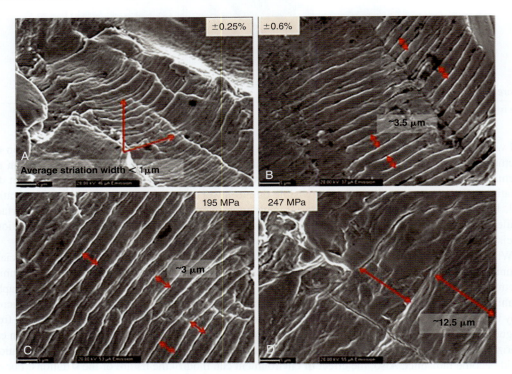

Figure 10.36
Comparison of striated fracture surface in 321 stainless steel specimens tested at 360°C under (a) and (b) strain-controlled fatigue testing (at ±0.28% and ±0.6%, respectively); (c) and (d) stress-controlled fatigue testing (at 195 MPa and 247 MPa, respectively).[49]

---

[48]Beachen CD. Microscopic fracture processes. In: Liebowitz H, ed. *Fracture an Advanced Treatise*. Vol. 60. New York: Academic Press; 1968:311.
[49]Prasad Reddy GV, Dinesh PM, Sandhya R, Laha K, Jayakumar T. Behavior of 321 stainless steel under engineering stress and strain controlled fatigue. *Int J Fatigue*. 2016;92:272-280.

Figure 10.37
(a) Image of the rail ends bolted together, (b) fatigue cracks initiated from bolt holes, (c) fractograph of the fracture surface, and (d) fatigue striations on the fractured surface.[50,51]

10.6 Design aspects

In laboratory fatigue testing, a specific stress amplitude is used until the specimen fractures. However, in service, variations in stress amplitudes, mean stresses, and loading frequency could complicate the situation.

Cumulative damage theories are many and include those proposed by Marco-Starkey, Henry, Gatts, Corter-Dolan, Marin, and Manson double linear. However, the Palmgren-Miner rule[52,53] is the one widely used method for evaluating residual fatigue life and thus that method helps in fatigue life prediction. The underlying premise of this approach is the assumption that fatigue damage created by each loading pattern is accumulated in the material, and fatigue life will be exhausted progressively.

For the general case of multiple stress amplitudes, damage is defined as a function of total life. For example, if stress amplitude of σ_1 is cycled for n_1 cycles and fatigue life at σ_1 is N_1, then the damage (D_1) for σ_1 is given by

$$D_1 = \frac{n_1}{N_1}.$$

---

[50]Zhang, P., Li, J., Zhao, Y. et al. Crack propagation analysis and fatigue life assessment of high-strength bolts based on fracture mechanics. Sci Rep 13, 14567 (2023).
[51]Galya V. Duncheva, Jordan T. Maximov, A new approach to enhancement of fatigue life of rail-end-bolt holes, Engineering Failure Analysis, Volume 29, 2013, Pages 167-179.
[52]Palmgren A. Die Lebensdauer von Kugellagern. Zeitschrift des Vereins Deutscher Ingenieure. 1924;68:339-341.
[53]Miner MA. Cumulative damage in fatigue. J Appl Mech. 1945;12:159-164.

And if $n_1 = N_1$, then failure should occur.

In reality, however, the component would go through different types of loading history with variable stress amplitudes. Fig. 10.38 (a) shows a variable stress-amplitude loading history comprised of stress cycles with three different stress amplitudes, whereas Fig. 10.38 (b) presents the corresponding S-N curve. Note that it does not depend on the order in which the stress cycles are imposed on the component. Generically, the total damage that can cause failure under variable stress amplitudes can be expressed as a linear summation of all the damage:

$$D_{total} = \frac{n_1}{N_1} + \frac{n_2}{N_2} + \frac{n_3}{N_3} + \cdots \frac{n_{i-1}}{N_{i-1}} + \frac{n_i}{n_i} = \sum \frac{n_j}{N_j} = 1. \tag{10.18}$$

Fatigue failure would be expected if the total sum of fractional lives becomes unity or more.

The rule discussed above has various issues with the simplistic linear summation without regard to the mechanisms involved under different stress amplitude ranges. Also, this rule cannot take into

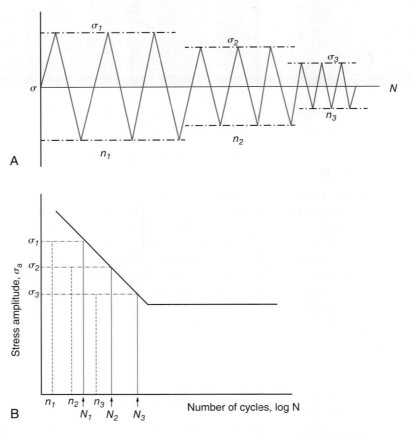

Figure 10.38

(a) Variable amplitude stress cycles showing three different stress blocks, and (b) corresponding S-N curves.

account why tensile overload applied to notched or precracked specimens helps to decrease the crack growth rate while compressive overload has an opposite effect. Also, the assumption that loading sequence does not have any bearing may not be right.

10.6.1 Fatigue design approaches

Since the 19th century, considerable literature sources have been presented to establish the methods and procedures of mechanical design against fatigue failure. In modern times, more refinement of fatigue design has been implemented. Here we would limit ourselves to introduction of the design philosophies without getting into details. Fatigue design approaches need to be followed for numerous applications.

10.6.1.1 *Infinite-life design*

Since this design approach evolved early in the development of fatigue testing methodologies, it is regarded as the oldest fatigue design philosophy. Design stresses are kept well below the fatigue limit (at some fraction of fatigue strength of the material) under the assumption that infinite life can be achieved under low-stress conditions. This basically uses the data from the stress-life approach; that is, it is based on traditional *S-N* curves. The assumption is that the starting material is free from flaws. Over the years, this design philosophy has been replaced by other advanced methodologies. However, the approach is quite simple to use and can be particularly useful in cases where regular inspections are not feasible or desired. It is still applicable for components in the elastic range, that is, HCF. Since it is a conservative design approach, it results in heavier section of components.

10.6.1.2 *Safe-life design*

This design approach is based on the consideration that the component is essentially defect-free at the start but over time develops a critical crack size that eventually leads to failure. Therefore the assumption is based on the finite life of the component. The methodology takes into account service conditions for components exposed to greater loads that can lead to plastic strains. Under such conditions, deformation is described in terms of strain. Thus the approach resulted in the development of fatigue assessment techniques that gave a mathematical description of strain (ε) versus number of cycles (*N*). Some of this design approach appears in pressure vessels, bearings, and jet turbine design.

10.6.1.3 *Fail-safe design*

This design philosophy is rooted in the assumption that fatigue cracks can be detected and repaired before they lead to eventual component failure. This approach is used mainly in the aircraft industry, where high safety margins cannot be used because of greater weight, and it also cannot risk too low a safety margin for fear of accidents. Generally, this type of design involves many loading paths and crack-stoppers in the structure itself. If a primary load path fails, an alternate load path can pitch in and prevent structure failure. This design philosophy relies heavily on the strict regime of detection and inspection of cracks and must follow relevant regulatory certifications.

10.6.1.4 *Damage-tolerant design*

The damage-tolerant design is the newest design philosophy and can be considered an extension of the fail-safe design approach discussed above. This approach is based on the concept of fracture mechanics, which states that the components contain cracks and thus can be used to evaluate crack propagation rates. The necessary mitigation method can be implemented safely by repairing the component or by discontinuing its use before failure occurs. In this approach, crack detection and monitoring are important before critical crack dimension is reached. Materials that have high fracture toughness and slow fatigue crack growth rate should be used in damage-tolerant designs. The efficacy of this design approach depends largely on the use of a suitable non-destructive evaluation technique to detect and monitor cracks in the damage-prone areas of the whole system.

Despite much advancement in fatigue design approaches, fatigue-related failures do still occur quite frequently. The relatively frequent failures are perhaps due to the difficulty in predicting the variable service conditions and the lack of robust microstructure-based fatigue models. Thus full-component system testing should be performed before the system is put to real-world use. Also, the simulated tests should be run in the context of test conditions that are truly representative of actual service conditions. However, although sometimes accelerated testing would be needed to save time and money, various other factors may be missed under those test conditions, which can significantly influence the overall fatigue life.

10.6.2 Use of Ashby charts in materials selection and design

This section discusses Ashby charts (or material property chart) to emphasize the importance of material classes on fatigue properties. "A material property chart is a diagram with one or a combination of material properties as its axes on which the fields are occupied by each material class and of the individual members within each material class are plotted."[54] Fig. 10.39 presents a plot of threshold stress intensity factor (ΔK_{th}) against the endurance limit (σ_e). The chart reveals that, for materials with low ΔK_{th} and high σ_e, the fatigue property is crack-growth limited (extrinsic fatigue). On the other hand, a high value of ΔK_{th} and low value of σ_e imply that fatigue is crack initiation limited (intrinsic fatigue). While the difference between these two different behaviors is difficult to discern, close inspection of the Ashby chart (Fig. 10.39) reveals the difference in fatigue behavior among the materials classes. Engineering ceramics have low ΔK_{th} and higher σ_e. This means that engineering ceramics have already pre-existing flaws in the form of pores and cracks that result in low ΔK_{th}, but fatigue life is dominated mainly by crack propagation. On the other hand, metals that are tougher (copper, mild steel) or polymers (nylons) can create cracks inside the material, cracks which then propagate to failure under fatigue. However, metals (magnesium) and polymers (PMMA) that are less tough are susceptible to fatigue failure by propagation of pre-existing cracks. This observation can be further substantiated by considering the materials property chart in terms of fatigue threshold versus fracture toughness (Fig. 10.40).

In **Fig. 10.40**, diagonal contours represent three different $\Delta K_{th}/K_{Ic}$ ratios. For tough steels and titanium alloys, this ratio starts from 0.03, whereas for certain ceramics and polymers, it starts from 0.8. Broadly, the ratio falls between the range of 0.05 and 0.5, which is quite narrow indeed. Generally, not much advantage is gained for designing against fatigue for materials that have the $\Delta K_{th}/K_{Ic}$ ratios greater than 0.7. For those materials, designing using the fracture toughness approach should be sufficient. However, sometimes the mechanistic factors that increase fracture toughness of a ceramic may

---

[54]Fleck NA, Kang KJ, Ashby MF. The cyclic properties of engineering materials. *Acta Metall Mater.* 1994;42:365-381.

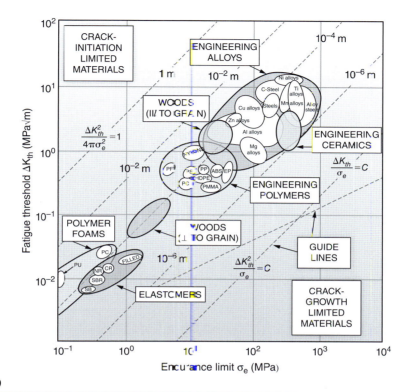

Figure 10.39

Material property chart of fatigue threshold (for stress ratio, $R = 0$) versus endurance limit ($R = -1$).[55]

in fact decrease fatigue resistance, for example, transformation toughening, and crack bridging do not really help the situation. Thus toughened ceramics would need design against fatigue as the $\Delta K_{th}/K_{Ic}$ ratio becomes smaller.

10.6.3 Geometrical stress concentrations

Fatigue cracking often starts from stress concentration sites during fatigue loading. A brief prior discussion of microscale stress concentration sites also mentioned that these stress concentration sites could be present as holes, fillets, threads, or other geometrical discontinuities (i.e., notches). Stress concentration sites such as surface roughness can be created by the manufacturing process. Local stress is given by multiplying applied stress with the stress concentration factor (K_t):

$$\sigma_{notch} = K_t \times \sigma_{applied} \tag{10.19}$$

For brittle materials, when σ_{notch} reaches $\sigma_{fracture}$, failure occurs. Regarding yielding for ductile materials, unfortunately, the expression of stress concentration applies to elastic material, not material

---

[55]Fleck NA, Kang KJ, Ashby MF. The cyclic properties of engineering materials. *Acta Metall Mater.* 1994;42:365-381.

Chapter 10 Fatigue behavior of materials: fatigue limiting design

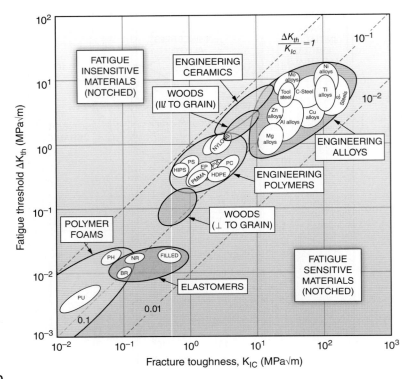

Figure 10.40

Material Property Chart (threshold stress intensity range, ΔK_{th}, versus fracture toughness, K_{Ic} [all data at stress ratio of $R = 0$]).[56]

with plasticity. The influence of geometrical stress concentration sites on fatigue can be studied by running fatigue experiments on notched samples. There are different ways in which a notch under uniaxial loading can affect the behavior: by increasing the local stress level at the notch; it can develop a stress gradient over the distance between the notch and the sample center; and a triaxial state of stress can develop at the notch root. Taking into account the adverse effect of notches on fatigue strength, the terms *fatigue strength reduction factor* or *fatigue notch factor* (K_f) were coined:

$$K_f = \frac{S_{El}}{S^*_{El}} \tag{10.20}$$

where S_{El} is the endurance limit of the unnotched sample and S^*_{El} is the endurance limit of the notched sample. The K_f value can vary with type of notch and notch geometry, material type, loading type, and stress applied.

---

[56]Fleck NA, Kang KJ, Ashby MF. The cyclic properties of engineering materials. *Acta Metall Mater*. 1994;42:365–381.

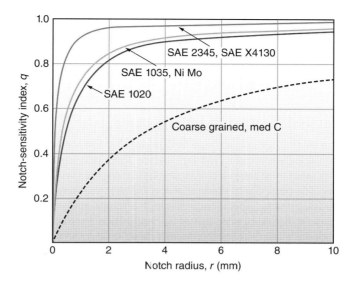

Figure 10.41

The variation of notch sensitivity factor against the notch radius for steels with different grain sizes and hardness. The lowest line is dotted because of insufficient data.[57]

Notched fatigue data are also reported using the notch sensitivity factor, q, which is defined as

$$q = \frac{K_f - 1}{K_t - 1}. \tag{10.21}$$

The above relationship compares theoretical stress-concentration factor, K_t, to fatigue strength reduction factor, K_f. The two limiting values of q are, (a) a material that does not exhibit any adverse effect due to a notch has $q = 0$ and (b) the other that has $q = 1$ experiences a reduction in fatigue up to the theoretical value. The q depends on material type and notch root radius. Fig. 10.41 shows the variation of q with respect to the notch radius for steels with different grain sizes. The topmost curve represents a quenched and tempered steel with a fine grain size. The middle two curves represent normalized steels with intermediate grain size. The lowest curve in the figure is constructed for a very coarse-grained medium carbon steel. The figure also reveals that geometric notches affect the fatigue resistance of high-strength materials more than the low-strength materials, largely because the former generally have limited ductility and thus less capacity for crack tip blunting.

Peterson's method[58] found the following expression of q as

$$q = \frac{1}{\left(1 + a/r\right)} \tag{10.22}$$

where a is a materials constant and r is the radius of notch.

---

[57]Peterson RE. *Relation Between Life Testing and Conventional Tests of Materials*. American Society for Testing Materials Bulletin; 1945:9-16.
[58]Peterson RE. Notch sensitivity. In: Sines G, Waisman JL, eds. *Metal Fatigue*. McGraw-Hill; 1959:293-306.

For steels, a can be of the following form:

$$a = \left[\frac{300}{S_{UTS}}\right]^{1.8} \times 10^{-3} \text{ inches}$$

where S_{UTS} is the ultimate tensile strength in ksi.

10.6.4 Avoidance/minimization of fatigue damage

A potential way of minimizing fatigue damage and improving fatigue life of components includes introducing compressive residual stress in the top surface layer of components. The success of this approach is due to the fact that most fatigue cracks originate from the surface and/or near-surface areas of the parts/components. The presence of compressive residual stress may help suppress initiation and propagation of such fatigue cracks, specifically for high-strength materials. Moderate compressive stress at the surface increases life (shot peening); it is more difficult to nucleate a crack when the local stress state opposes crack opening. On the other hand, tensile residual stresses are deemed detrimental to fatigue resistance. For larger stress amplitudes, that is, at lower fatigue lives, cyclic stresses can relax compressive residual stress generated by surface treatments, especially in low-strength materials.

One technique to introduce residual compressive stress in the surface layer is shot peening. In this technique, copious small-diameter (0.1–1 mm) hard steel balls are made to impinge on the component surface, creating localized plastic deformation. Fig. 10.42 shows a schematic residual stress profile in a shot-peened component with residual compressive stress at the surface region. The diameter of the shots, shot velocity, and process duration are all important process parameters. While the effect of shot peening is largely beneficial, care should be taken not to allow too long a time, as shot peening may create cracks at the slip bands and also promote additional cracking because of the resulting excessive

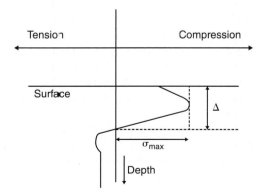

Figure 10.42

A schematic profile of residual stress distribution in a shot-peened component. The compressive residual stress is present up to the depth represented by Δ.[59]

---

[59]Gallitelli D, Boyer V, Gelineau M, et al. Simulation of shot peening: from process parameters to residual stress fields in a structure. *CR Mecanique*, 2016;344(4-5):355-374.

surface roughness. Also, this roughness needs to be removed for wear-intensive applications, but processes used to remove such roughness also remove the majority of the surface layer under the beneficial compressive residual stress. In shot peening, compressive residual stresses are limited to small depth, typically within 0.25 mm in soft metals (e.g., aluminum alloys) but even less in harder metals (steels, nickel-based superalloys). Nonetheless, shot peening is used extensively for automobile applications such as gears, chassis, various shafts, and valve springs. It is also used in the aircraft industry.

An alternative surface treatment technology, laser shock processing (also known as laser peening), can create residual compressive stresses to greater depths (can be up to 2.5 mm) using high-power, Q-switched laser pulses. The ability of a pulsed laser beam to generate shock waves using a short-pulsed laser beam was first recognized and explored during the early 1960s. Subsequent studies have established the fundamentals and conditions for enhancing the amplitude of stress shock waves, making it possible to induce plastic deformation on metal target surfaces. Follow-on work on prototype facilities and initial feasibility studies performed at Battelle Columbus Laboratories in the United States and other efforts worldwide have enabled laser peening to emerge as a promising alternative to traditional shot-peening processes. Laser peening has also been instrumental in enhancing the fatigue behavior of titanium alloys that are used as turbine compressors and Inconel-type, nickel-based superalloys used in the hot sections of jet turbine engines. Preliminary tests of laser-peened blades have confirmed that 10% to 40% improvement in metal fatigue strength can be obtained, as well as significant increase in fatigue failure resistance of fan blades caused by foreign object damage (FOD).

In Fig. 10.43, fatigue properties (in terms of maximum stress vs. number of cycles) of notched 7075-T7351 Al alloy specimens (exposed to shot peening and laser peening) evaluated from bending fatigue tests are compared with those of untreated specimens. Laser peening leads to almost 22%

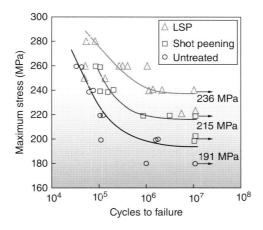

Figure 10.43

Plots of maximum stress versus number of cycles for untreated, shot-peened, and laser shock–processed (LSP) 7075-T7351 Al alloy (notched) specimens.[60]

---

[60]Peyre P, Fabbro R, Merrien P, Lieurade HP. Laser shock processing of aluminum alloys. Application to high cycle fatigue behaviour. *Materials Science and Engineering*. 1996;A210:102-113.

improvement in fatigue strength at 10^7 cycles, while shot peening provided 11% increase, as compared to the untreated specimens. The greater improvement in laser-peened surfaces can be explained because of the greater depth of the residual compressive stress fields, as discussed earlier.

Surface hardening heat treatment techniques such as carburizing and nitriding applied to certain steels involve volume change in the surface layer during phase transformation that can incorporate residual compressive stress on the surface and thus can improve fatigue resistance in the process.

10.7 Key takeaways

In this chapter, we highlighted various essential concepts of fatigue property fundamentals and elements of materials design principles against fatigue. Fatigue is an intriguing dynamic material property. HCF lives are best described by fatigue testing under stress-controlled mode (within the elastic regime), whereas LCF properties are evaluated using strain-controlled fatigue testing under stresses high enough where gross plasticity is involved in addition to elastic deformation. Fatigue data are often expressed in terms of S-N curves from which the concepts of endurance/fatigue limits are evaluated. The importance of Coffin-Manson for presenting LCF information was also discussed, as were the characteristics of cyclic stress-strain curves. While precracked specimens are not used in the aforementioned fatigue tests, there are tests that involve precracked specimens and that are used for fatigue crack propagation studies. Most often Paris law is used to understand the Stage II regime of the fatigue crack growth regime and has been useful in calculating fatigue lives. Microstructural factors affect fatigue property and are essential to the design of materials against fatigue. Examination of fatigue fracture surface is a great way to understand the causes of fatigue failure. While there are many tools, the fatigue fracture surface can be studied most effectively by scanning electron microscopy because of its high depth of field. Finally, different practical means of avoiding or minimizing fatigue failure involve avoidance of stress concentration sites in the mechanical design of components as well as pursuing specific surface engineering treatments/processes (e.g., shot peening, laser peening, carburizing) that introduce residual compressive stresses in the surface layer.

10.8 Exercises

1. A cylindrical metallic alloy rod is put under a cyclic uniaxial loading condition with a maximum load of 200 kN in tension and a minimum of 75 kN in compression. The metal has the following mechanical properties: ultimate tensile stress = 500 MPa; yield stress = 250 MPa; and endurance limit = 200 MPa. Predict the rod diameter that will give infinite fatigue life if the safety factor during the design of rod was given 2.
2. What are the different parts/segments of total fatigue life?
3. Define R value for fatigue testing. What are the typical values of R used?
4. Why does the fatigue data scatter increase at lower stresses?
5. What is the typical range of fatigue ratio (= fatigue strength/UTS)?

6. Describe Bauschinger effect.
7. How can fatigue life be increased through residual stress? Mention one method.
8. What are the fatigue crack growth regimes? Briefly describe each regime.
9. How does grain refinement influence crack initiation life?
10. What is the influence of constituent particles on fatigue life?
11. What are the ways to enhance fatigue life through microstructural engineering?
12. (a) How does the deformation-induced transformation influence the fatigue deformation? (b) What will happen if there is volume change during the transformation?
13. What are the typical values of fatigue endurance limit for different engineering alloys? Do a Google search and create a table. Do you see a correlation between strength and fatigue limit? HINT: Try both the yield strength and UTS.
14. What is the range of number of cycles to failure for LCF and HCF?
15. What are the different design approaches to fatigue?
16. Is heterogeneous microstructure good for fatigue?
17. Is deformation-induced twinning (TWIP effect) beneficial for fatigue life? Take a material that exhibits TWIP and compare with the table you built for Problem number (13).

Further readings

Courtney TH. *Mechanical Behavior of Materials*. 2rd ed. McGraw-Hill Publisher; 2000.
Dieter GE. *Mechanical Metallurgy, SI Metric Edition*. McGraw-Hill Publisher; 2001.
Laird C. *The Influence of Metallurgical Structure on the Mechanisms of Fatigue Crack Propagation, Fatigue Crack Propagation (ASTM Special Publication No. 415)*. Philadelphia, PA, USA: 1967:131-168.
Suresh S. *Fatigue of Materials*. 2nd ed. Cambridge University Press; 2001.

CHAPTER 11

High-temperature deformation of materials: creep-limiting design

Chapter outline

11.1 Classification of creep **303**
    11.1.1 Logarithmic creep 303
    11.1.2 Recovery-based creep...304
    11.1.3 Diffusional creep 305
11.2 Role of diffusion during creep deformation **305**
11.3 High-temperature deformation constitutive relationships **306**
    11.3.1 Dislocation glide 308
    11.3.2 Power law creep...309
        11.3.2.1 Weertman model for dislocation creep 310
    11.3.3 Diffusional creep 312
    11.3.4 Harper-Dorn creep 318
    11.3.5 Grain boundary sliding and superplasticity..319
    11.3.6 Summary of creep mechanisms 325
    11.3.7 Creep in particle-containing alloys 326
    11.3.8 Creep deformation mechanism maps 330
        11.3.8.1 Independent and sequential processes...330
        11.3.8.2 Ashby's deformation mechanism map..332
11.4 Creep damage mechanisms **333**
    11.4.1 Damage by loss of external section 336
    11.4.2 Creep damage by loss of internal section 336
    11.4.3 Damage by microstructural degradation 337
    11.4.4 Damage by gaseous-environmental attack 337
11.5 Creep fracture mechanism maps **337**
11.6 Creep life–related relationships **338**
11.7 Design of creep-resistant materials... **343**
11.8 Exercises **345**

Learning objectives

- Know about the basic definition of creep.
- Learn about the change in dislocation-obstacle interaction physics at elevated temperatures.
- Learn about various creep mechanisms.
- Learn about the temperature and stress range when creep deformation becomes important.
- Introduce design approaches for creep-limiting design.

Chapter 11 High-temperature deformation of materials: creep-limiting design

The preceding chapters have not dealt with how the mechanical behavior of materials can be affected at elevated temperatures. That aspect of materials behavior is important, especially for components that need to serve at higher temperatures or for enhanced formability to shape components. Design of materials for high-temperature applications can also require understanding of various other properties such as corrosion/oxidation resistance or the physical property of density. For example, to design lightweight high-temperature components, we can use the following Ashby chart (Fig. 11.1), which shows the variation of creep strength at 950°C for different materials as a function of density.

Different high-temperature deformation design approaches based on engineering importance include but are not limited to the following:

a) Deflection-limited designs
- Turbine blades
- Reactor assemblies

b) Stress relaxation
- Any bolted assembly
- Polymer gaskets

c) Time and/or strain to fracture or rupture

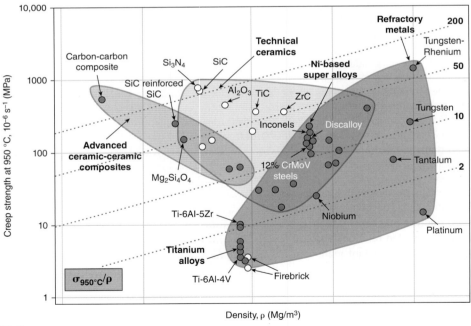

Figure 11.1

An Ashby property chart showing the strength of some important selected materials at a temperature of 950°C and a strain-rate of 10^{-6} s^{-1}, plotted against density.[1]

---

[1]Ashby MF. *Materials Selection in Mechanical Design*. 5th ed. Butterworth-Heinemann; 2016.

d) Hot working (formability related)
- Extrusion
- Hot rolling
- Forging
- Superplastic forming

For categories (a) to (c), the materials need to have microstructural design to resist deformation at elevated temperatures by reducing dislocation- and diffusion-related deformation mechanisms. On the contrary, the purpose of hot working is quite different. In this case, we want to accumulate more plastic deformation; that is, the goal is enhanced formability. Note the inherent contradiction in developing high-temperature components! Deformation-based forming or shaping of high-temperature alloys is quite challenging. In this book, we are focused on the creep-resistant material design, but the fundamentals carry over to high-temperature formability.

In engineering design, most often, designers are concerned with creep and other associated effects. The basic definition of creep is *time-dependent plastic deformation of materials*. Generally, metallic materials start showing the tendency to creep significantly at temperatures of 0.3 to $0.4T_m$ or above (T_m—melting temperature in K), whereas ceramics would need about 0.4 to $0.5T_m$ or above. Note that all the scientific discussion is done on a normalized basis with T_m and unit of K is used, whereas a practicing engineer may use °C or °F depending on the country. Students are encouraged to use the SI unit of K to avoid confusion, even though some examples in this chapter include other units because of historical context. In polymeric materials, the term used to signify a creep-like phenomenon is viscoplasticity.

11.1 Classification of creep

Depending on the range of homologous temperature used, creep can be classified into three broad categories. Before we delve deeply into discussing the individual creep categories, we need to define the concept of homologous temperature (T_H). T_H is the ratio of the actual temperature of interest (T) and the material's melting point (with both temperatures being in K), that is, $T_H = T/T_m$. Note that T_H does not have any unit; that is, it is dimensionless. So, all materials can be discussed in the range of 0 to 1, 0 being 0 K, where all atomic movement ceases to exist, and 1 being the melting temperature for metals and ceramics where the crystal structure is lost.

11.1.1 Logarithmic creep

Logarithmic creep occurs at lower homologous temperatures, typically $0.05 < T_H < 0.3$. The material subjected to this kind of creep starts to strain harden and continues the strain hardening process until the creep rate becomes essentially zero or approaches zero. This can be viewed in two different ways. One mechanistic interpretation is that the starting material has significant dislocation density. The applied stress is not high enough to activate dislocation sources. So, if the recovery process (i.e., softening) is absent because the diffusion rate is insignificant, the existing dislocations will move finite distance and then stop. This kind of creep is sometimes termed a type of exhaustion creep behavior (see creep curve schematic in Fig. 11.2 [a]). Note that generally instantaneous deformation (ε_o) in the

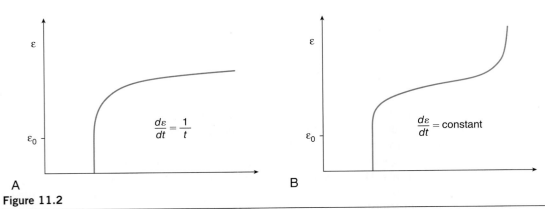

Figure 11.2
Two types of creep curves: (a) a curve that has continuous decrease of creep rates, applicable for logarithmic creep and for recovery creep with no steady state, and (b) a curve that has three stages, applicable for dislocation-based recovery creep with steady state and for diffusional creep.

creep specimen (under tensile loading) consists mainly of elastic deformation. The second mechanistic interpretation is that the applied stress causes some dislocation sources to get activated. These sources then create a pile of dislocations. The back stresses generated by the pileup stop further generation of dislocations, and in the absence of recovery processes, the creep strain saturates. Logarithmic creep behavior is given by the following equation:

$$\varepsilon = \varepsilon_0 + \alpha \ln(1 + \beta t) \qquad (11.1)$$

where α and β are constants, and t is the time.

11.1.2 Recovery-based creep

At higher homologous temperatures, most creep curves are based on recovery-based creep. Between the homologous temperatures of 0.2 to 0.5, the creep curve looks much like Fig. 11.2 [a] without any steady state. The appearance of the creep curves for limited recovery rate is similar to the logarithmic creep, because the strain rate continues to reduce with time. However, the distinction will be in the magnitude of the creep strain, which will increase with increasing temperature. For homologous temperatures higher than that (i.e., $0.5 < T_H < 1.0$), the creep curves take up a typical shape with three distinct stages (primary, secondary or steady state, and tertiary) of creep (Fig. 11.2 [b]). In very few cases, an inverse primary stage can be observed, but to keep the discussion generic, we will ignore that in this chapter.

Most creep curves represent the competition between recovery and strain hardening effects. In the primary creep stage (Stage I), strain hardening dominates over recovery; that is, the rate of strain hardening is greater than the rate of recovery. In this stage of creep, the creep rate continues to decrease. Following the primary stage, creep deformation enters the steady state creep regime (Stage II). In this stage, the microstructure enters a dynamic state where the rate of recovery equals the rate of strain hardening, and thus the creep rate remains constant. However, as time progresses, the specimen enters an accelerated creep stage where damage mechanisms (voids, microcracking/cracking, necking) and other microstructural degradation processes begin to dominate. This last stage of creep prior to fracture is known as the tertiary creep stage.

We can use the Orowan-Bailey equation to describe this process mathematically. This equation can explain the competition between the strain hardening and recovery that occurs during the primary creep stage before reaching the steady state creep stage. If we express true stress during high-temperature deformation as a function of true strain and time, we can write

$$\sigma = \sigma(\varepsilon, t)$$

$$\text{or, } d\sigma = \left(\frac{d\sigma}{dt}\right)_\varepsilon dt + \left(\frac{d\sigma}{d\varepsilon}\right)_t d\varepsilon$$

As creep tests are performed at constant stress or at least it is constant in the low strain regime of the constant load tests, for constant true stress, $d\sigma = 0$. Therefore, the expression becomes

$$\left(\frac{d\sigma}{dt}\right)_\varepsilon dt + \left(\frac{d\sigma}{d\varepsilon}\right)_t d\varepsilon = 0$$

After arranging the above equation, we get

$$\frac{d\varepsilon}{dt} = \dot{\varepsilon} = -\frac{\left(\frac{d\sigma}{dt}\right)_\varepsilon}{\left(\frac{d\sigma}{d\varepsilon}\right)_t} = \frac{r}{h} \quad (11.2)$$

where r represents the recovery rate and h represents the strain hardening rate. When r and h are in balance, we get steady-state creep stage. The extent of this stage is material and microstructure dependent (in addition to stress and temperature). Early onset of fracture mechanisms can limit this stage, and in that case the material can directly transition from primary stage to tertiary stage.

11.1.3 Diffusional creep

Generally, diffusion creep occurs at higher homologous temperatures (>0.6), which will be discussed further in later sections. The creep curve is much like the one shown in Fig. 11.2 [b]. The classic diffusion theories do not predict the primary stage, and this is touched upon in the later section.

11.2 Role of diffusion during creep deformation

Dorn and coworkers[2] collected creep and diffusion data of several pure metals and demonstrated that the temperature dependence (termed as activation energy) of steady-state creep rate and diffusion are essentially the same. Additional data confirmed that the temperature dependence of creep deformation is similar to the lattice self-diffusion in pure metals.[2] Later, this phenomenon was shown to be applicable also for solid solution alloys. Just as temperature dependence of diffusion is described in terms of activation energy

---

[2]Sherby OD, Orr RL, Dorn JE. Creep correlations of metals at elevated temperatures. *Trans Am Inst Min Metall Eng.* 1954;200:71-80.

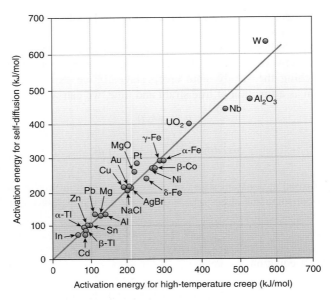

Figure 11.3

Correlation between lattice self-diffusion activation energy and creep activation energy.[3]

for diffusion, the rate of high-temperature creep, being a thermally activated process, is also described by creep activation energy. Such a correlation, as shown in Fig. 11.3, confirms the close correlation between the activation energies of high-temperature creep and those of lattice self-diffusion. However, later we will see that the activation energy of creep may not always be that of lattice self-diffusion but may depend on the dominant diffusion paths (such as grain boundaries and dislocation core).

11.3 High-temperature deformation constitutive relationships

In creep deformation, an equation to describe the steady-state creep rate as a function of stress and other parameters is helpful. Note that all the mechanistic analysis of creep deformation mechanisms is based on the steady-state creep rate or the minimum creep rate for a given combination of stress, temperature, and microstructure. For most of the analysis, it is assumed that the microstructure is constant, and this may not be correct for many alloys because of the microstructural evolution due to long-term temperature exposure and deformation strain. So, as you go through this chapter, keep the discussion of microstructures in engineering materials (**Chapter 4**) in mind and think of their thermal stability. The simplest of these equations is known as Norton's law:

$$\dot{\varepsilon}_{ss} = C\sigma^n \tag{11.3}$$

where $\dot{\varepsilon}_{ss}$ is the steady state creep rate, σ is true stress, C is a temperature- and microstructure-dependent constant, and n is the stress exponent. Eq. (11.3) clearly shows the dependence of steady-state

---

[3]Sherby OD, Burke PM. Mechanical behavior of crystalline solids at elevated temperature. *Prog Mater Sci.* 1967;13:323-390.

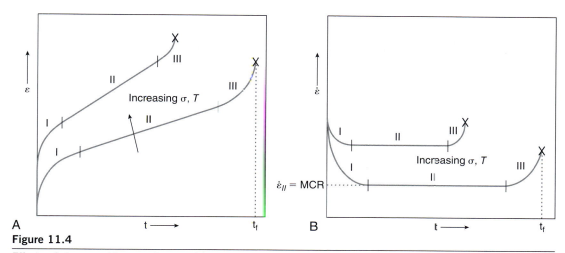

Figure 11.4

Effects of stress and temperature on (a) creep curves and (b) curves of strain rate versus time.

creep rate on stress (Fig. 11.4). One drawback of the Norton equation is that C as a function of temperature and microstructural parameters is not readily apparent. In this regard, the Bird-Mukherjee-Dorn (BMD) equation, which can be described as an expanded form of Norton's equation, is useful. The BMD equation can describe the general form of constitutive equation governing diffusion-assisted high-temperature deformation. That means creep and superplasticity can be described by this equation. The general form of the BMD equation is

$$\dot{\varepsilon}_{ss} = A \frac{DEb}{kT} \left(\frac{\sigma}{E}\right)^n \left(\frac{b}{d}\right)^p \tag{11.4}$$

where E is the elastic modulus of the material, b is the Burgers vector, k is the Boltzmann constant, T is temperature, d is grain diameter, and p is the inverse grain size exponent. The terms $\dot{\varepsilon}_{ss}$, σ, and n have the same meaning as described in relation to Norton's law. D (diffusivity) is the most important temperature-dependent term in the BMD equation and is described by the following equation:

$$D = D_o \exp\left(-\frac{Q}{RT}\right) \tag{11.5}$$

where D_o is the pre-exponential factor, R is the universal gas constant (8.314 J/mol K), and Q is the appropriate activation energy. Actually, the reason for increase in creep rate with increasing temperature (Fig. 11.4) can be explained from Eqs. (11.4) and (11.5).

Each creep deformation micromechanism comes with unique combinations of n, p, Q, and A values. A standard method of identifying creep mechanisms has been to determine n, p, Q, and A to establish the rate-controlling micromechanisms during creep by comparing these parametric dependencies with the predicted values from the standard creep models (to be discussed in the following sections). Sometimes, comparing the experimentally determined A values with the A values in the standard models becomes difficult because of the lack of understanding of how A is affected by various microstructural and material composition factors.

11.3.1 Dislocation glide

Generally, at homologous temperatures less than 0.3 and/or high stresses, creep in metallic materials occurs through time-dependent dislocation movement, which causes the accumulation of strain with little or no help from diffusion. The failure occurs eventually through ductile rupture as a result of geometric instability such as necking. This kind of dislocation glide-related creep is the analogous version of low-temperature dislocation glide; it is just that because of elevated temperatures, thermally activated dislocation glide is at play. Under these conditions, grain boundaries and other interfaces still act as sources of strengthening. The activation barrier under such conditions would be lowered in the direction of the dislocation movement (Fig. 11.5). For thermally activated jump of dislocations, the following constitutive equation is appropriate:

$$\dot{\varepsilon} = C' \sinh\left(\frac{a\tau}{RT}\right)\exp\left(-\frac{Q}{RT}\right) \quad (11.6)$$

where τ is shear stress applied to the dislocation, and C' and a are constants. At low-stress regime, the sinh function essentially becomes a linear function, and at higher stress it approximates to an exponential behavior. This kind of equation can be used to describe the creep behavior of certain polymers where unfolding of links within the molecular chains can be the rate-controlling micromechanism for time-dependent permanent deformation.

Creep that occurs in the upper range of homologous temperature can be broadly divided into two types of processes: (a) dislocation-based and (b) diffusion-based flow processes. The creep micromechanisms change as functions of temperature and stress.

Concept revision: *What happens to dislocation-based strengthening mechanisms at high temperatures?* In **Chapter 8**, you learned about strengthening mechanisms. The fundamental physics was that dislocations are repelled by interfaces and elastically hard particles. Let us take these two obstacles separately, that is, dislocation/grain boundary interaction and dislocation/particle interaction.

Effect of temperature on grain boundary strengthening: The Hall-Petch strengthening in **Chapter 8** was based on grain boundaries being obstacle to the dislocation motion. This concept was accepted

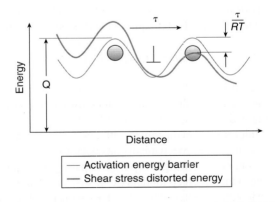

Figure 11.5

Schematic variation of activation energy as a function of dislocation position: under no shear stress applied to the dislocation, and under the influence of shear stress.

without modification till the end of the 1970s. In early 1980s, Srolovitz et al.[4,5] introduced the concept of diffusional relaxation of interaction forces at the interfaces. In parallel, in situ transmission electron microscopy evidence showed that the dislocation enters the high angle grain boundaries and gets absorbed. Although a detailed discussion is not possible here, for curious students, a general direction of discussion is mentioned. The 1980s was also the time when researchers started investigating the grain boundary structure in detail. An understanding developed that the grain boundary structure consists of "intrinsic" and "extrinsic" grain boundary dislocations (GBDs). A simple way to visualize the difference between these two types of GBDs is to think of intrinsic GBDs as the dislocations necessary to define the misorientation between the adjoining grains/crystals. The tilt and twist angles that define the misorientation can be described as a set of intrinsic GBDs. On the other hand, extrinsic GBDs are formed by absorption of lattice dislocations at the grain boundaries. Consider an edge dislocation (an extra half plane with Burgers vector b) moving toward the grain boundary. As this extra half-plane merges in the grain boundary, the lattice spacing on two sides of the boundary is disturbed. There is an additional strain component on one side of the boundary. In transmission electron micrographs, the contrast for extrinsic GBDs is quite prominent and it is spaced non-periodically in the grain boundaries. The intrinsic GBDs give patterned contrast. So, the steps involved in the absorption of lattice dislocations in the grain boundaries include, (1) approach of dislocation at the interface, and (2) dislocation core spreading at the interface. It is now easy to conceptualize that such a process will have strain rate–dependent kinetics. The rate-controlling step in this process will be rate at which the core can spread by a diffusion-assisted process. Based on the microstep assumptions in this process, one can develop a series of theoretical models! Regardless of the details, one thing is obvious: *the grain boundaries no longer lead to increase of strength at high temperatures.*

Effect of temperature on particle strengthening: The particle strengthening from non-shearable hard particles in **Chapter 8** was based on "repulsive" interaction of dislocation and particle. Similar to the discussion above, as the dislocation approaches an incoherent particle, it gets attracted to the interface because of reduction in the core energy (see Mishra et al.[6] [1993] for the concept of lattice dislocation dissociation into interfacial dislocations at the matrix/particle interface). This concept is further discussed in **Section 11.3.7**. It is important to keep in mind that this also leads to decrease in the magnitude of strengthening calculated from the Orowan mechanism. The overall drop in the level of strengthening will again depend on the assumptions in the micromechanism of the models.

11.3.2 Power law creep

At higher stresses and homologous temperatures >0.4, dislocation creep follows the power law relationship, with the steady state creep rate being directly proportional to σ^n. The n value generally varies from 3 to 7 for dislocation-based creep mechanisms. Sometimes, the observed stress exponent can go even higher for engineering alloys, which may have complicated compositions and microstructures.

---

[4]Srolovitz DJ, Petkovic-Luton RA, Luton MJ. Edge dislocation-circular inclusion interactions at elevated temperatures. *Acta Metall.* 1983;31(12):2151-2159.
[5]Srolovitz DJ, Luton MJ, Petkovic-Luton R, Barnett DM, Nix WD. Diffusionally modified dislocation-particle elastic interactions. *Acta Metall.* 1984;32(7):1079-1088.
[6]Mishra RS, Nandy TK, Greenwood GW. The threshold stress for creep controlled by dislocation-particle interaction. *Phil Mag A.* 1994;69(6):1097-1109.

All these creep mechanisms are referred to as **power law creep**. The micromechanisms involved during creep are essentially a combination of dislocation glide and dislocation climb. Depending on which process is rate controlling, the power-law creep mechanism becomes viscous glide or climb controlled.

11.3.2.1 Weertman model for dislocation creep

Different versions of dislocation creep models have been put forward. Dislocation creep can be thought of as involving three basic steps: (1) creation of dislocations, (2) movement of dislocations, and (3) subsequent annihilation. Variation in models depends on which of these steps become rate controlling. During transient creep, the dislocation sources produce and send out loops (Fig. 11.6 [a]). Soon, material work hardens and sources are subjected to back stress, both of which hinder further production of loops. Dislocation climb occurs at the head of the pileup, leading to annihilation and thus producing recovery. When two dislocations annihilate each other, another glide dislocation is generated from each source, as illustrated in Fig. 11.6 [b]. In this process, the majority of creep strain is still created by the glide dislocation, but the overall creep rate is climb controlled. Here "L" is the distance between the dislocation source and h is the height between the climb planes that the dislocation must climb for annihilation. For climb control condition, the glide velocity (v_{glide}) should be larger than the climb velocity (v_{climb}) as given by

$$v_{glide} \geq (L/h)\, v_{climb} \tag{11.7}$$

The time to climb the distance h is $\Delta t = h/v_{climb}$. Let us consider N as the number of dislocation sources per unit volume and L as the glide distance or average loop radius. We can write the number of dislocations per source as constant multiplied by L/h. So, the mobile dislocation density can be written as

$$\rho_m = NL(L/h) = NL^2/h \tag{11.8}$$

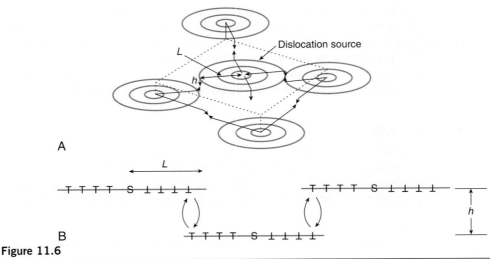

Figure 11.6

Weertman's pillbox model for creep deformation controlled by dislocation annihilation by climb. Note the expansion of loops shown in 'A' and the climb annihilation in 'B'[7]

---

[7]Weertman J. Theory of steady-state creep based on dislocation climb. *J Appl Phys*. 1955;26:1213.

From Orowan equation, we can write

$$\dot{\gamma} = \rho_m b v_{glide} \qquad (11.9)$$

By substituting ρ_m and v_{glide} from Eqs. (11.8) and (11.7) into Eq. (11.9), we can write

$$\dot{\gamma} = \frac{NL^2}{h} b \, A' \frac{\tau}{\tau} v_{climb} = A' \frac{NL^3 b}{h^2} v_{climb} \qquad (11.10)$$

where A' is a constant. Also, we can write "L" in terms of N and h derived from the source of each volume ($\pi L^2 h N =$ constant):

$$L = (\pi N h)^{-1/2} \qquad (11.11)$$

By substituting Eq. (11.11) for L into Eq. (11.10), we get

$$\dot{\varepsilon} = \frac{A'' \times v_{climb}}{h^{3.5} N^{1/2}} \qquad (11.12)$$

where A'' is a constant. The climb velocity is given by the following relationship:

$$v_{climb} = \frac{2 D_L \sigma \Omega}{kT}$$

Note that depending on the stress dependence of h and N, one can get different model for creep. Assuming, $h \propto \frac{1}{\sigma}$ and N being independent of stress, we can write:

$$\dot{\varepsilon} = A_C \, \sigma^{3.5} \frac{\Omega}{kT} D_L = A_C \sigma^{4.5} \frac{\Omega}{kT} D_L \qquad (11.13)$$

So, Weertman's analysis shows that edge dislocation climb-controlled creep rate would have a stress exponent of 4.5.

Now, note that in the preceding derivation, N was considered to be independent of stress. If we consider $N \propto \sigma^2$, based on $\rho \propto \sigma^2$ for Frank-Read source, then the stress exponent in Eq. (11.13) will reduce to 3.5. In both cases the dislocation climbs by lattice diffusion, so the activation energy associated with the climb-controlled creep is the same as the activation energy for lattice diffusion. Additionally, since the dislocation source density is assumed to be independent of grain size, there is no microstructural term in Eq. (11.13).

Later, Weertman assumed that Frank-Read sources create dislocation loops on parallel slip planes wherein the leading edge components of the dislocation loops climb and are annihilated. This is commonly referred to as Weertman pillbox model and does not require physical obstacles, such as Lomer-Cottrell barriers, as are encountered in FCC metals. Based on a large body of experimental observations, it is common to assign $n = 5$ for climb-controlled dislocation creep.

When dislocation climb creep occurs at lower homologous temperatures, the measured activation energy may become less than that of lattice diffusion, and the stress exponent rises (from $n = 5$ to 7). In that case, dislocation core diffusion becomes predominant. So, creep activation energy becomes the activation energy of core diffusion (or pipe diffusion) for lattice atoms. The exact determination of activation energy for dislocation core diffusion is not possible; however, it is considered to be close to the activation energy for grain boundary diffusion. Given that in this case the dislocation cores provide the primary diffusion path for atoms, dislocation density is an important factor. Since dislocation density is directly proportional to the square of stress ($\rho \propto \sigma^2$), we can write the power law creep equation for low-temperature dislocation climb in the following form:

$$\dot{\varepsilon} = \frac{A_{dcl} D_c E b}{kT}\left(\frac{\sigma}{E}\right)^{5+2} \tag{11.14}$$

where D_c is the activation energy for dislocation core diffusion and A_{dcl} is the corresponding dimensionless constant. Here the stress exponent observed in high-temperature climb is 5, so adding 2 to 5 (i.e., 7) would give the stress exponent for low-temperature climb. As the low-temperature climb is essentially a type of dislocation creep, it also does not have any grain size dependence ($p = 0$). However, we need to be cautious on that, in that if analysis of creep data of a material gives a stress exponent of 7, we should not automatically assume that the associated creep mechanism must then be the low-temperature climb, because the stress exponents of high-temperature climb can vary widely, from 4.5 to 7 (sometimes 8 or higher). Therefore, to identify low-temperature dislocation climb, a key determinant is the creep activation energy, which should be distinctly lower compared to the activation energy of lattice diffusion and should be consistent with that of dislocation core diffusion. A similar postulation is that the viscous glide process occurring at low homologous temperature would be controlled by solute diffusion through dislocation cores, and the stress exponent (n) would be $3 + 2 = 5$.

One way of examining creep mechanisms in solid solutions is to classify creep behaviors into two types: Class-M (metal type) or Class-2 and Class-A (alloy type) or Class-1. The creep behavior of some pure metals and some alloys can be considered Class-M type, exhibiting a stress exponent of 5 (generally remains within 4–7). On the other hand, the creep behavior of many solid solution alloys termed Class-A shows stress exponents close to 3. In such cases, solute atoms are dragged by mobile dislocations, thereby making the glide progress sufficiently slower than the dislocation climb under a range of creep conditions. The activation energy for creep essentially equals the activation energy for solute diffusion or Darken diffusivity. The climb-controlled regime may reappear as the stress decreases. However, at a higher stress, dislocations locked by the solutes break away. The breakaway stress can be calculated from the following equation[8]:

$$\sigma_b = \frac{W_m^2 c_o}{2^\beta k T b^3} \tag{11.15}$$

where c_o is the solute concentration; W_m is the binding energy for solute atom/dislocation interaction; β is typically 2 to 4 depending on the shape of the solute atmosphere; and k, T, and b represent the usual meanings as defined previously.

11.3.3 Diffusional creep

At lower stress levels, dislocation creation becomes difficult. The applied stresses are not large enough to operate the conventional dislocation sources. In such cases, at high enough homologous temperatures, vacancy flow through the bulk/lattice diffusion process or along grain boundaries becomes an important creep mechanism known as diffusional creep. Diffusional creep is a type of Newtonian viscous process with the stress exponent value of 1 ($n = 1$). While one form of dislocation creep

---

[8]Murty KL. Viscous creep in Pb-9Sn alloy. *Mater Sci Eng*. 1974;14:169.

(known as Harper-Dorn creep[9]) also exhibits linear stress dependency of strain rate, it does not have a grain size dependency of creep rate (i.e., $p = 0$). On the other hand, diffusional creep is strongly grain size dependent ($p = 2$ or 3). We will briefly describe the characteristics of Harper-Dorn creep in a later section. Of the two kinds of diffusional creep, one is Nabarro-Herring (N-H) creep[10,11] and the other is Coble creep,[12] depending on the predominant path of vacancy flow. Fig. 11.7 shows the N-H creep and Coble creep mechanisms schematically in a square grain. Of course, even though only one grain is shown here, in reality the process involves multiple grains in polycrystalline materials. If the vacancy flow takes place predominantly through the lattice, it is called N-H creep, whereas Coble creep occurs because of the predominance of vacancy diffusion along the grain boundaries.

Note that under applied stress, the grain boundaries oriented perpendicular to the stress experience tensile stress and create a higher concentration of vacancies (Fig. 11.7). Conversely, grain boundaries parallel to the applied stress (under compression) experience a reduced vacancy concentration. Thus, a vacancy concentration gradient is set up with vacancy flow occurring from the high vacancy concentration region (i.e., from the perpendicular grain boundaries to the lateral grain boundaries). The arrows on the solid lines in Fig. 11.7 represent the atom or mass transport (opposite to the vacancy flow), whereas the dotted lines show the vacancy flow (opposite to the atom movement). Here diffusion of vacancies creates strain and actually leads to an elongated shape of grains.

The following mathematical treatment of diffusional creep can allow calculation of creep rate from the steady-state vacancy concentration gradient (basically the driving force). Here consider Q_f as the

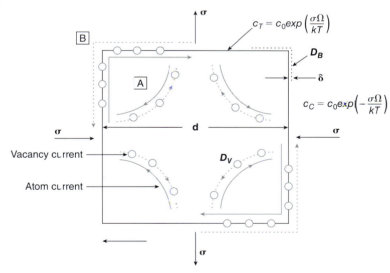

Figure 11.7

Schematic of (a) Nabarro-Herring creep and (b) Coble creep mechanisms.

---

[9]Harper J, Dorn JE. Viscous creep in aluminum near its melting temperatures. *Acta Metall.* 1957;5:654-665.
[10]Nabarro FRN. Deformation of crystals by the motion of single ions. *Report Conf. on Strength of Solids*. London: The Physical Society; 1948:75.
[11]Herring C. Diffusional Viscosity of a Polycrystalline Solid. *J Appl Phys.* 1950;21(5):437-445.
[12]Coble RL. A model for boundary diffusion controlled creep in polycrystalline materials. *J Appl Phys.* 1963;34(6):1679-1682.

vacancy formation energy and $\sigma\Omega$ as the extra work required to form the vacancy where Ω is the atomic volume and σ is the applied stress. Then we can write the following for the high-concentration vacancy region (tensile region) and the low-concentration vacancy region (compressive region):

$$N_v^{(T)} = A \exp\left(-\frac{Q_f}{kT}\right) \exp\left(\frac{\sigma\Omega}{kT}\right) \tag{11.16}$$

$$N_v^{(C)} = A \exp\left(-\frac{Q_f}{kT}\right) \exp\left(-\frac{\sigma\Omega}{kT}\right) \tag{11.17}$$

where k is the Boltzmann constant, and T is the temperature in K.

Considering steady state of vacancy flow, according to Fick's first law we can write the vacancy flux (J_v) as follows:

$$J_v = -D_v \frac{dN_v}{dx} \tag{11.18}$$

where D_v is the vacancy diffusivity, N_v is the vacancy concentration, and x diffusion distance. Thus and $\frac{dN_v}{dx}$ can be written as

$$\frac{\Delta N_v}{\Delta x} = \frac{N_v^C - N_v^T}{d_G} \tag{11.19}$$

In the above equation, diffusion distance (Δx) is replaced by grain size (d_G; note that some times we have used symbol d for grain size). Thus Eq. (11.19) can be rewritten with the help of Eqs. (11.17) and (11.18) as

$$J_v = -D_v \frac{N_v^C - N_v^T}{d_G} = -D_v \frac{N_v^C - N_v^T}{d_G} = -D_{0v} \exp\left(-\frac{Q_m}{kT}\right) \frac{\exp\left(-\frac{Q_f}{kT}\right)\left\{\exp\left(-\frac{\sigma\Omega}{kT}\right) - \exp\left(\frac{\sigma\Omega}{kT}\right)\right\}}{d_G} \tag{11.20}$$

Given that, $D_{0v} \exp\left(-\frac{Q_m}{kT}\right) \exp\left(-\frac{Q_f}{kT}\right)$ is actually considered the lattice diffusivity (D_L). To further simplify the Eq. (11.20), we will use Taylor's exponential series expansion $e^y = 1 + y + \frac{y^2}{2!} + \frac{y^3}{3!} + \cdots \frac{y^n}{n!}$.
If "y" is a small number, e^y can be approximated as $(1 + y)$, as the higher-order terms become very small and hence negligible. Following this rationale, we can write

$$\exp\left(-\frac{\sigma\Omega}{kT}\right) - \exp\left(\frac{\sigma\Omega}{kT}\right) = \left(1 - \frac{\sigma\Omega}{kT}\right) - \left(1 + \frac{\sigma\Omega}{kT}\right) = -\frac{2\sigma\Omega}{kT}$$

Thus, the simple form of Eq. (11.20) is given by

$$J_v = \frac{D_L}{d_G} \frac{2\sigma\Omega}{kT} \tag{11.21}$$

Let us consider volume (V) of the grain in three-dimension as a cube (Fig. 11.8), with dl being an infinitesimal change in length of the cube due to diffusional creep.

Due to diffusional creep, the rate of volume change can be given by

$$\frac{dV}{dt} = J_v A = J_v d_G^2 = \frac{D_L}{d_G} \frac{2\sigma\Omega}{kT} d_G^2 = D_L d_G \frac{2\sigma\Omega}{kT}$$

$$dV = d_G^2 dl, \text{ or,}$$

11.3 High-temperature deformation constitutive relationships

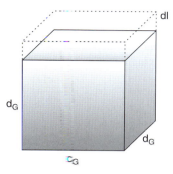

Figure 11.8
A simple geometry of diffusional creep model.

$$\frac{dV}{dt} = d_G^2 \frac{dl}{dt} \tag{11.22}$$

The creep rate ($\dot{\varepsilon}$) can be written as

$$\dot{\varepsilon} = \frac{1}{d_G}\frac{dl}{dt}$$

But from Eq. (11.22),

$$\frac{dl}{dt} = \frac{1}{d_G^2}\frac{dV}{dt}$$

Therefore,

$$\dot{\varepsilon} = \frac{1}{d_G}\frac{1}{d_G^2}\frac{dV}{dt} = \frac{1}{d_G^3}D_L\frac{2\sigma\Omega}{kT} = \frac{2D_L d_G \sigma \Omega}{d_G^2 kT}$$

Thus we can write the constitutive equation for the N-H creep as

$$\dot{\varepsilon}_{N-H} = \frac{2D_L}{d_G^2}\left(\frac{\sigma\Omega}{kT}\right) = A_{N-H}\frac{D_L}{d_G^2}\left(\frac{\sigma\Omega}{kT}\right) \tag{11.23}$$

In Eq. (11.23) for the N-H creep, the steady-state creep rate depends on stress linearly (i.e., $n = 1$) and on grain size through an inverse square relationship (i.e., $p = 2$). While N-H creep is observed in many types of materials, this mechanism is particularly important in ceramics under low stress and elevated temperatures.

On the other hand, in the flux equation, if A is replaced by $d_G \times \delta$, where δ is grain boundary width, we get the equation for Coble creep, which is shown below:

$$\dot{\varepsilon}_{Coble} = A_{Coble}\left(\frac{\delta D_{GB}}{d_G^3}\right)\left(\frac{\sigma\Omega}{kT}\right) \tag{11.24}$$

where D_{GB} is grain boundary diffusivity and A_{Coble} is a constant. An important note is that Coble creep has a stronger grain size dependence of creep rate ($p = 3$) compared to N-H creep ($p = 2$), even though the stress exponents for both N-H creep and Coble creep are the same. Note that the activation energy of diffusion for this process is associated with grain boundaries, which is roughly 1/2 to 2/3 of lattice diffusion. Thus, Coble creep predominates at temperatures lower than N-H creep and for finer grain sizes, whereas N-H creep predominates at higher temperatures and larger grain sizes. N-H creep and Coble creep can operate in parallel, that is, independent of each other. Thus, the overall creep rate by diffusional creep is given by the following equation:

$$\dot{\varepsilon}_{eff} = A_{eff} \frac{D_{eff}}{d_G^2 kT}\left(\frac{\sigma \Omega}{kT}\right) \tag{11.25}$$

where D_{eff} is the effective diffusivity is expressed by

$$D_{eff} = D_L\left(1 + \frac{\pi \delta D_{GB}}{d_G D_L}\right) \tag{11.26}$$

For Eqs. (11.25) and (11.26) to be valid, the basic underlying assumptions are that grain boundaries are considered perfect sources and sinks of vacancies, and that the initial dislocation density in the material is low. One interesting observation of the effect of diffusional creep on microstructure has been reported to be the creation of "denuded zones" in particle-containing alloys on the grain boundaries normal to the tensile stress. We can refer to Fig. 11.9 [a] for a simple schematic of denuded zone formation as described by Burton and Greenwood.[13] Note the particle collection at the parallel grain boundaries to the tensile stress direction, while areas devoid of particles in the surrounding regions of the normal grain boundaries are created as denuded zones. We can see in Fig. 11.9 (b) an actual example of denuded zones formed following diffusion creep in a Mg-0.5Zr alloy.[14]

> *Food for thought*: Why does it take a significant time to ask conceptual questions? Students are encouraged to think about the timeline of evolution of creep theories. Let us look at diffusional creep. Nabarro-Herring diffusional creep mechanism was proposed in 1950.[15] Coble (1963) modified the diffusional creep theory to include the grain boundary as potential diffusion path. Why did it take more than a decade to think about this possibility? Those of you who are more curious about this topic, try to link the assumptions with progression of our understanding. In this book, we have started with the context of why mechanical behavior is important and we reviewed the microstructure of engineering materials before jumping into the theories. Can such an approach prepare you better for asking pertinent questions about the gap between experimental observations and theoretical models?

- In 1950, there was no understanding of grain boundary structure. So, the assumption that grain boundaries are perfect sink and source of vacancies was made. As the understanding of grain boundary structure evolved, interface-controlled diffusional creep theory was developed in 1980s.
- The initial diffusional creep theories do not predict primary creep or threshold stress. As researchers gathered experimental creep curves showing primary creep, it took decades to address this fundamentally.
- Similarly, analysis of creep rate-stress data showed increase in stress exponent in the regime below $n = 1$. This was interpreted as existence of threshold stress. A conceptual model for this was proposed by Mishra et al.[16] in 1989.

---

[13] Burton B, Greenwood GW. The limits of the linear relation between stress and strain rate in the creep of copper and copper-zinc alloys. *Acta Metall.* 1970;18(12):1237-1242.
[14] Vickers W, Greenfield P. Diffusion creep in magnesium alloys. *J Nucl Mater.* 1967;24(3).
[15] Nabarro FRN. Deformation of crystals by the motion of single ions. *Report Conf. on Strength of Solids*. London: The Physical Society; 1948:75.

11.3 High-temperature deformation constitutive relationships

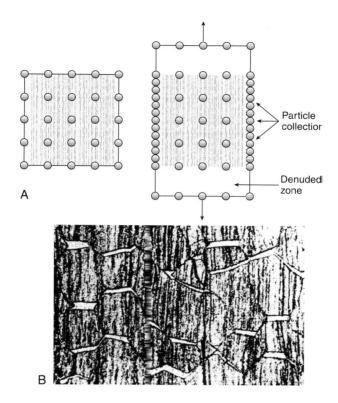

Figure 11.9

(a) A schematic representation of formation of denuded zone; *(left)* a square grain containing particles before creep deformation; *(right)* the same grain after creep deformation. (b) An optical micrograph of a crept Mg-0.5Zr alloy showing "denuded zones."

- Note that during diffusional creep, as grains become elongated or change shape, the aggregate of grains still need to be together, and the accommodation mechanism is grain boundary sliding. Both Nabarro-Herring and Coble creep[17] models assume that the accommodation is going to happen without any problem. We will highlight implication of this aspect further in a later conceptual box.

Concept enhancer: What if the basic assumptions do not hold true? As emphasized in the previous concept box, there are key assumptions in diffusional creep theories. Note the position of the grain boundary in Fig. 11.9 (a) schematic and think of this in general terms. First, as the grain boundaries act as source and sink of vacancies, they "*migrate*." In fact, if you consider the current understanding of grain boundary structure, all the grain boundary–related processes such as grain boundary

---

[16]Mishra RS, Jones H, Greenwood GW. On the threshold stress for diffusional creep in pure metals. *Phil Mag A*. 1989;60:581.

[17]Coble RL. A model for boundary diffusion controlled creep in polycrystalline materials. *J Appl Phys*. 1963;34(6):1679-1682.

sliding and grain boundary migration are linked with motion of GBDs. Starting from Nabarro-Herring, none of the researchers working on diffusional creep raised the question, "what if the grain boundaries are pinned by particles and cannot move or migrate?" What happens to Eq. (11.22)? Because the creep rate is defined as rate of plating of atoms on the tensile boundaries, if the boundaries are locked, should not the entire process cease? In such cases, will diffusional creep be completely suppressed?

We highlight two *Nature* papers published in 1998 and 2016 to illustrate examples of evolution of understanding based on key experimental observations. In the 1980s, Gleiter[18] started the field of nanocrystalline materials. There was significant excitement about the impact of nanocrystalline grain structure on diffusional creep and grain boundary sliding. A key paper by Chokshi et al.[19] highlighted the importance of diffusional creep in deformation of nanocrystalline copper at room temperature. This has become one of the most highly cited papers and has major implications on how we think about diffusional creep.

Key concept: Refinement of grains to nanocrystalline range in pure metals or solid solutions also makes the microstructure thermally unstable. So, are the observations of creep mechanisms impaired by microstructural change during creep? Many researchers have raised the issue of concurrent grain growth on diffusional creep and grain boundary sliding.

Implications: The kinetics of diffusional processes are likely to be enhanced because of grain boundary migration in nanostructured alloys. The *Nature* paper by McFadden et al.[20] (1998) indeed reported superplasticity at temperatures as low as $0.35T_m$. Note that the conventional wisdom at that time was that diffusion-assisted deformation mechanisms start operating above $0.5T_m$. So, as you work on new alloys and/or microstructure, always ask what the intrinsic assumptions are and how valid they are. The second *Nature* paper by Darling et al.[21] (2016) had just the opposite finding. In this case, the nanocrystalline Cu-Ta alloy was highly creep resistant. Note that the Cu-Ta alloy is an immiscible system. This made this nanostructure highly thermally stable. The stability of second phase depends on solubility in the matrix and its diffusivity. For an immiscible element both of these are very low. In hindsight, it is obvious that an immiscible system would be creep resistant, but why did researchers not think of this for over 40 years of active research on diffusional creep? As you work through this last chapter, keep making connections between the mechanisms and microstructure of engineering materials. Question the assumptions behind the creep models, and how it impacts the applicability of a particular model for your system of interest.

11.3.4 Harper-Dorn creep

As noted earlier, another creep mechanism has Newtonian viscous flow nature (i.e., $n = 1$) but is not a type of diffusional creep. Harper and Dorn[22] reported on the evidence for the existence of a type of dislocation-based Newtonian viscous creep, now referred to as the Harper-Dorn (H-D) creep. They performed creep experiments on high-purity (99.95%) metals at a homologous temperature of 0.99, that is, very close to the melting temperature. Even though N-H creep has identical stress exponent and

---

[18]Gleiter H. Nanocrystalline materials. *Prog Mater Sci*. 1989;33(4):223-315.
[19]Chokshi AH, Rosen A, Karch J, Gleiter H. On the validity of Hall-Petch relationship in nanocrystalline materials. *Scr Met*. 1989;23(10):1679-1683.
[20]McFadden S, Mishra R, Valiev R, et al. Low-temperature superplasticity in nanostructured nickel and metal alloys. *Nature*. 1999;398:684-686.
[21]Darling K, Rajagopalan M, Komarasamy M, et al. Extreme creep resistance in a microstructurally stable nanocrystalline alloy. *Nature*. 2016;537:378–381.
[22]Harper J, Dorn JE. Viscous creep in aluminum near its melting temperatures. *Acta Metall*. 1957;5:654-665.

creep activation energy consistent with lattice diffusion, H-D creep has a grain size exponent of zero ($p = 0$), meaning no grain size dependence, and experimental creep rates about 1400 times greater than the theoretical predictions of N-H creep. The steady-state strain rate is expressed as

$$\dot{\varepsilon}_{H-D} = A_{H-D}\left(\frac{\sigma b D_L}{kT}\right) \qquad (11.27)$$

The H-D creep is generally observed in large-grained pure metals. Even though the first study on H-D creep was done at very high homologous temperature, creep studies have been performed on other metals including alpha-zirconium, alpha-titanium, beta-cobalt, and alpha-iron in the homologous temperature range of 0.3 to 0.6 and grain size of about 500 μm.

11.3.5 Grain boundary sliding and superplasticity

Grain boundary sliding (GBS) occurs under creep conditions as strain accommodation mechanism, which is believed to promote grain boundary crack nucleation. Note that during diffusional creep deformation, grain boundary sliding is operating at a faster rate and is assumed to accommodate the geometrical shape change. However, when the grain size becomes smaller and temperatures are high ($T_h > 0.4$), grains tend to slide past each other along the grain boundary under applied (resolved) shear stress, and grain boundary sliding becomes a predominant mode of high-temperature creep mechanism instead of just being an auxiliary process of accommodation or flaw nucleation. It should be noted that grain boundary sliding cannot continue without an associated accommodation process, because stress concentrations created at the grain corners due to GBS open up flaws, as illustrated in Fig. 11.10. Two types of accommodation processes have been noted. One is called Lifshitz sliding, which occurs with the accommodation process provided by diffusional creep. In fact, the

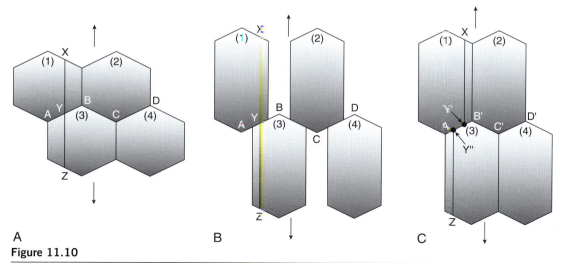

Figure 11.10

Schematic of Lifshitz sliding mechanism steps (A, B, C).[23]

[23]Evans AG, Langdon TG. Structural ceramics. *Prog Mater Sci.* 1976;21:17.

diffusion creep process requires Lifshitz sliding to take place simultaneously. As a result of Lifshitz sliding, elongation occurs in the grain structure (Fig. 11.10). The other one is called Rachinger sliding, which is accommodated by dislocation-based processes such as dislocation climb-glide process. Rachinger sliding is expected to maintain the grain shape and structure through grain rearrangement.

The hallmark of the grain boundary sliding process is that the stress exponent associated with this process is generally 2. Note that grain boundary sliding is also related to superplastic deformation. One of the earliest models was proposed by Ashby and Verrall[24]:

$$\dot{\varepsilon} = \left(\frac{100\Omega}{kTd_G^2}\right)\left(\sigma - \frac{0.72\gamma}{d_G}\right)D_L\left(1 + \frac{3.3\delta D_{GB}}{d_G D_L}\right) \quad (11.28)$$

where γ is the grain boundary surface energy, $(0.72\ \gamma/d_G)$, the minimum stress to induce grain sliding (i.e., threshold stress), D_{GB} is grain boundary diffusivity, D_L is lattice diffusivity, δ is grain boundary width, and Ω, σ, d_G, k, and T all have their usual meanings. The threshold stress involving grain boundary surface energy is very small and can usually be neglected. The strain rate is directly proportional to stress, with a stress exponent of 1.

The Ashby-Verrall model involves diffusional flow as an accommodation process. The model predicts larger elongations than do the diffusional creep models. Diffusional flow occurs through a grain-switching mechanism; the process is shown schematically in Fig. 11.11 at various stages of deformation.

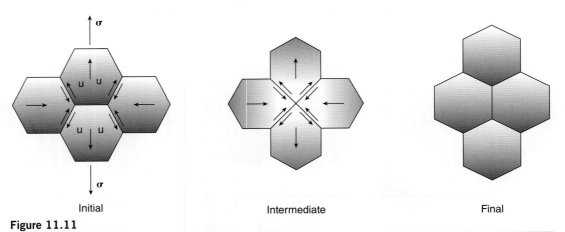

Figure 11.11

A schematic illustration of grain switching event as described by Ashby and Verrall (1973).[24]

---

[24]Ashby MF, Verrall RA. Diffusion accommodated flow and superplasticity. *Acta Metall.* 1973;21:149-163.

11.3 High-temperature deformation constitutive relationships

In superplasticity, most models of grain boundary sliding accommodated by diffusional flow or dislocation slip are indicative of n of 2 and p of 2. The Ball-Hutchison model[25] is described by

$$\dot{\varepsilon} = \frac{200 L_{GB} G b}{kT}\left(\frac{\sigma}{G}\right)^2 \left(\frac{b}{d_G}\right)^2 \tag{11.29}$$

where all terms have been defined in prior sections. As shown in Fig. 11.12, the model is based on the premise that when grains slide under a state of shear stress, stress concentration develops at the triple points. Stress concentration leads to the creation of dislocations, which then move to the opposite grain boundaries and start piling up against them. As a result, back stress is developed and hinders further dislocation generation. Back stress is relieved when the lead dislocation climbs into the grain boundary and further deformation becomes plausible. Similar models have also been developed by Langdon,[26] Mukherjee,[27] and Arieli-Mukherjee.[28] On the other hand, Gifkins[29] came up with a core-mantle model where grain interior is considered "core," and the areas adjacent to the grain boundary are considered "mantle." In Gifkins' model, the entire deformation is assumed to occur in the mantle region.

The grain boundary sliding mechanisms/models discussed above are essentially the same ones for superplastic deformation. Superplastic deformation is generally defined as the ability of a material to exhibit larger-than-usual tensile elongations prior to failure. This behavior is attributed mainly to grain boundary sliding. Among various kinds of superplasticity, fine-grain superplasticity or structural superplasticity draws the most interest. *For metal, superplasticity must involve a minimum tensile elongation to failure of 200%, whereas for ceramics it is 100%!* Let us observe the superplastically deformed tensile specimens of a friction stir-processed Al-8.9Zn-2.6Mg-0.09Sc (wt%) alloy (original

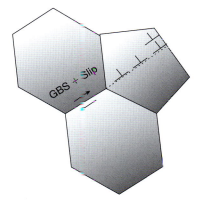

Figure 11.12

Schematic illustration of the Ball-Hutchison model.[25]

---

[25]Ball A, Hutchison MM. Superplasticity in the aluminum-zinc eutectoid. *J Met Sci.* 1969;3(1):1-7.
[26]Langdon TG. A unified approach to grain boundary sliding in creep and superplasticity. *Acta Metall Mater.* 1994:42;2437.
[27]Mukherjee AK. Role of grain boundaries in superplastic deformation. In: Walter JL, et al., eds. *Grain Boundaries in Engineering Materials.* Claitor Publishing; 1975:93-105.
[28]Arieli A, Mukherjee AK. A model for the rate-controlling mechanism in superplasticity. *Mater Sci Eng* A. 1980;45:61.
[29]Gifkins RC. Grain boundary sliding and its accommodation during creep and superplasticity. *Metall Trans A.* 1976;7:1225.

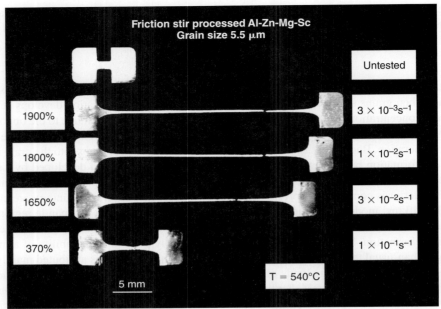

Figure 11.13

Examples of enhanced superplasticity in a friction stir-processed Al-8.9Zn-2.6Mg-0.09Sc alloy. Friction stir processing is a unique thermomechanical processing technique based on the principles of friction stir welding (Charit and Mishra, unpublished work).

average grain size of 5.5 μm) (Fig. 11.13). The gage section appears to be uniform, with no sign of typical localized necking. So, it is essentially extreme diffuse necking, which means superplastic deformation is highly resistant to necking, and that is why strain rate sensitivity ($m = 1/n$, i.e., reciprocal of stress exponent) needs to be high for superplasticity (at least 0.3, but typically 0.5 or more). In such cases, ultimate failure occurs by internal cavitation. The higher the "m" value, the more enhanced the superplasticity. Note the huge elongation to fracture (up to 1900%) obtained in one of the specimens tested at an initial strain rate of 3×10^{-3} s^{-1} and temperature of 540°C. The m values were close to 0.7 for the specimens exhibiting enhanced superplasticity. Even the elongations of the specimens tested at higher strain rates (10^{-2} s^{-1} and 3×10^{-2} s^{-1}) are truly remarkable. However, instances of extreme elongations (way out of the mainstream superplasticity) were reported in some cases, for example, as a record 5500% obtained in aluminum bronze. A fairly large number of studies have been reported for a variety of materials showing large superplasticity, even in some ceramics (Nieh et al.[30])! Traditionally, superplasticity occurs at homologous temperatures of 0.4 or beyond and at lower strain rates (10^{-5} to 10^{-3} s^{-1}).

The contribution of grain boundary sliding to superplasticity and creep has been examined experimentally by the fiducial scratch marker offset method. As grain boundary sliding results in shearing of

---

[30]Nieh TG, McNally CM, Wadsworth J. Superplasticity in intermetallic alloys and ceramics. *JOM*. 1989;41:31-35.

11.3 High-temperature deformation constitutive relationships

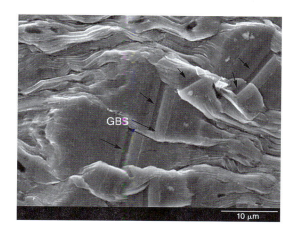

Figure 11.14

A secondary electron scanning electron microscopy (SEM) image of the surface topography of an Al-Zn-Mg-Sc alloy specimen deformed at a temperature of 480°C and a strain rate of 10^{-2} s^{-1} (Charit and Mishra, unpublished work).

the scratch markers, a shear offset is created. A reliable statistical measurement of such shear offsets enables determination of grain boundary sliding contribution to the overall deformation. Fig. 11.14 shows the topography of an Al-Zn-Mg-Sc alloy deformed superplastically. In the figure, the individual arrows indicate the configurations of one scratch that has sheared/rotated because of superplastic deformation. The image also confirms that grain boundary sliding is accompanied by grain rotation and rearrangement.

Superplastic deformation is used for commercial applications known as superplastic forming (SPF), which allows fabrication of components with complex design and near-net shape relatively cost-effectively. SPF has been recognized as an enabling technology for building unitized structures.

Superplasticity is intrinsically related to microstructural characteristics. Hence, to obtain a superplastic material, we need to pay special attention to microstructural design. Microstructural requirements for superplasticity are the following:

(i) **Fine grain size and grain shape** (mean linear intercept grain size of 10 μm or less for metals; for ceramics, further finer, typically submicron). Given that most grain boundary sliding models show strong grain size dependence ($p = 2-3$) of strain rate, a natural outcome is that grain size is important for superplasticity. Finer grain sizes mean more grain boundary areas and thus more opportunities for substantial grain boundary sliding. For superplastic material, the preference is equiaxial grain shape, as grain boundaries can experience shear stresses easily to promote grain boundary sliding and associated grain rotation.

With the possibility of advanced processing techniques capable of producing ultrafine-grained and nanocrystalline materials, superplasticity can be achieved at much higher strain rates or much lower temperatures than usual. This is apparent from the constitutive equations. High strain rate superplasticity (HSRS) is generally defined as superplasticity achieved at strain rates of 10^{-2} s^{-1} or more and can be commercially attractive. Fig. 11.15 illustrates that with refinement in grain size from 15 μm to

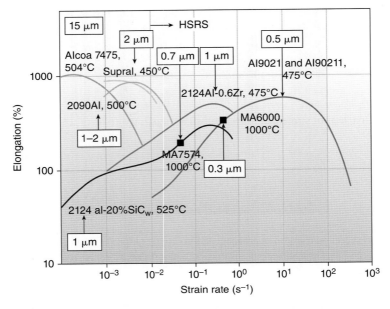

Figure 11.15

Increase in optimum superplastic strain rate with decreasing grain size.[31]

0.5 μm, we can see orders of magnitude increase in optimum superplastic strain rates. We can consider this as a natural extension of the inverse relationship between strain rate and grain size $(\dot{\varepsilon} \propto d_g^{-p})$. But the challenge is how to produce the fine, stable grain structure. Dynamic grain growth in superplastic materials is known to affect superplastic ductility by reducing the extent of grain boundary sliding. In recent times, developments involving equal channel angular pressing, high-pressure torsion, accumulative roll bonding, multiaxial forging, friction stir processing, and other advanced processing techniques have shown promise for extensive grain refinement and thus enhanced superplasticity.

(ii) **Fine, stable dispersoids** to pin grain boundaries are crucial because superplasticity obtained at elevated temperatures can make the microstructure unstable and lead to considerable grain growth and loss of the fine grain size required for superplasticity. Thus if we have fine, stable (resistance to coarsening and dissolution) dispersoids in the microstructure, they can resist grain growth by imparting a particle pinning effect (known as Zener pinning) on the grain boundaries. For instance, Al_3Sc dispersoids impart resistance to grain growth in an Sc-containing Al alloy. Also, fine eutectic/eutectoid structures are inherently superplastic. A classic superplastic material is Zn-22Al eutectoid. Single-phase materials (or pure metals) rarely show any superplastic behavior due to lack of second phase particles in sufficient volume fraction to hinder the grain growth.

---

[31] Sherby OD, Wadsworth J. Superplasticity—Recent advances and future directions. *Prog Mater Sci.* 1989;33:169.

A key criterion to remember is that fine second phase particles are important. If large particles are present at the grain boundaries, strain accommodation will be difficult, as the diffusion distance will be longer. Stress concentration around the particle-matrix interface may lead to nucleation of cavities, and premature cavitation can lower superplastic elongations significantly.

In ceramics, either Zener pinning or solute segregation (impurity drag) effects can help develop superplasticity. For example, in 3 mol% yttria-stabilized tetragonal zirconia (3YTZ), yttrium ions move to grain boundaries and create solute drag, thus keeping the grain size small. For instance, F. Wakai et al.[32] first demonstrated evidence of superplasticity (over 120%) in a 3YTZ ceramic at 1450°C and a strain rate of 5.5×10^{-4} s^{-1}. Since then, many ceramics and ceramic composites have exhibited enhanced superplastic behavior, including 2500% elongation in a ceramic composite (40 3YTZ – 30MgAl$_2$O$_4$ – 30Al$_2$O$_3$) at 1650°C under a strain rate of 8×10^{-2} s^{-1}.[33]

(iii) The importance of **grain boundary/interface character** to superplasticity can hardly be overemphasized. Random high angle grain boundaries are known to have a greater tendency for sliding than the special grain boundaries.[34] Also, the nature of grain boundaries can control the failure mechanisms via which they can control the superplastic elongation. Thus emerged the grain boundary engineering (GBE) concept of developing superplastic material. The advent of advanced characterization tools promoted understanding of the grain boundary structure, and this factor will become increasingly important in the design of superplastic microstructures.

11.3.6 Summary of creep mechanisms

So far, we have discussed the details of various creep mechanisms. Accordingly, Table 11.1 summarizes all that we have covered so far in terms of important creep parametric dependencies. The para-

Table 11.1 Parametric dependencies of different diffusion-assisted creep mechanisms:
$$\dot{\varepsilon} = \frac{ADEb}{kT}\left(\frac{\sigma}{E}\right)^n \left(\frac{b}{d_G}\right)^p$$

Creep mechanism	n	P	D	A
Harper-Dorn	1	0	D_L	3×10^{-10}
Coble	1	3	D_{GB}	100
Nabarro-Herring	1	2	D_L	12
Grain boundary sliding	2	2	D_{GB}	200
Viscous dislocation glide or solute drag creep	3	0	D_S	6
High-temperature dislocation climb	5	0	D_L	6×10^7
Low-temperature dislocation climb	7	0	D_C	2×10^8

D_C, dislocation pipe diffusivity; D_{GB}, grain boundary diffusivity; D_L, solute diffusivity; D_S, dislocation core diffusivity.

---

[32]Wakai F, Sakaguchi S, Matsuno Y. Superplasticity of yttria-stabilized polycrystals. *Adv Ceram Mater*. 1986;1(3):259-263.
[33]Kim BN, Hiraga K, Sakka Y. High strain rate superplastic ceramic. *Nature*. 2001;413:288.
[34]Watanabe T, Kimura S, Karashima S. The effect of a grain boundary structural transformation on sliding in <10–10> tilt zinc bicrystals. *Philos Mag*. 1984;49:845-864.

metric dependency values listed are the most generic. Changes in numbers are possible; for example, A value is present only in model values and may change in real conditions based on the microstructure because the microstructural parameters can vary widely. Also, as mentioned earlier, the stress exponent value of high-temperature climb can vary from 4 to 7. Furthermore, descriptions of many creep mechanisms do not fall directly under the conventional creep mechanisms. So, while Table 11.1 is instructional, use and interpretation with proper context are important.

11.3.7 Creep in particle-containing alloys

Particle-containing alloys are common in many engineering applications at high temperatures. In **Chapter 8**, we discussed precipitation strengthening and dispersion strengthening for low temperatures. But at higher temperatures, particles can provide a strengthening effect so long as the particles remain stable. In many cases, analysis of creep data in particle-containing alloys (especially the dispersion-strengthened ones) shows unusual stress exponents and higher activation energy than that of the relevant lattice self-diffusion. Threshold stress (σ_o) arising out of the dislocation-particle interactions is thought to account for such an apparent discrepancy.

The following constitutive equation developed for dispersion-strengthened alloys can be used to describe creep deformation in the oxide dispersion-strengthened (ODS) steels:

$$\dot{\varepsilon} = \frac{AD_L Gb}{kT}\left(\frac{\sigma - \sigma_o}{G}\right)^n \tag{11.30}$$

where A is a dimensionless constant, D_L is lattice diffusion coefficient, G is shear modulus, b is Burgers vector, k is Boltzmann's constant, T is absolute temperature, σ is applied stress, σ_o is threshold stress ($\sigma - \sigma_o = \sigma_e$ is called the effective stress), and n is the true stress exponent.

There is a difference between the physics of the particle strengthening effect at low temperatures and at high temperatures. The attachment of dislocation to the particles at high temperatures changes the problem from ***approach side at lower temperatures to departure side at high temperatures***. While such a phenomenon has been observed and analyzed since the 1980s, the physics of such interaction has not yet been fully sorted out.

At low temperatures, the Orowan strengthening mechanism considers the dislocations being repelled by elastically hard particles. Note that although we discussed another way of particle strengthening accomplished by particle shearing, that mechanism will not be discussed here. The discussion here will be on dispersion-strengthened alloys that exhibit remarkable creep strength at elevated temperatures.

At higher temperatures, the diffusional processes become important, and the climb by-pass process can surmount the particle obstacle. Several models were proposed to explain the high-temperature dislocation-particle interaction. However, most of those models were proposed based on the basic concept of repulsive interaction between particles and dislocations. These models focused largely on how the line energy of dislocations changes during the climb by-pass event. Two types of possible climb processes, local climb and general climb, were considered, as depicted in Fig. 11.16.

We can see that because of the geometry involved, less atomic diffusion is needed in local climb, but line length increase associated with local climb is higher than that in general climb. The debate on which type of climb would be favorable continued until the early 1980s. Then, the concept began

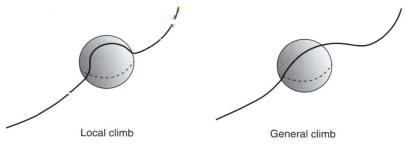

Figure 11.16

Geometry of local climb and general climb of an edge dislocation.[35]

to change with both theoretical and experimental work. Srolovitz and coworkers[36] were the first to develop a model that could explain the transition from repulsive interaction to attractive interaction at elevated temperatures. Srolovitz's model assumed that the stress field associated with dislocation-particle interaction can undergo diffusional relaxation. As a result, the dislocation moving on the glide plane containing the particle experiences an attractive pull toward the particle. When the dislocation enters the particle-matrix interface, the core of the dislocation relaxes and thus results in pinning of the dislocation at the particle. Srolovitz et al.[37] developed an expression for the stress needed to depin the dislocation from the particle:

$$\frac{\sigma_o}{G} = \frac{b}{2\pi(1-\nu)\lambda}\left[\frac{\pi^2}{12} + \ln\left(\frac{r}{r_o}\right)\right] \tag{11.31}$$

where G is shear modulus of matrix, b is Burgers vector of dislocation, ν is Poisson's ratio, λ is inter-particle spacing, r is particle radius, and r_o is the inner cutoff radius. This model appeared to give a value of threshold stress similar to that predicted by Orowan strengthening.

Nardone et al.[38] and Arzt and Wilkinson[39] have proposed models based on the assumption that the elastic energy of the dislocation line is lower at the particle-matrix interface. However, no physical mechanism was proposed to account for the specific reduction in energy. The expression for threshold stress proposed by Arzt and Wilkinson is given by

$$\frac{\sigma_c}{G} = \frac{b}{\lambda}(1 - k_A^2) \tag{11.32}$$

where k_A is a relaxation parameter and other parameters have already been defined in previous sections, which is defined as the ratio of dislocation line energy in the interface and matrix. For $k_A = 1$, no detachment is possible. The precise reason for such relaxation was not explained.

---

[35]Mishra RS. Dislocation-particle interaction at elevated temperatures. *JOM*. 2009;61:52-55.
[36]Srolovitz D, et al. On dislocation-incoherent particle interaction at high temperatures. *Scr Met*. 1982;16(12):1401-1406.
[37]Srolovitz et al. Edge dislocation circular inclusion interactions at elevated temperatures. *Acta Metall*. 1983;31(12):2151-2159.
[38]Nardone VC, Matejczyk DE, Tien JK. The threshold stress and departure side pinning of dislocations by dispersoids. *Acta Metall*. 1984;32(9):1509.
[39]Arzt E, Wilkinson DS. Threshold stresses for dislocation climb over hard particles: the effect of an attractive interaction. *Acta Metall*. 1986;34(10):1893-1898.

Mishra et al.[6] put forth a theory to explain the reason for relaxation and developed a new model for threshold stress. They stated that the elastic energy associated with dislocations is lowered by dissociation of lattice dislocations (recall Frank's rule) at the matrix-particle interface and their conversion to interfacial dislocations. The corresponding threshold stress because of such dislocation dissociation reaction can be given by

$$\frac{\sigma_o}{G} = A\frac{b}{r}\exp\left(B\frac{r}{\lambda}\right) \qquad (11.33)$$

where A and B are the parameters that depend on the specific nature of dislocation dissociation reactions.

Fig. 11.17 (a) shows a transmission electron microscopy image showing evidence of departure side pinning in a dispersion-strengthened platinum alloy. Also, Fig. 11.17 (b) displays multiple particle-dislocation and dislocation-dislocation interactions in a rapid solidification-processed (RSP) Al-5wt%Ti alloy. Note that such interactions have not been modeled and may have major impact on the kinetics of dislocation motion. In complex microstructure of high temperature alloys, such complexities should be expected. The students are again encouraged to think about the approach of modeling simple features and reality of complex microstructures in engineering alloys.

Creep studies performed on the ODS alloys show consistently higher than normal stress exponent values, which are higher than predicted for metallic alloys (Table 11.1). This behavior is sometimes

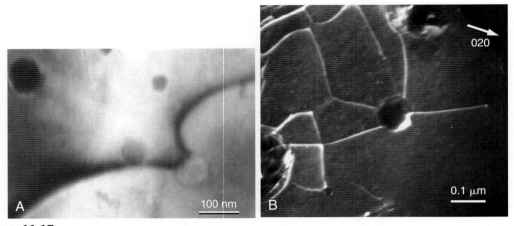

Figure 11.17

(a) A bright field transmission electron micrograph showing departure-side pinning due to attractive dislocation-particle interaction in a dispersion-strengthened platinum alloy.[40] (b) A dark field transmission electron micrograph showing multiple dislocation-particle and dislocation-dislocation interactions in a fine-grained RSP Al-5wt.% Ti alloy crept at 623 K and 25 MPa to a strain of 2%.

---

[40]Heilmaier M, Reppich B. In: Mishra RS, Mukherjee AK, Murty KL, eds. *Creep Behavior of Advanced Materials for the 21st Century*. TMS; 1999:503.

typical of various ODS alloys and is generally attributed to threshold stress. It also leads to an overestimation of creep activation energy. This behavior is dependent on the precise nature of interaction between the oxide particles and mobile dislocations. Fig. 11.18 shows steady-state creep rates versus stress in a double-logarithmic plot for three different yttria-containing ODS steels (12Y1, 12YW, 12YWT) exhibiting high "n" (slope) values.[41]

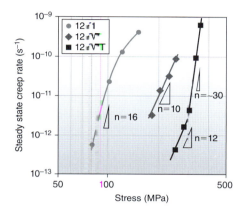

Figure 11.18

Steady-state creep rate as a function of stress in three different ODS steels at 700°C.[4]

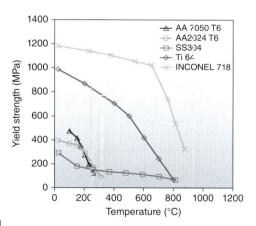

Supplementary Figure S11.1

Yield strength versus temperature for multiple alloys.[42-45]

---

[41]Modified from I. Kim, B.Y. Choi, C.Y. Kang, T. Okuda, P. Maziasz, K. Miyahara, ISIJ International, 43 (2003) 1640.
[42]Kaufman JG. *Properties of Aluminum Alloys: Tensile, Creep and Fatigue Data at High and Low Temperatures*. ASM International; 1999.
[43]Nikulin I, Kaibyshev R, Skorobogatykh V. High temperature properties of an austenitic stainless steel. *J Phys Conf Ser.* 2010;240:012071.
[44]Thomas A, El-Wahabi M, Cabrera JM, Prado JM. High temperature deformation of Inconel 718. *J Mat Process Tech.* 2006;177(1-3):469-472.
[45]Ivanov S et al. Effect of elevated temperatures on the mechanical properties of a direct laser deposited Ti-6Al-4V. *Materials (MDPI)*. 2021;14(21):6432.

Concept builder—Transition in dislocation-particle interaction with temperature and loss of strengthening. Students are encouraged to look at this historically and ponder over the timeline. Consider the fact that while the theory of dislocations developed in the late 1930s, dislocations were directly imaged by transmission electron microscopy only in the 1950s. When images of high-temperature deformed or crept materials were published in the 1960s, why did researchers not raise *the basic question that dislocations appear to be attached to the particles*? Sometimes we see what we expect to see! With the understanding of shift in physics of dislocation-particle interaction, it is important to consider the impact of this on strengthening mechanisms. Let us take a quick look at some of the strengthening mechanisms covered in **Chapter 8**.

- Solid solution strengthening—At high temperatures, the solutes start to diffuse and its contribution reduces to solute-drag effect.
- Grain boundary (gb) strengthening—Hall-Petch strengthening is very effective strengthening contributor for many elements. The physics of dislocation-gb interaction is similar to dislocation-particle interaction. At elevated temperatures, dislocations are attracted to grain boundaries! So, the repulsive interaction that led to dislocation pileup at grain boundaries no longer applies above a critical temperature. Therefore, the associated gb strengthening disappears. The mechanism of this dislocation-gb interaction is based on absorption of dislocations at the interface. This occurs through spreading of the lattice dislocation core and results in gb "extrinsic" dislocations. Note that twin boundaries will have different impact. Can you build up an argument to justify this difference? Why the structure of gb is important for dislocation-gb interaction?
- Dispersion strengthening—The magnitude of strengthening reduces from Orowan strengthening to threshold stress for dislocation-particle interaction, which is significantly smaller.
- Composite strengthening (**Chapter 6**)—Note that this strengthening mechanism is based on elastic mismatch and associated load transfer. This mechanism is only dependent on efficacy of load transfer to the stiffer phase. So, it is unaffected by the nature of dislocation-obstacle interaction.

NET IMPACT—What is the overall impact on strength? Take a look at some strength versus temperature data in this box figure (Supplementary Fig. 11.1). It is not difficult to appreciate the impact of the change in dislocation-obstacle interaction physics as a function of temperature. This is very important for development of high-temperature materials.

11.3.8 Creep deformation mechanism maps

Before we discuss deformation mechanism maps, let us understand independent and sequential mechanisms during creep deformation.

11.3.8.1 Independent and sequential processes

Creep mechanisms generally do not operate in isolation. In most cases, multiple mechanisms operate simultaneously, independent of one another during creep processes. This combination of processes is called independent or parallel mechanisms. The net effect of the mechanisms is linearly additive. In other words, we can add the creep rate of individual creep mechanisms ($\dot{\varepsilon}_i$) to obtain the overall creep rate ($\dot{\varepsilon}$):

$$\dot{\varepsilon} = \sum \dot{\varepsilon}_i. \tag{11.34}$$

Two mechanisms operating in parallel are presented schematically **in** Fig. 11.19. The double logarithmic plot of stress versus creep rate shows that Mechanism-1 acts in parallel or simultaneously with Mechanism-2. In such a scenario, the faster mechanism will dominate the creep deformation process as the rate-controlling mechanism, except for the transition region of the involved mechanisms, where both mechanisms would contribute comparably. Diffusional creep and dislocation creep are considered independent mechanisms.

In some instances, creep mechanisms occur in sequence or series. The overall creep rate in terms of individual creep rates is given in the following expression:

$$\frac{1}{\dot{\varepsilon}} = \sum \frac{1}{\dot{\varepsilon}_i} \tag{11.35}$$

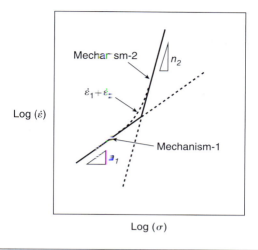

Figure 11.19
Schematic showing creep rate versus stress on a log-log plot showing parallel creep mechanisms.

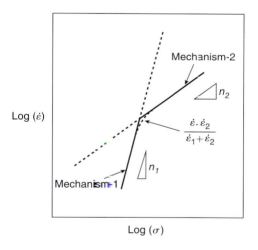

Figure 11.20
Schematic of creep rate versus stress (log-log) plot showing sequential creep mechanisms.

If Mechanism-1 and Mechanism-2 act in series, the double logarithmic plot of stress and creep rate would appear as described in Fig. 11.20. In such a case, the slower mechanism is the rate-controlling process. However, the total extent of creep deformation is not necessarily controlled by the slower process. In fact, in the case of glide-climb creep, the viscous dislocation glide mechanism operates in sequence with the dislocation climb mechanism, with the former providing most of the creep strain compared to the latter. The glide-climb creep was discussed at length in a prior section of this chapter.

11.3.8.2 Ashby's deformation mechanism map

The concept of a deformation mechanism map was first proposed by Ashby[46] to describe various creep mechanisms of importance as a function of temperature, stress, and grain size. Later, this very concept was extended to fracture, wear, and sintering. We discussed in the prior section how more than one creep mechanism can operate, depending on temperature and stress for a given material and microstructure. Nevertheless, only a single creep mechanism has a dominant effect under a certain range of temperature and stress. Thus, a deformation mechanism map can show the predominance of different creep mechanisms in different spaces. Originally, the axes of a deformation mechanism map were defined as follows: Y-axis is the tensile stress normalized by shear modulus, $\frac{\sigma}{G}$, and the X-axis is the temperature normalized by the melting point, $\left(\frac{T}{T_m}\right)$. An example map (Fig. 11.21) is for pure silver with grain size of 32 μm and critical strain rate of 10^{-8} s^{-1}. Let us look at the map a bit more closely. At low homologous temperatures and stresses, the material would undergo elastic deformation, shown as an elastic regime. However, in later modifications, it has been replaced by Coble diffusion creep. For example, the transition between the Coble creep and Nabarro-Herring creep can be obtained by

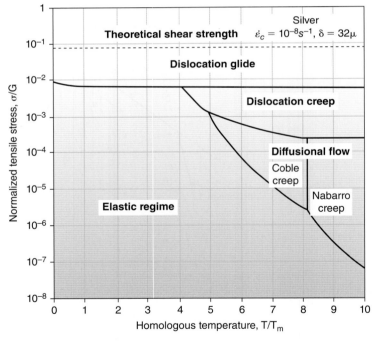

Figure 11.21

Deformation mechanism map for pure silver.[46]

---

[46] Ashby MF. A first report on deformation-mechanism maps. *Acta Metall.* 1972;20(2972):887-897.

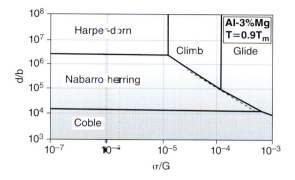

Figure 11.22

Deformation mechanism map in the space of normalized grain size and normalized stress as described by Mohamed and Langdon.[18]

equating the steady-state creep rates of the two mechanisms using the respective constitutive equations discussed earlier.

Mohamed and Langdon[47] showed a deformation mechanism map as a function of grain size. We have noted previously the importance of grain size on creep mechanisms. This type of deformation mechanism map is constructed as a plot of normalized grain size (d/b) versus normalized stress (σ/G) under a fixed temperature. Fig. 11.22 shows such a deformation mechanism map for an Al-3wt%Mg alloy at a high homologous temperature. From the figure, we can observe that Coble creep is operative at a smaller grain size, and Nabarro-Herring creep and Harper-Dorn creep mechanisms become favorable as grain size increases. Dislocation creep mechanisms, climb, and glide occur at larger grain sizes and higher stresses.

Mishra[48] (1992) extended the concept of the deformation mechanism map to dispersion-strengthened materials. This map shows conceptual transition in mechanisms as we change the particle size. Fig. 11.23 shows an example of such a map for aluminum alloys and composites. Note that this depicts the operating space for dispersion-strengthened alloys and composites.

11.4 Creep damage mechanisms

Resistance to both creep deformation and fracture are important to the design of high-temperature structural components. Generally, the stress at which fracture happens in 10^5 h can be used to estimate the allowable stress. However, as completing such tests takes a long time, we need to depend on short-term data based on parameters such as Larson-Miller parameter or other parameters (to be discussed later). However, we do not advise extrapolating short-term data to long-term, as the micromechanism by which fracture happens can change. So, the estimation of allowable stress or life should be based on sound physics-based mechanisms, not just empirical parameters. Successfully doing so would avoid

---

[47]Mohamed FA, Langdon TG. Deformation mechanism maps for solid solution alloys. *Scr Metall*. 1975;9:137-140.
[48]Mishra RS. Dislocation creep mechanism map for particle strengthened materials. *Scr Met*. 1992 26(2):309-313.

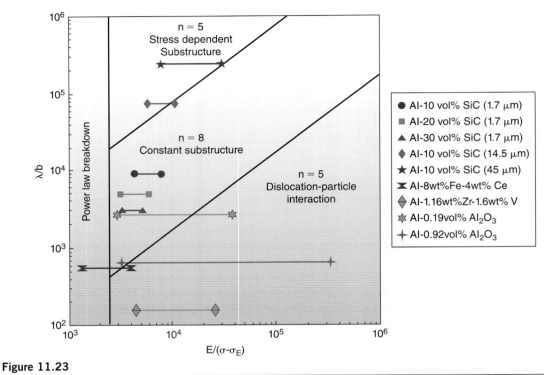

Figure 11.23

Deformation mechanism map for dispersion-strengthened materials.[49]

the danger of *blind extrapolation*. In short, empirical extrapolation of short-term data will not consider the ever-increasing effects of microstructural instability and degradation under long-term service and might lead to erroneous life predictions.

We discussed a few low-temperature, fracture mechanism–related issues in **Chapter 8**. Here we consider specifically the high-temperature creep fracture mechanisms. Different fracture modes can be encountered during creep depending on creep test parameters, primarily the applied stress and temperature. Ashby and Dyson[49] defined the term creep damage as follows: "… a term coined by engineers to describe the material degradation which gives rise to the acceleration of creep rate known as tertiary creep." Previous continuum mechanics approaches described creep damage, where the importance of microstructure was not taken into account and was lumped into a generic damage term, without understanding the individual micromechanisms at play. Ashby and Dyson merged these two together in their seminal work and categorized creep damage into various types (Fig. 11.24).

---

[49] Ashby MF, Dyson BF. Creep damage mechanics and micromechanisms. In: Rama Rao P, Raju KN, Knott JF, Taplin DMR, eds. *The 6th International Conference on Fracture (Fracture 84)*. Vol. 1. Elsevier. 1984:3-30.

11.4 Creep damage mechanisms

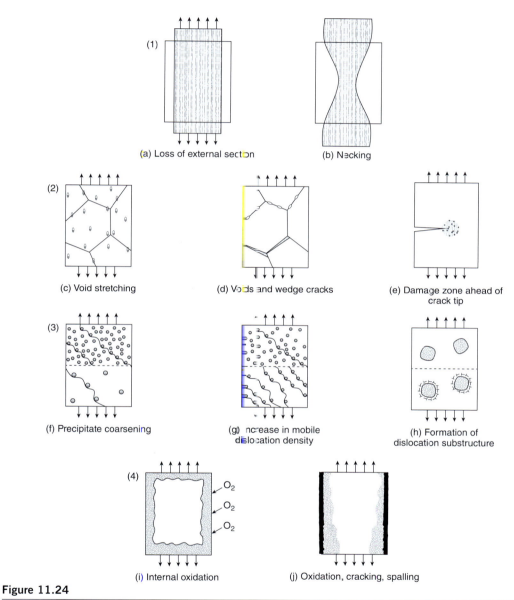

Figure 11.24

Various forms of creep damage: (1) loss of external section, (2) damage by loss of internal section by creep cavitation or cracking, (3) creep damage by microstructural degradation, and (4) creep damage by environmental chemical interaction (internal or external).[49]

11.4.1 Damage by loss of external section

This kind of damage occurs by external necking. In tensile creep under constant tensile loading, the cross-section gradually decreases with accumulation of strain, and the stress progressively increases, thereby enhancing the creep rate. If there is no other intervening damage mechanism, the section could neck down essentially to a point, that is, associated reduction in area being close to 100%. This kind of situation may appear at high homologous temperatures (above 0.8) in pure metals and solid solution alloys. This type of failure is often referred to as ductile rupture.

11.4.2 Creep damage by loss of internal section

The intergranular fracture could happen in two distinct ways:

(i) Creep cavitation: The creation of cavities (hole-like features or voids) during creep often occurs at the grain boundaries that lie perpendicular to the maximum principal tensile stress (Fig. 11.25 [a]). The cavities essentially reduce the internal cross-section and lead to acceleration of creep deformation, thus leading to further damage growth. At low stresses, the cavities may remain void like, but at higher stresses, they link up (i.e., cavity coalescence) to create grain boundary cracks. This type of crack sometimes is referred to as an *r*-type crack. The cavity shape can take different forms based on temperature, stress, and spacing. Creep cavitation has been observed in a wide variety of materials undergoing creep, in materials such as pure metals, commercial alloys like nickel-based superalloys, and ferritic and austenitic steels. However, in some circumstances more than one creep damage mechanism could be at play. Cavities form on the grain boundaries in the vicinity of second-phase particles in a crept specimen (Fig. 11.25 [b]).

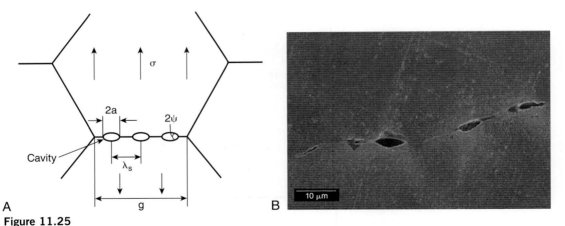

Figure 11.25

(a) A schematic of creep cavitation (r-type cracking) on a transverse grain boundary.[50] (b) View of cavities in copper crept at an applied stress of 20 MPa and temperature of 550°C to a strain of about 0.04 (still within Stage-II or steady-state creep deformation region).[51]

---

[50]Kassner ME, Hayes TA. Creep cavitation in metals. *Inter J Plast*. 2003;19:1715-1748.
[51]Evans HE. *Mechanisms of Creep Fracture*. Elsevier Applied Science; 1984.

(ii) Wedge crack formation at the grain boundary triple junctions as a result of stress concentration created by unaccommodated grain boundary sliding.

11.4.3 Damage by microstructural degradation

(i) Many high-temperature engineering alloys are hardened by second-phase particles, including various carbides in ferritic and austenitic stainless steels, γ' in nickel based superalloys, and the like. During long durations of creep, thermally induced/stress-assisted coarsening of precipitates can take place, as can change of nature of particle (e.g., coherency). In either case, creep rate is accelerated. In extreme cases, if the temperature is very high, precipitates can just dissolve.

(ii) Substructure-induced acceleration of creep. In this case, it refers to the substructure formation in some particle-containing alloys, which increases mobile dislocation density as creep rate increases. At higher stresses, a cellular structure can form and lead to greater dislocation annihilations at those boundaries (rate of recovery) and thus accelerated creep rate. Dynamic subgrain coarsening can also be considered a creep-damage mechanism.

11.4.4 Damage by gaseous-environmental attack

This kind of damage can occur through two processes: damage by internal oxidation and damage via failure of external oxides.

When exposed to oxidizing atmospheres, metallic materials without protective oxide film or scale have been noted to creep faster and thus lead to reduction in creep life. Species like oxygen have ingressed in the material and start reacting with certain alloying elements within the material, with resultant regions of internal oxidation. Also, gases can form bubbles at the grain boundaries or cavities can form on those internally oxidized regions, hastening creep failure. On the other hand, creep can disturb the stability of the protective surface oxide films by straining them during creep deformation. So, perturbance in the external oxide film may lead to further creep damage effects by directly attacking the film disruption sites and thus lead to spalling or further aggravation of the creep effects.

11.5 Creep fracture mechanism maps

Like deformation mechanism maps, Ashby et al.[52] developed fracture mechanism maps for various types of materials. In their original version of fracture mechanism maps, the normalized tensile stress was included on the Y-axis and the homologous temperature on the X-axis (see Fig. 11.26, which is a calculated fracture mechanism map of a solution-annealed and aged Cu-Cr-Zr alloy[59]). The domains shown in the map correspond to the fracture mechanism dominant in the stress-temperature space. These types of maps can provide helpful guidelines for understanding damage evolution as a function of service conditions and making relevant life assessments. Creep fracture is generally considered intergranular in nature. While intergranular creep fracture covers a wide region, there is a fracture domain involving a

---

[52]Ashby MF, Gandhi C, Taplin DMR. Fracture mechanism maps and their construction for FCC metals and alloys. *Acta Metall.* 1972;20:887-897.

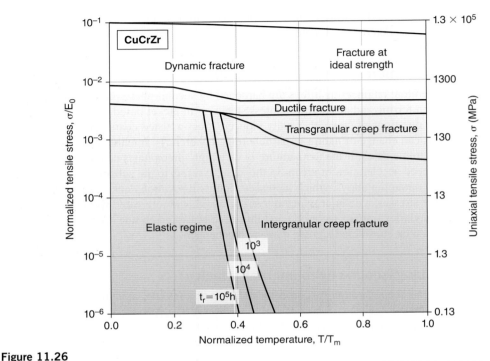

Figure 11.26

Fracture mechanism map of a CuCrZr alloy.[53]

transgranular creep fracture region above it. Another way of plotting fracture mechanism maps is to plot tensile stress against failure time with constant temperature lines superimposed on the plot.

11.6 Creep life–related relationships

The common design methodology largely uses the stress level to design high-temperature components that would not cause failure or accumulation of large plastic strains resulting in loss of dimensional stability of the component under service. Generally, the design philosophy is that the material should have sufficient creep strength determined from the following criteria. (1) *Based on rupture life*: Creep rupture strength is determined by the applied stress necessary to cause fracture in 10^5 h or 2×10^5 h or any other specified time period. (2) *Based on specified strain*: This refers to creep resistance that is defined as the applied stress level at which an engineering strain of 0.1% (or 0.2% or 0.5%) is accumulated.

While simple short-term laboratory creep tests can be used to compare extrapolated creep behavior of different materials under equivalent stress and temperature, there is no surety that these materials will compare the same way in real service conditions in the long term.

---

[53]Li M, Zinkle SJ. Fracture mechanism maps in unirradiated and irradiated metals and alloys. *J Nucl Mater*. 2007;361:192-205.

We can measure time to rupture at a variety of stress levels and temperatures. Usually, the tests are performed at temperatures and stresses well above the expected engineering applications to be completed within a short period of time (could be months!). Then the short-term data are used to extrapolate creep behavior for longer terms at lower temperatures and stresses.

While characterizing the creep curve is useful for mechanistic understanding, often rupture time (t_R) is needed for design of engineered structures. Stress rupture tests are similar to creep tests except that creep strains are not monitored during the test, tests are carried out to fracture, and times to fracture are noted at various temperatures and stresses. The stress-time to rupture data are plotted (Fig. 11.27). The basic information obtained in the stress-rupture test can be used for design of high-temperature components in power plants where design life is the criterion.

Larson-Miller parameter (LMP) is a creep life prediction parameter that was derived by treating the creep rate as the inverse of rupture life (t_R). The parameter was derived as follows:

$$Rate = \frac{1}{t_R} = A\exp\left(-\frac{Q}{kT}\right)$$

$$\ln\left(\frac{1}{t_R}\right) = \ln(A) - \left(\frac{Q}{kT}\right)$$

$$\frac{Q}{kT} = \ln(A) + \ln(t_R)$$

$$\frac{Q}{k} = T[\ln(A) + \ln(t_R)] = T[C + \ln(t_R)]$$

where Q is activation energy for the creep process, k is Boltzmann's constant, T is absolute temperature, and A and C are constants.

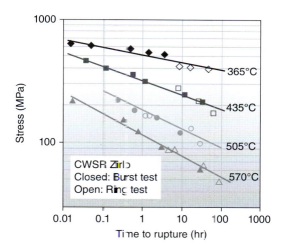

Figure 11.27

Stress versus rupture time for a niobium-bearing zirconium alloy (Zirlo™) at different temperatures.[54]

[54]Seok CS, Marple B, Song YJ, Gollapudi S, Charit I, Murty KL. High temperature deformation characteristics of Zirlo™ tubing via ring-creep and burst tests. *Nucl Eng Des.* 2011;241(3):599-602.

Chapter 11 High-temperature deformation of materials: creep-limiting design

This is commonly written as the following:

$$\frac{Q}{R} = T[C + \ln(t_R)] = LMP \tag{11.36}$$

where Q/R is usually treated as a materials parameter that is constant. The right-hand side of the equation involving temperature and rupture life is called the Larson-Miller parameter. We should note that T is given in K, t_R is rupture life in hours, and C is generally taken as 20, which, however, can vary. C is best determined experimentally. We can plot t_R versus $1/T$ in a semilogarithmic plot. The lines for a fixed stress should extrapolate back to C.

Fig. 11.28 shows a plot depicting variation of stress against LMP for nuclear-grade zirconium alloys (Zirlo and Zircaloy-4). Note that even though LMP can be used to extrapolate data from short stress-rupture data, extrapolation to long term could be problematic, as the microstructure would undergo degradation and thus would change the failure mechanisms.

Monkman-Grant relationship follows from the fact that time to rupture decreases with stress or minimum creep rate:

$$\dot{\varepsilon}_s^m \cdot t_R = K_1 \tag{11.37}$$

where $\dot{\varepsilon}_s$ is the steady state creep rate and m_{MG} is a factor that is typically 1 or close to 1. However, Dobes and Milicka[55] modified Eq. (11.37) to obtain a better correlation of rupture time with steady-state or minimum creep rate by the following relationship:

$$\dot{\varepsilon}_s^{m'} \cdot \frac{t_R}{\varepsilon_f} = K_2 \tag{11.38}$$

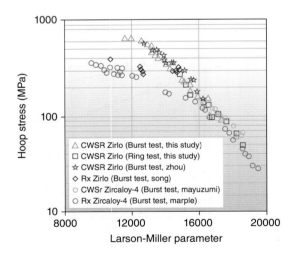

Figure 11.28

Stress versus LMP for two zirconium alloys (Zirlo[TM] and Zircaloy-4) under two conditions (recrystallized, or Rx, and cold-worked stress relieved, or CWSR).[25]

---

[55]Dobes F, Milicka K. Origin of internal stress in creep of long range ordered beta brass. *Met Sci.* 1976;10:382-384.

11.6 Creep life–related relationships

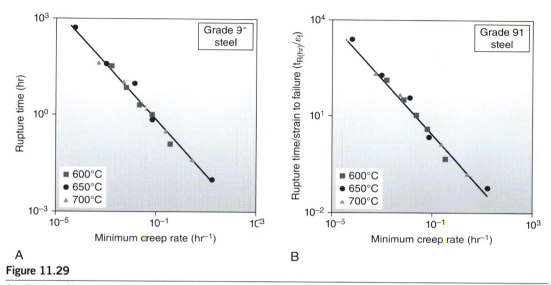

Figure 11.29

Application of Monkman-Grant relationship: (a) original and (b) modified form, for stress-rupture of Grade 91 steel.[56]

where ε_f is fracture strain (based on elongation to fracture), and K_2 and m' are constants. Fig. 11.29 shows the plots of the Monkman-Grant, using both original and modified Monkman-Grant relations, for modified 9Cr-1Mo steel (Grade 91) (Figs. 11.29 [a] and [b]). These relations are useful in predicting creep-rupture life for service stresses under relatively low stresses and temperatures using data obtained in short-term laboratory creep tests carried out under high stresses and temperatures. After stress and temperature variations of minimum creep rates are measured for a material and are determined from laboratory tests, we can predict first the secondary creep rate at the service temperature and stress, following which the application of Monkman-Grant relationship can be used to predict the rupture life in-service.

Another useful parameter known as the Zener-Hollomon parameter (Z) is applicable to a given applied stress:

$$Z = \dot{\varepsilon} e^{Q/RT} \qquad (11.39)$$

Fig. 11.30 shows such a correlation for alpha-iron as a single curve describing all the stages of creep, that is, the primary, steady-state, and tertiary stages for different temperatures under a fixed stress of about 27.6 MPa.

Sherby-Dorn parameter (P_{SD}) involves normalization with temperature-compensated time (t):

$$\varepsilon_{SL} = t\, e^{-Q/RT} \qquad (11.40)$$

The creep curves at three different temperatures at a constant stress (13.8 MPa) merge into a single curve when plotted strain versus compensated time for aluminum (Fig. 11.31).

Other life-prediction parameters include Manson-Haferd, Orr-Sherby-Dorn, and White-LeMay.

---

[56]Shrestha T, Basirat M, Charit I, Potirniche GP, Rink EK. Creep rupture behavior of Grade 91 steel. *Mater Sci Eng A.* 2013;565:382-391.

Chapter 11 High-temperature deformation of materials: creep-limiting design

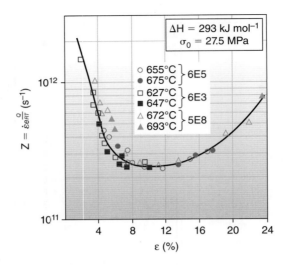

Figure 11.30

Zener-Holloman parameter against time at various temperatures at a constant stress of 27.6 MPa in alpha-iron.[57]

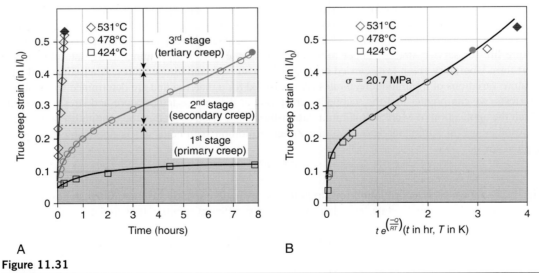

Figure 11.31

(a) Creep curves at three different temperatures at an applied stress of 20.7 MPa (3000 psi), and (b) all data coalesce into a single curve when plotted against Sherby-Dorn parameter.[58]

---

[57]Murty KL. Thesis MS. Cornell University; 1967.
[58]Sherby OD, Dorn JE. Creep correlations in alpha solid solutions of aluminum. *Trans AIME*. 1952;194:959-964.

11.7 Design of creep-resistant materials

In the preceding sections, we discussed various creep deformation and fracture mechanisms. We must also recognize the vastness of creep literature that could not be fully covered in this chapter. However, hopefully we have gained enough basic understanding of creep to consider important aspects for design of creep-resistant materials.

a) **Crystal structure and melting point:** As the creep rate is directly proportional to diffusivity, close-packed structures such as FCC and HCP lattice structures tend to reduce the creep rate. Examples: Steels based on austenitic structure (FCC), nickel-based matrix (FCC) for superalloys, and cobalt-based superalloys (HCP). Having a higher melting point would definitely help. However, choosing a higher-melting-point metal is not always possible. Of course, there is also interest in developing a creep-resistant alloy in the category of base element, for example, a creep-resistant aluminum alloy. In such a case, obviously we do not have the choice of melting temperature or crystal structure, and we move to other alloy design principles.

b) **Grain size and grain boundary character:** Grain boundary sliding occurs at high temperatures. Grain boundary sliding is promoted in fine-grained microstructures. So, creep-resistant alloys with fine-grain microstructures with large numbers of grain boundaries would undergo a higher rate of deformation and have greater chances of grain boundary cavity or crack nucleation. On the other hand, large grain size can reduce creep rate. Note that a single crystal is essentially a single grain! Interestingly, single-crystal nickel superalloys with superior creep resistance are commonly used to make jet turbine engine blades. In Fig. 11.32, the left blade contains equiaxed grains, the middle one has long directionally solidified grains, and the right one is a single-crystal turbine blade.

Evidence in the literature is sufficient that certain coincident site lattice boundaries (CSLBs) with Σ value ≤ 29 (i.e., special CSLBs) have characteristics such as less susceptibility to impurity segregation, lower diffusivity, and better resistance to grain boundary sliding.[59] Watanabe's[60] contributions to the understanding of grain boundary character distribution led to the development of the GBE concept. In the context of creep, higher numbers of special CSLBs in the microstructure lead to greater resistance to intergranular degradation against fracture and cavitation, among many other beneficial characteristics. On the other hand, random grain boundaries tend to be prone to grain boundary sliding, which is not beneficial for a creep-resistant material but is good for a superplastic material!

c) **High-temperature corrosion/oxidation:** We can develop alloys with good intrinsic oxidation and corrosion resistance by adding Cr, Al, and/or Si. These elements provide stable, protective oxide scales that can survive at elevated temperatures.

d) **Particles:** The most effective method of strengthening is particle strengthening. Generally, large numbers of stable particles with high volume fraction provide greater particle strengthening. If the particle coarsens via Ostwald ripening mechanism or dissolves, it will have a negative effect on creep resistance. Particles also help to resist grain boundary migration and grain boundary sliding. Note the concept box that highlights Cu-Ta system. Immiscible alloys can form a subcategory for particle strengthening.

---

[59] Watanabe T. Grain boundary design and control for high temperature materials. *Mater Sci Eng A*. 1993;166:11-28.
[60] Watanabe T. An approach to grain boundary design for strong and ductile polycrystals. *Res Mech*. 1984;11(1):47-84.

344   Chapter 11 High-temperature deformation of materials: creep-limiting design

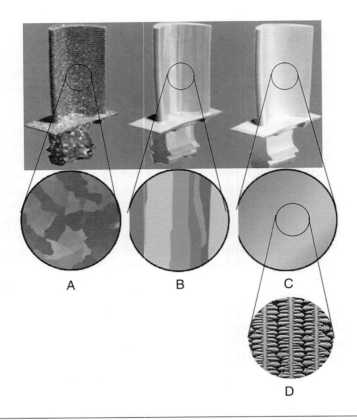

Figure 11.32

Turbine blade castings—*(left)* equiaxed grains, *(middle)* directionally solidified large grain size, and *(right)* single crystal turbine blade.

Dai, H., Gebelin, J., Newell, M., Reed, R.C., D'Souza, N., Brown, P.D., & Dong, H. (2008). Grain Selection During Solidification in Spiral Grain Selector. Superalloys, 367-374, TMS 2008, Proceedings of the 11th International Symposium on Superalloys, 14-18 September 2008. Copyright 2008 The Minerals, Metals & Materials Society. Used with permission.

- e) **Substitutional solutes** affect the recovery rate more than hardening. The solutes that reduce overall diffusivity of alloys are preferred for creep resistance. For example, Kanthal APMT™ (Fe-22Cr-5Al-3Mo) provides higher strength and creep resistance compared to Kanthal APM™ (Fe-22Cr-5Al), largely because molybdenum imparts significant solid solution strengthening even at elevated temperatures. Such an approach is also used for a number of nickel-based superalloys.
- f) **Lower stacking fault energy:** The separation of dissociated partials is a very important concept. Partial dislocations have smaller Burgers vector and cannot participate in the non-conservative climb motion of edge dislocations. Therefore, *the partial dislocations cannot climb without constriction*. This reduces climb recovery and can be a very good principle to strengthen the matrix, before combining other concepts.
- g) **Composite strengthening:** As mentioned earlier, this concept is based on load transfer and its use can be quite synergistic. Note that the dislocation creep relationships that we covered

are power law with stress exponent being 3 or higher. The implication is that any reduction in matrix stress level will have a very high impact on lowering of creep rate. This mechanism is independent of other mechanisms and will lead to synergistic outcome.

11.8 Exercises

1. In a laboratory creep experiment at 1200°C, a steady-state creep rate of 0.2% per hour is obtained and the specimen failed at 1500 hours. The creep mechanism for this alloy is known to be dislocation climb with an activation energy of 25 kcal/mol. If the service temperature of the alloy is 600°C, evaluate the service life using Larson-Miller parameter.
2. Enhanced higher-temperature strength reduces high-temperature formability! Then how do we process very strong high-strength alloys?
3. Explain diffusional creep with the help of a sketch. What is the grain size dependence for one of the mechanisms?
4. What are the steps involved with drawing a creep deformation mechanism map?
5. Write stress exponent and grain size dependence for three creep mechanisms in a tabular form.
6. How would you strengthen a material against creep deformation? Mention at least three concepts using a hypothetical alloy.
7. How does the dislocation-particle interaction change at elevated temperatures? How can you use threshold stress to increase high-temperature strength?
8. Suppose a polycrystalline alloy with a very fine grain size. Assume that due to some faulty processing the material becomes a single grain material (i.e., a single crystal). Explain how the creep resistance of the material will be affected. Discuss in terms of the effect of microstructure and mechanisms.
9. **Open-ended question**—Create a hypothetical microstructure that uses all the high-temperature strengthening mechanisms you can think of.
10. In the previous chapters, we considered TRIP and TWIP effects as strengthening mechanisms. What happens to TWIP and TRIP mechanisms as you increase temperature? Consider (a) low-, (b) medium-, and (c) high-temperature regions for this answer.

    Qualify the following statements as "True" or "False." Give adequate reasons for supporting your choice.
11. Homologous temperature is the ratio of a temperature of interest to the melting temperature of a metal (both temperatures given in °F).
12. Strain hardening exponent increases with temperature.
13. Grain boundary sliding is the main mechanism for superplasticity and typically occurs in fine-grained materials.

14. Strain rate sensitivity during superplastic deformation remains very small, typically less than 0.05.
15. Stacking fault energy does not depend on temperature.
16. Creep resistance of fine-grained materials is generally lower than the coarse-grained material.

Further readings

Bailey, R. W. Note on the softening of strain hardening metals and its relation to creep. *J. Inst. Metals* 35 (1926): 27.

Kassner ME, Perez-Prado MT. *Fundamentals of Creep in Metals and Alloys*. 2nd ed. Elsevier Science; 2008.

Murty KL, ed. *Materials' Aging and Degradation in Light Water Reactors: Mechanisms and Management*. Woodhead Publishing; 2013.

Penny RK, Marriott DL. *Design for Creep*. 2nd ed. Cornwall: Chapman & Hall; 1995.

Tin S. Modeling of creep. In: Furrer DU, Semiatin SL, eds. *Fundamentals of Modeling for Metals Processing*. Vol 22A. ASM International; 2009:400-407.

Orowan, E. Creep in Metals. *J. West Scotland Iron & Steel Inst*. 54 (1946): 45.

Index

Note: Page numbers followed by "*f*" indicate figures, "*t*" indicate tables, and "*b*" indicate boxes.

A

A-basis, 21
ABCDEF plane, 131
Accelerated Insertion of Materials (AIM) program, 36–38
Activation energy, 305–306
Advanced aircraft, 14, 14f
AFM. *see* Atomic force microscopy
AIJF plane, 131
Alloys
   aluminum, 51–52, 51f, 52t, 53f, 54f
   M54, 41–43, 41f
   titanium, 55, 57f
Al-rich Al-Cu alloy system, 181–182, 182f, 183f, 184f
Al-SiC metal matrix composite, 35–36
Alternating stress, 254
Aluminum alloys, 51–52, 51f, 52t, 53f, 54f
Amorphous materials, 55–56
Amplitude ratio, 254
Anisotropy, in elastic property, 110–112, 111f, 111t
"Annealing phenomena," 157
Antiphase boundary (APB), 166, 172–174, 172t
Applied plastic strain, 270
Aramid fiber-reinforced polymer (AFRP) composites, 11
Artificial aging, 181–182
Arzt and Wilkinson model, 327
Ashby charts, 292–293, 293f, 294f, 302, 302f
Ashby-Orowan equation, 178, 178f
Ashby's basic mechanical design framework, 21–30
   creep limiting design, 29–30, 30f
   fatigue-limited design, 28–29
   stiffness-limited design, 22–23, 22f, 23f
   strength-limited design, 23–25, 24f, 26t
   toughness-limited design, 25–28, 27f, 28t
Ashby's deformation mechanism map, 332–333, 332f, 333f, 334f
Ashby's property charts, 5–6, 6f
Ashby-Verrall model, 320, 320f
Atom, definition of, 48
Atomic bond strength, 94–95
Atomic force microscopy (AFM), 68–69
Atomic spacing, 22
Automobile, 13, 13f
Average values, 20
Axes, transformation of, 102–105, 103f, 103t, 104f

B

Back stress, 321
Ball-Hutchison model, 321, 321f
Basquin's equation, 258
Bauschinger effect, 261–265, 262f, 263f, 264f, 265f
B-basis, 21
Beach marks, 285–286
Biomimetics, 11
Bird-Mukherjee-Dorn (BMD) equation, 306–307
BMD. *see* Bird-Mukherjee-Dorn
Body-centered cubic (BCC), 7–8, 128–129, 128f, 130f
   Hall-Petch parameters for, 160t
Boeing 787 plane, 14, 14f, 114
Boltzmann constant, 314
Breakaway stress, 312
Brinell hardness, test of, 65, 66f
Brittle fracture, 190, 190f, 192f
Broad categories, of materials, 6–11, 7f
   ceramics, 8–9, 9f, 10f, 12t
   composites, 11
   metals/alloys, 7–8, 12t
   polymers, 10–11, 10f, 12t
Brown-Embury analysis, for tensile ductility, 191, 194f
Bulk modulus, 105–108
Burgers vector, 132

C

Calibration constant, 277
Carbon fiber-reinforced polymer (CFRP) composites, 11
Carbon steel, strain hardening exponent and strength coefficients for, 75t
Carbon-carbon composites, 11
Cartesian coordinates, 97–98, 103f
Castings, 281
Cauchy stress, 97
Cauchy stress tensor, 99
Cementite (Fe_3C), 142–143
Ceramic matrix composites (CMCs), 11
Ceramics, 8–9, 9f, 10f, 12t, 58–60, 59f
   elastic moduli range in, 92–93, 93f
   toughening of, 244–247, 244t, 246f, 247f, 248f
CGHD plane, 131
Charpy V-notch (CVN) test, 215–216, 216f
Chemical strengthening, 174, 174f
Clamshell marks, 285–286
Cleavage fracture, 198
Close-packed directions, 130, 130f
Close-packed planes, 130, 130f
CMOD. *see* Crack mouth opening displacement
Coarse particles, in fracture toughness, 242
Coble creep, 312–313, 313f
Coffin-Manson relationship, 265–266
Coherency hardening, 169, 170f
Cold working, 155, 156f
Comet airplanes, 252, 253f
Compact test specimen, for fracture toughness, 231–232
"Complex concentrated alloys" (CCA), 138
Composite strengthening, 344–345
Composites, 11
   Al-SiC metal matrix, 35–36
   elastic moduli range in, 93f
   fiber-reinforced, 119–122, 119t, 120f
   high stiffness, design of, 114–122, 114t, 115f

347

Composites *(Continued)*
  particulate-reinforced, 116–119, 116f, 117f
  polymer-based, 56–58, 58f
Compression tests, 4
Constant strain approach, to particulate-reinforced composites, 118, 118f
Constant stress and strain amplitude testing, 252–268, 254f
  strain life approach in, 265–268, 266f, 267f, 267t
  strain-controlled fatigue testing and Bauschinger effect in, 261–265, 262f, 263f, 264f, 265f
  stress-life fatigue (*S-N* curve) in, 255–261, 255f, 256f, 257f, 259f, 259t, 260f, 261f
Constant stress approach, to particulate-reinforced composites, 119
Constituent particles/inclusions, unintended consequence of, 241–244, 242f, 243b, 243f
Continuous object, 92
Continuum theories, 92
Contracted notation, 110, 111f, 111t
Copper, strain hardening exponent and strength coefficients for, 75t
Corrosion, high-temperature, 343
Corrosion-resistant martensitic landing-gear steel, properties of, 39–41, 40f
Crack bridging, 245, 246f, 247f
Crack deflection, 245, 245f
Crack deformation, 224, 225f
Crack growth models, 273–276, 275f
Crack initiation, 258, 269–273, 269f, 270f, 272f
Crack length, 240
Crack mouth opening displacement (CMOD), 232
Crack propagation, 258, 268
Crack size, 276, 277f
Crack tip opening displacement (CTOD), 236–237, 236f
  experimental determination of, 237, 237f
Creep, 303
  classification of, 303–305
  damage mechanisms, 333–337, 335f
    by gaseous-environmental attack, 337
    by loss of external section, 336
    by loss of internal section, 336–337, 336f
    by microstructural degradation, 337
  diffusional, 304f, 305
  logarithmic, 303–304, 304f
  in particle-containing alloys, 326–330, 327f, 328f, 329f
  recovery-based, 304–305, 304f
Creep activation energy, 312
Creep cavitation, 336, 336f
Creep deformation, 326
  mechanism maps, 330–333
    Ashby's deformation mechanism map in, 332–333, 332f, 333f, 334f
    independent and sequential processes during, 330–331, 331f
  role of diffusion during, 305–306, 306f
Creep fracture mechanism maps, 337–338, 338f
Creep life-related relationships, 338–341, 339f, 340f, 341f, 342f

Creep limiting design, 29–30, 30f, 301–346, 345–346b
  creep damage mechanisms in, 333–337, 335f
  creep fracture mechanism maps in, 337–338, 338f
  creep life-related relationships in, 338–341, 339f, 340f, 341f 342f
  of creep-resistant materials, 343–346, 344f
  high-temperature deformation constitutive relationships, 306–333, 307f
    creep deformation mechanism maps in, 330–333
      Ashby's deformation mechanism map in, 332–333, 332f, 333f, 334f
      independent and sequential processes during, 330–331, 331f
    creep in particle-containing alloys, 326–330, 327f, 328f, 329f, 330b
    creep mechanisms, summary of, 325–326, 325t
    diffusional creep in, 312–318, 313f, 315f, 317f
    dislocation glide in, 308–309, 308f
    grain boundary sliding in, 319–325, 319f, 320f
    Harper-Dorn creep in, 318–319
    power law creep in, 309–312
      Weertman model for dislocation creep, 310–312, 310f
    superplasticity in, 319–325, 321f, 322f, 323f, 324f
  role of diffusion during creep deformation, 305–306
Creep tests, 84–85, 86f, 87f
Creep-limited design approach, 4
Critical energy release rate, 4
Critical resolved shear stress (CRSS), 142
Critical stress intensity factor, 224
Cross-slip, 155
Crystalline materials, general dislocation theory for, 127–148, 143–147b
  crystals and defects in, 127–131, 128f
    Miller and Miller-Bravais indices in, 130–131, 130f
    theoretical strength of, 129–130, 129f
  dislocation generation in, 139–142
    interfacial dislocation sources, 141–142
    intragranular dislocation sources, 139–141, 140f
  dislocation movement in, 135–136, 135f
  dislocation reactions in, 136–139, 136f, 138f
  dislocations in, types of, 132–134, 132f, 133f, 133t
  geometrically necessary dislocations in, 142–143
Crystalline silica (cristobalite), 10f
Crystallographic fracture, 190, 190f
Crystals
  defects and, 127–131, 128f
  structure of, 48, 343
  theoretical strength of, 129–130, 129f
CTOD. *see* Crack tip opening displacement
Cup-cone type fracture, 192f, 193f
Cybersteel, Olson's flying, 39–43, 40f, 41f, 42f
Cyclic loading, 3
  in stress-controlled fatigue testing, 253
Cyclic softening, 265
Cyclic stress-strain curve, 263, 265f

D

da Vinci's sketches, 1–3, 2f
Damage, defined, 289

Damage tolerance, 4
Damage-tolerant design approach, 26–28, 291–292
  based on assumption of flaws, 240–241, 241f
  to fatigue-limited design, 29
Darken diffusivity, 312
DBTT. *see* Ductile-to-brittle-transition temperature
Deflection-limited designs, 302
Deformation-induced transformation, 138, 139b
Deformation-induced twinning, 138, 139b
Denuded zones, 316, 317f
Design allowables, concept of, 19–21, 20f, 21b
Deterministic microstructure, 35–36
Deviatoric strain tensor, 102
Diffusional creep, 304f, 305, 312–318, 313f, 315f, 317f
Dilatation, 100
Dislocation creep
  creation of, 310
  movement of, 310
  subsequent annihilation of, 310
  Weertman model for, 310–312, 310f
Dislocation density, 134
Dislocation generation, in strength limiting design, 188
Dislocation glide, 308–309, 308f
Dislocation pileup model, of grain boundary strengthening, 158–161, 158f, 160f, 160t
Dislocation theory, for crystalline materials, 127–148, 143–147b
  crystals and defects in, 127–131, 128f
    Miller and Miller-Bravais indices in, 130–131, 131f
    theoretical strength of, 129–130, 129f
  dislocation generation in, 139–142
    interfacial dislocation sources, 141–142
    intragranular dislocation sources, 139–141, 140f
  dislocation movement in, 135–136, 135f
  dislocation reactions in, 136–139, 136f, 138f
  dislocations in, types of, 132–134, 132f, 133f, 133t
  geometrically necessary dislocations in, 142–143
Dislocation-based plasticity theory, influence of, 86–87
Dislocation-dislocation interactions, 135f
Dislocation-particle interaction, 167
Dispersion strengthening, 185
Dispersoids
  definition of, 49
  in superplasticity, 324
Distortion, 100
Ductile fracture, 191–197, 192f, 193f, 194f, 194t, 195f, 196f, 197f, 198t
Ductile rupture, 336
Ductile striations, 287
Ductile-to-brittle-transition temperature (DBTT), 215–215
Ductility, 73–74
Dynamic deformation, 3
Dynamic grain growth, 323–324
Dynamics, 100

E
"Easy glide" regime, 151
EBSD. *see* Electron backscatter diffraction
Edge dislocations, 132, 132f, 133f, 133t
Elastic constants, 106t

Elastic deformation, 4, 105
Elastic energy, storage and transfer of, 110b
Elastic interaction, 163
Elastic modulus, 92–93, 93f, 106t
  of particulate composites, 118f
  resilience and, 110t
Elastic properties, 92–93, 93f
Elastic response, of materials, 91–126
  discussion on, 92–96, 93f, 94f, 95f, 96f
  elasticity theory in, development of, 96–114, 104f
    anisotropy in, 110–112, 111f, 111t
    axes in, transformation of, 102–105, 103f, 103t, 104f
    bulk modulus in, 105–108
    elastic modulus in, 106t
    Hooke's law in, 105–108, 106t, 107f
    Poisson's ratio in, 105–108, 106t
    polycrystal in, elastic behavior of, 112–114, 113t
    resilience in, 108–110, 109f, 110b, 110t
    shear modulus in, 105–108, 106t
    strain in, 100–102, 100f
    stress in, 97–100, 97f, 98f, 99f
    Young's modulus in, 105–108
  for high stiffness composite materials, 114–122, 114t, 115f
    fiber-reinforced composites in, 119–122, 119t, 120f
    particulate-reinforced composites in, 116–119, 116f, 117f
Elastic strain, 263
Elastic strain energy, 108
Elasticity theory, development of, 96–114, 104f
  anisotropy in, 110–112, 111f, 111t
  axes in, transformation of, 102–105, 103f, 103t, 104f
  bulk modulus in, 105–108
  elastic modulus in, 106t
  Hooke's law in, 105–108, 106t, 107f
  Poisson's ratio in, 105–108, 106t
  polycrystal in, elastic behavior of, 112–114, 113t
  resilience in, 108–110, 109f, 110b, 110t
  shear modulus in, 105–108, 106t
  strain in, 100–102, 100f
  stress in, 97–100, 97f, 98f, 99f
  Young's modulus in, 105–108
Elastic-plastic fracture mechanics, 235–240
  crack tip opening displacement in, 236–237, 236f
    experimental determination of, 237, 237f
  J-integral in, 238–239, 238f
    experimental determination of, 239–240, 239f
Electrical interaction, 166
Electron backscatter diffraction (EBSD), 284
"Electron sea," 7–8
Electronic gadgets, 17f
Endurance limit, 255–256, 257t
Engineering shear strain, 101
Engineering stress-strain curves, 72–73, 73f, 74f
Engineering structural materials, microstructural elements in, 47–62, 60b
  metallic materials, 49–55
    aluminum alloys, 51–52, 51f, 52t, 53f, 54f
    nickel-based superalloys, 55, 56f
    steels, 49–51, 50f

Engineering structural materials, microstructural elements in *(Continued)*
    titanium alloys, 55, 57f
    nonmetallic materials, 55–60
        amorphous materials, 55–56
        ceramics, 58–60, 59f
        polymer-based composites, 56–58, 58f
        polymers, 56–58, 58f
Engineering systems, materials and, 12–18
    advanced aircraft, 14, 14f
    automobile, 13, 13f
    miscellaneous examples of, 15–18, 16f, 17f
    power plant, 14, 15f
External necking, 336
Extrinsic toughening mechanisms, 247, 249b
Extrusions, in slip steps, 269, 269f

F

Face-centered cubic (FCC), 7–8, 128–129, 128f, 130f
    Hall-Petch parameters for, 160t
Fail-safe design approach, 29, 291
Failure mechanisms, in strength limiting design, 188–198, 189f, 190f
    low-temperature tensile fracture in, 191–198
        cleavage fracture, 198
        ductile fracture, 191–197, 192f, 193f, 194f, 194t, 195f, 196f, 197f, 198t
Far-field stress, 228–229
Far-for-equilibrium processing, of materials, for microstructural tailoring, 185b
Fatigue
    stress-life, 255–261, 255f, 256f, 257f, 259f, 259t, 260f, 261f
    test of, 79–81, 81f
Fatigue damage, 262
Fatigue deformation, influence of microstructure, 278–285, 278f, 279f, 280f, 281f, 282f, 283f, 284f
Fatigue ductility coefficient, 266
Fatigue failure, 290
Fatigue fracture, 285–287, 286f, 287f, 288f, 289f
Fatigue limit, 255–256
Fatigue limiting design, 28–29, 251–300, 298–299b
    avoidance/minimization of fatigue damage in, 296–298, 296f, 297f
    constant stress and strain amplitude testing in, 252–268, 252t, 254f, 258b
        strain life approach in, 265–268, 266f, 267f, 267t
        strain-controlled fatigue testing and Bauschinger effect in, 261–265, 262f, 263f, 264f, 265f
        stress-life fatigue (*S-N* curve) in, 255–261, 255f, 256f, 257f, 259f, 259t, 260f, 261f
    crack growth models, 273–276, 275f
    crack initiation and growth in, 268–276, 269f, 270f, 272f
    design approaches, 291–292
        Ashby charts, 292–293, 293f, 294f
        damage-tolerant, 29, 291–292
        fail-safe, 29, 291
        geometrical stress concentrations, 293–296, 295f
        infinite-life, 28–29, 291

Fatigue limiting design *(Continued)*
        safe-life, 29, 291
    design aspects in, 289–298
    fatigue deformation and influence of microstructure in, 278–285, 278f, 279f, 280f, 281f, 282f, 283f, 284f
    fatigue fracture, 285–287, 286f, 287f, 288f, 289f
    life prediction in, 276–277, 277f
Fatigue ratio, 278
Fatigue resistance, 278
Fatigue striations, 286–287, 288f
Fatigue testing, 4, 5f
Fatigue zone, 285, 286f
FATT. *see* Fracture appearance transition temperature
Ferrite (α-Fe), 142–143
Ferrium® S53 (AMS 5922) corrosion-resistant landing gear steel, 39–41, 41f, 42f
Fiber-reinforced composites, 119–122, 119t, 120f
Fiducial scratch marker offset method, 322–323
Fine grain size, in superplasticity, 323
Fine particle strengthening, 167–187
    associated work hardening behavior, 186
    further topics on, 179–187, 180f, 181f, 182f, 183f, 184f, 185b, 186f, 187b
    particle cutting or shearing mechanism in, 167–174, 168f, 169f, 170f, 171f, 172f, 172t, 173f, 174f
    particle hardening via strong barriers in, 174–179, 175f, 176f, 177f, 178f, 179b
    transition from particle cutting to Orowan looping in, 179
Flow stress, strain rate *vs.*, 77, 78f
Flying cybersteel, 39–43, 40f, 41f, 42f
Fracture appearance transition temperature (FATT), 218
Fracture surface, during impact testing, 218, 219f, 220f
Fracture toughness, 4, 5f, 25, 27f, 224–227, 226f, 227f
    testing, 81–83, 82f, 83f, 84f, 85f, 231–235
        for ductile materials, 233–235, 234f, 235f
        for less ductile materials, 231–233, 231f, 232f, 233t
Fracture transition plastic (FTP), 218
Frank-Read source, 139, 140–141b, 140f, 159–161
Frank's rule, 328
Friction stir processing, 322f
Friction stir-based additive manufacturing (FSAM), 185
FTP. *see* Fracture transition plastic

G

GBS. *see* Grain boundary sliding
Geometrical stress concentrations, 293–296, 295f
Glass fiber, compositions and properties of, 119t
Glass fiber-reinforced polymer (GFRP) composites, 11
Glide-climb creep, 331
Grain, definition of, 48
Grain boundary, 48, 284
    cavitation, 284
    character, 343
    diffusivity, 315–316
    migration, 317–318
Grain boundary sliding, 317–325, 319f, 320f, 343

Grain boundary strengthening, 157–161
  dislocation pileup model of, 158–161, 158f, 160f, 160t
  effect of temperature, 308–309
  Li's general model for Hall-Petch relationship in, 161
Grain boundary/interface character, in superplasticity, 325
Grain growth, dynamic, 323–324
Grain shape, in superplasticity, 323
Grain size, 199, 281–283, 283f, 284f, 343
  in transition temperature, 220–222, 221f
"Guaranteed minimum value," 20

H

Haigh diagrams, 259, 260f
Hall-Petch grain size effect, 180
Hall-Petch parameters, 159–161, 160t
Hall-Petch relation, 159–161, 160f
Hall-Petch relationship, Li's general model for, 161
Hall-Petch strengthening, 157, 308–309
Hardness, test of, 4, 5f, 64–71
  macrohardness, 65–66, 65t, 66f
  microhardness, 66–68, 67f, 68f
  nanoindentation, 68–71, 69f, 70f, 70t
Harper-Dorn creep, 312–313, 318–319
HCF. see High cycle fatigue
Hertz's theory, of nanoindentation, 70
Hexagonal close packed (HCP), 128–129, 128f
  Hall-Petch parameters for, 160t
High cycle fatigue (HCF), 4, 258, 261
High stiffness composite materials, design of, 114–122, 114t, 115f
High strain rate superplasticity (HSRS), 323–324, 324f
High-temperature deformation constitutive relationships, 306–333, 307f
  creep deformation mechanism maps in, 330–333
    Ashby's deformation mechanism map in, 332–333, 332f, 333f, 334f
    independent and sequential processes during, 330–331, 331f
  creep in particle-containing alloys, 326–330, 327f, 328f, 329f, 330b
  creep mechanisms, summary of, 325–326, 325t
  diffusional creep in, 312–318, 313f, 315f, 317f
  dislocation glide in, 308–309, 308f
  grain boundary sliding in, 319–325, 319f, 320f
  Harper-Dorn creep in, 318–319
  power law creep in, 309–312
    Weertman model for dislocation creep, 310–312, 310f
  superplasticity in, 319–325, 321f, 322f, 323f, 324f
HIP. see Hot isostatic pressing
Hollomon's equation, 74–77, 75f, 75t, 76f, 77t, 78f, 214
Homogeneous object, 92
Hooke's law, 92, 105–108, 106t, 107f
Hot isostatic pressing (HIP), 281
Hot tear, 43
Hot working, 303
HSRS. see High strain rate superplasticity
Hydrostatic component, of stress tensor, 100
Hydrostatic strain tensor, 102
Hyperplanes, 1–3

I

Impact energy, 216–218, 217f
Impact toughness, 215–224
  transition temperature approach to fracture control of, 215–218, 216f, 217f, 218t, 219f, 220f
Impurity content, 242
Incremental step method, 263
Independent process, during creep deformation, 330–331, 331f
Infinite-life design, 28–29, 291
Instantaneous zone, 285, 286f
Intended microstructural component, 34f
Interfaces, definition of, 48
Interfacial dislocation sources, 141–142
Intergranular fracture, 336–337
Interphase interfaces, 49
Interstitial atoms, 162, 162f
Intragranular dislocation sources, 139–141, 140f
Intrinsic creep strength-limited design, 29–30
Intrinsic toughening mechanisms, 247, 249b
Intrusions, in slip steps, 269, 269f
"Inverse Hall-Petch" effect, 159–161
Irwin, George, 227–228
Isotropic hardening, 205–206, 205f
Isotropic material, 92
Izod impact testing, 215–216, 217f

J

J-integral, 238–239, 238f
  experimental determination of, 239–240, 239f
Jog, 135–136, 135f

K

Kinematic hardening, 205–206, 205f
Kinks, 135–136, 135f
Knoop indenter, microhardness testing with, 67–68, 67f, 68f
Kocks model, 180

L

Large-scale plasticity, in strength limiting design, 188
Larson-Miller parameter (LMP), 333–334, 339, 340
Laser peening, 297
Laser powder-based additive manufacturing (LPBAM), 185
Laser shock processing, 297
LCF. see Low cycle fatigue
LEFM. see Linear elastic fracture mechanics
Life prediction, 276–277, 277f
Lifshitz sliding mechanism, 319–320
Light microscopy, for fatigue striations, 287
Linear elastic behavior, 92, 239
Linear elastic fracture mechanics (LEFM), 224, 235–236
Li's general model, for Hall-Petch relationship, 161
Load-displacement curves, 232, 232f
Logarithmic creep, 303–304, 304f
Lomer-Cottrell barriers, 311
Long-range order interaction, 166
Low cycle fatigue (LCF), 258
Lower stacking fault energy, 344

Low-temperature tensile fracture, 191–198
   cleavage fracture, 198
   ductile fracture, 191–197, 192f, 193f, 194f, 194t, 195f, 196f, 197f, 198t

M

M54 alloys, 41–43, 41f
Macrohardness, test of, 65–66, 65t, 66f
Macromolecules, 10
Martensite, 164
Materials
   broad categories of, 6–11, 7f
      ceramics, 8–9, 9f, 10f, 12t
      composites, 11
      metals/alloys, 7–8, 12t
      polymers, 10–11, 10f, 12t
   engineering systems and, 12–18
      advanced aircraft, 14, 14f
      automobile, 13, 13f
      miscellaneous examples of, 15–18, 16f, 17f
      power plant, 14, 15f
   fatigue behavior of, 251–300, 298–299b
      avoidance/minimization of fatigue damage in, 296–298, 296f, 297f
      constant stress and strain amplitude testing in, 252–268, 252t, 254f, 258b
         strain life approach in, 265–268, 266f, 267f, 267t
         strain-controlled fatigue testing and Bauschinger effect in, 261–265, 262f, 263f, 264f, 265f
         stress-life fatigue (S-N curve) in, 255–261, 255f, 256f, 257f, 259f, 259t, 260f, 261f
      crack growth models, 273–276, 275f
      crack initiation and growth in, 268–276, 269f, 270f, 272f
      design approaches, 291–292
         Ashby charts, 292–293, 293f, 294f
         damage-tolerant, 291–292
         fail-safe, 291
         geometrical stress concentrations, 293–296, 295f
         infinite-life, 291
         safe-life, 291
      design aspects in, 289–298
      fatigue deformation and influence of microstructure in, 278–285, 278f, 279f, 280f, 281f, 282f, 283f, 284f
      fatigue fracture, 285–287, 286f, 287f, 288f, 289f
      life prediction in, 276–277, 277f
   mechanical response of, 1–18, 2f, 17–18b
      Ashby's property charts, 5–6, 6f
      introduction, 3–6, 3f, 5f
   toughness of, 213–250, 249–250b
      damage-tolerant design approach based on assumption of flaws in, 240–241, 241f
      definitions of, 213–227
         fracture toughness, 224–227, 226f, 227f
         impact toughness, 215–224
         tensile toughness, 214–215, 215f
      elastic-plastic fracture mechanics in, 235–240
         crack tip opening displacement in, 236–237, 236f
         J-integral in, 238–239, 238f

Materials (Continued)
   fracture toughness testing in, 231–235
      for ductile materials, 233–235, 234f, 235f
      for less ductile materials, 231–233, 231f, 232f, 233t
   stress intensity and role of mechanics in, 227–230, 228f, 229f, 230b
   toughening of ceramics in, 244–247, 244t, 246f, 247f, 248f
   unintended consequence of constituent particles/inclusions in, 241–244, 242f, 243b, 243f
Materials Genome Initiative, 36–38, 37f
"Materials science," 3
Mean stress, 254, 260f
Mechanical anisotropy, in tensile testing, 78–79, 80f
"Mechanical Behavior of Materials," 36–38
Mechanical design framework, Ashby's, 21–30
   creep limiting design, 29–30, 30f
   fatigue-limited design, 28–29
   stiffness-limited design, 22–23, 22f, 23t
   strength-limited design, 23–25, 24f, 26t
   toughness-limited design, 25–28, 27f, 28t
Mechanical response, of materials, 1–18, 2f
   Ashby's property charts, 5–6, 6f
   introduction, 3–6, 3f, 5f
Mechanical tests, simple, 63–90, 88–89b
   creep tests, 84–85, 86f, 87f
   fatigue testing, 79–81, 81f
   fracture toughness testing, 81–83, 82f, 83f, 84f, 85f
   hardness testing, 64–71
      macrohardness, 65–66, 65t, 66f
      microhardness, 66–68, 67f, 68f
      nanoindentation, 68–71, 69f, 70f, 70t
   mechanics and dislocation-based plasticity theory, 86–87
   methods for, 64
   tensile testing, 71–78, 72f, 73f, 74f
      Hollomon's equation of, 74–77, 75f, 75t, 76f, 77t, 78f
      mechanical anisotropy in, 78–79, 80f
      temperature in, 78, 79f
Mechanics, influence of, 86–87
Melting point, 95, 343
Metal matrix composites (MMCs), 11
Metallic materials, 4, 49–55
   aluminum alloys, 51–52, 51f, 52t, 53f, 54f
   nickel-based superalloys, 55, 56f
   steels, 49–51, 50f
   titanium alloys, 55, 57f
Metallic Materials Properties Development and Standardization (MMPDS), 20–21, 20f
Metallurgical factors, transition temperature and, 219
Metals/alloys, 7–8, 12t
   elastic moduli range in, 92–93, 93f
Microelectronics, 16–18
Microhardness, test of, 66–68, 67f, 68f
Microstructural distribution, consequent effects and, 198–200, 199f, 200f, 201f, 202f
Microstructural tailoring, nonequilibrium processing of materials for, 185b
Microstructure, 25
   definition of, 47–48

Microstructure *(Continued)*
  design of, Olson's systems approach to materials by, 35–43, 35f, 37f
  processing-microstructure-properties correlations, importance of, 43–46, 44f, 45f
  unintended, 33–34, 34f
Microvoid nucleation, 193–195
Miller and Miller-Bravais indices, in general dislocation theory, 130–131, 130f
Miller indexing method, 110
Misfit dislocation, 168f
Modeling strong dislocation barriers, 179
Modulus effect, 171
Modulus interaction, 163
"Modulus of resilience," 109
Modulus-strength chart, 206, 207f
Monkman-Grant relationship, 311
Monoclinic, 247
Monocrystalline metals, critical normal stresses in, 198t
Monomers, 10–11
Monotonic loading, 3
Multiaxial loading, effects of, on yielding, 202–206
  Tresca's yield criterion, 204
  von Mises criterion, 203–204
  yield locus, 204–206, 205f

N

Nabarro-Herring (N-H) creep, 312–313, 313f
  diffusional, 316–317
Nanocrystalline grain structure, 318
Nanoindentation, test of, 68–71, 69f, 70f, 70t
Natural aging, 181–182
NDT. *see* Nil ductility temperature
Nickel-based superalloys, 55, 56f
Nil ductility temperature (NDT), 218
Noncrystalline silica glass, 9, 10f
Nonequilibrium processing, of materials for microstructural tailoring, 185b
Nonmetallic materials, 55–60
  amorphous materials, 55–56
  ceramics, 58–60, 59f
  polymer-based composites, 56–58, 58f
  polymers, 56–58, 58f
Normal stress, 98–99
Norton's law, 306
Notch sensitivity factor, 295f

O

ODS. *see* Oxide dispersion-strengthened
OIM. *see* Orientation imaging microscopy
Olson's systems approach, to materials, 33–46, 34f, 45–46b
  flying cybersteel, 39–43, 40f, 41f, 42f
  by microstructural design, 35–43, 35f, 37f
  processing-microstructure-properties correlations, importance of, 43–46, 44f, 45f
Optical micrographs, of aluminum alloys, 51f, 52–54, 53f, 54f

Order hardening, 171
Orientation imaging microscopy (OIM), 48
  of aluminum alloys, 52–54, 53f
Orientations, transition temperature and, 222, 223f
Orowan bypassing, 174, 175f, 176f, 185
Orowan equation, 135
  derivation of, 176, 177f
  modification of, 178
Orowan looping, transition to, 179
Orowan strengthening mechanism, 326
Orowan-Bailey equation, 305
Oxidation, high-temperature, 343
Oxide dispersion-strengthened (ODS), 328–329

P

Palmgren-Miner rule, 289
Parabolic hardening stage, 151
Paris law, 271–272, 274
Partial dislocation, 136–137, 139b
Particle cutting, in fine particle strengthening, 167–174, 168f, 169f, 170f, 171f, 172f, 172t, 173f, 174f
Particle hardening, in fine particle strengthening, 174–179, 175f, 176f, 177f, 178f, 179b
Particle size, 199, 199f, 200f, 201f
Particle strengthening, 343
  effect of temperature on, 309
Particulate-reinforced composites, 116–119, 116f, 117f
Persistent slip bands (PSBs), 271
Peterson's method, 295
Plane strain fracture toughness testing procedure, 231–233, 231f, 232f, 233t
Plane stress, 104, 225–227
Plastic deformation, 3, 139, 155
Plastic strain, 263
Plastic zone
  defined, 225
  fracture toughness testing for, 233–235, 234f, 235f
Plastics. *see* Polymers
Poisson's ratio, 105–108, 106t
Polycrystal
  elastic behavior of, 112–114, 113t
  fine particle strengthening and, 179, 180f
  single crystal to, 152–157, 153f, 155f, 156f
Polyethylene, molecular structure of, 10f
Polymer-based composites, 56–58, 58f
Polymers, 10–11, 10f, 12t, 56–58, 58f
  elastic moduli range in, 93f
Pop-in behavior, 233
Powder metallurgy type method, 167
Power law creep, 309–312
Power plant, 14, 15f
Precipitate-free zones, 280
Precipitates, definition of, 49
Precipitation strengthening, 181, 182f
Probabilistic microstructure, 35–36
PSBs. *see* Persistent slip bands
Push-pull type test, 255, 255f
Pythagorean superposition, 187

Q

Quasistatic deformation, 3, 3f
QuesTek Innovations, 39–41, 39f

R

Rachinger sliding, 319–320
Rate of loading, 3
Recovery-based creep, 304–305, 304f
Resilience, 108–110, 109f, 110b, 110t
Reuss average method, 113
Rockwell hardness test, 65, 65t

S

Safe-life design, 29, 291
S-basis, 21
Scanning electron microscope (SEM), 48
    of aluminum alloys, 52–54, 53f, 54f
    of ceramics, 59f
    for fatigue striations, 287
Scatter, in fatigue data, 256–257
Schmid's law, 270–271
Screw dislocations, 132, 132f, 133f, 133t, 162
SEM. *see* Scanning electron microscope
Sequential process, during creep deformation, 330–331, 331f
SFE. *see* Stacking fault energy
Shakedown, in constant stress amplitude testing, 262–263
Shear fracture, 192f
Shear lip region, 225–227
Shear modulus, 22, 105–108, 106t
Shear offset, 322–323, 323f
Shear stress, 98–99, 99f
    dislocation and, 174
    Orowan relationship and, 174
Shearing mechanism, in fine particle strengthening, 167–174, 168f, 169f, 170f, 171f, 172f, 172t, 173f, 174f
Sherby-Dorn parameter, 341, 342f
Shockley partials, 136–137
Short-range order interaction, 166–167
Shot peening, 296, 296f
Silica glass (noncrystalline), 9, 10f
Silicate glasses, 9
Simple mechanical tests, 63–90, 88–89b
    creep tests, 84–85, 86f, 87f
    fatigue testing, 79–81, 81f
    fracture toughness testing, 81–83, 82f, 83f, 84f, 85f
    hardness testing, 64–71
        macrohardness, 65–66, 65t, 66f
        microhardness, 66–68, 67f, 68f
        nanoindentation, 68–71, 69f, 70f, 70t
    mechanics and dislocation-based plasticity theory, 86–87
    methods for, 64
    tensile testing, 71–78, 72f, 73f, 74f
        Hollomon's equation of, 74–77, 75f, 75t, 76f, 77t, 78f
        mechanical anisotropy in, 78–79, 80f
        temperature in, 78, 79f
Simple tensile test, 4
Simplest tests, 3

Single crystals
    to polycrystal, 152–157, 153f, 155f, 156f
    stress-strain curve of, 150–152, 151f, 152f
Single-phase polycrystals, 142
Sliding or in-plane shear mode, of crack deformation, 224, 225f
Slip activity, 271
Slip homogenization, 279–280
Slip localization, 280
S-N data, 4
Society of Automotive Engineers (SAE), 20–21
Solid solution strengthening, 161–167, 162f, 165f, 166f
Soluble constituents, in fracture toughness, 242
Solution heat treatment (SHT), 181–182
SPF. *see* Superplastic forming
Sports equipment, 15–16, 16f
Squeeze casting, 281, 282f
Srolovitz's model, 326–327
Stacking fault energy (SFE), 138, 171, 171f
Stacking faults, 137
Steady-state creep rate, 306
Steels, 49–51, 50f
    carbon, strain hardening exponent and strength coefficients for, 75t
    corrosion-resistant martensitic landing-gear, properties of, 39–41, 40f
    production process of, 43–46, 44f
Stiffness limiting design, 22–23, 22f, 23t, 91–126
    discussion on, 92–96, 93f, 94f, 95f, 96f
    elasticity theory in, development of, 96–114, 104f
        anisotropy in, 110–112, 111f, 111t
        axes in, transformation of, 102–105, 103f, 103t, 104f
        bulk modulus in, 105–108
        elastic modulus in, 106t
        Hooke's law in, 105–108, 106t, 107f
        Poisson's ratio in, 105–108, 106t
        polycrystal in, elastic behavior of, 112–114, 113t
        resilience in, 108–110, 109f, 110b, 110t
        shear modulus in, 105–108, 106t
        strain in, 100–102, 100f
        stress in, 97–100, 97f, 98f, 99f
        Young's modulus in, 105–108
    for high stiffness composite materials, 114–122, 114t, 115f
        fiber-reinforced composites in, 119–122, 119t, 120f
        particulate-reinforced composites in, 116–119, 116f, 117f
Strain, definition of, 100–102, 100f
Strain energy density, 108–110, 109f, 238
Strain hardening, 150–157
    single crystal to polycrystal, 152–157, 153f, 155f, 156f
    stress-strain curve of single crystals, 150–152, 151f, 152f
Strain life approach, 265–268, 266f, 267f, 267t
Strain rate, 77, 77t, 78f
Strain tensor, 102
Strain-controlled fatigue testing, 261–265, 262f, 263f, 264f, 265f
Strength, *versus* toughness, 243b
Strength limiting design, 23–25, 24f, 26t, 149–212, 208–211b
    dislocation generation in, 188

Strength limiting design *(Continued)*
  effect of multiaxial loading on yielding, 202–206
    Tresca's yield criterion, 204
    von Mises criterion, 203–204
    yield locus, 204–206, 205f
  failure mechanisms in, 188–198, 189f, 190f
    low-temperature tensile fracture in, 191–198
      cleavage fracture, 198
      ductile fracture, 191–197, 192f, 193f, 194f, 194t, 195f, 196f, 197f, 198t
    large-scale plasticity in, 188
  microstructural distribution and consequent effects in 198–200, 199f, 200f, 201f, 202f
  principles and examples of, 206, 207f
  strengthening mechanisms in engineering materials, 150–188
    fine particle strengthening in, 167–187
      further topics on, 179–187, 180f, 181f, 182f, 183f, 184f, 185b, 186f, 187b
      particle cutting or shearing mechanism in, 167–174, 168f, 169f, 170f, 171f, 172f, 172t, 173f, 174f
      particle hardening via strong barriers in, 174–179, 175f, 176f, 177f, 178f, 179b
      transition from particle cutting to Orowan looping in, 179
    grain boundary strengthening in, 157–161
      dislocation pileup model of, 158–161, 158f, 160f, 160t
      Li's general model for Hall-Petch relationship in, 161
    solid solution strengthening in, 161–167, 162f, 165f, 166f
    strain hardening/work hardening, 150–157
      single crystal to polycrystal, 152–157, 153f, 155f, 155f
      stress-strain curve of single crystals, 150–152, 151f, 152f
    superposition of, 187–188
Strengthening mechanisms, in engineering materials, 150–188
  fine particle strengthening in, 167–187
    further topics on, 179–187, 180f, 181f, 182f, 183f, 184f, 185b, 186f, 187b
    particle cutting or shearing mechanism in, 167–174, 168f, 169f, 170f, 171f, 172f, 172t, 173f, 174f
    particle hardening via strong barriers in, 174–179, 175f, 176f, 177f, 178f, 179b
    transition from particle cutting to Orowan looping in, 179
  grain boundary strengthening in, 157–161
    dislocation pileup model of, 158–161, 158f, 160f, 160t
    Li's general model for Hall-Petch relationship in, 151
  solid solution strengthening in, 161–167, 162f, 165f, 166f
  strain hardening/work hardening, 150–157
    single crystal to polycrystal, 152–157, 153f, 155f, 155f
    stress-strain curve of single crystals, 150–152, 151f, 152f
  superposition of, 187–188
Stress, definition of, 97–100, 97f, 98f, 99f
Stress amplitude, 254, 289, 290, 290f
  constant, 262–263, 264f
Stress concentration, 230, 230b
Stress cycles, 253, 254f
Stress intensity, 227–230, 228f, 229f, 230b
Stress intensity factor (K), 81–82, 83f, 224, 229f
Stress ratio, 254
Stress relaxation, 302
Stress triaxiality, 225

Stress-amplitude testing, 4, 5f
Stress-controlled fatigue testing, 253
Stress-life fatigue (*S-N* curve), 255–261, 255f, 256f, 257f, 259f, 259t, 260f, 261f
Stress-strain curves
  engineering, 72–73, 73f, 74f
  of single crystals, 150–152, 151f, 152f
  true, 74, 74f
Structural components, framework of five basic design approaches for, 19–32, 30b
  Ashby's basic mechanical design framework, 21–30
    creep limiting design, 29–30, 30f
    fatigue-limited design, 28–29
    stiffness-limited design, 22–23, 22f, 23t
    strength-limited design, 23–25, 24f, 26t
    toughness-limited design, 25–28, 27f, 28t
  concept of design allowables, 19–21, 20f, 21b
Substitutional atoms, 161–162, 162f
Substitutional solutes, 344
Substructure-induced acceleration, of creep, 337
Superalloys, nickel-based, 55, 56f
Super-dislocation, 172–174
Superplastic deformation, 323
Superplastic forming (SPF), 323
Superplasticity, 319–325, 321f, 322f, 323f, 324f
Superposition, of strengthening mechanisms, 187–188
Supersaturated solid solution (SSSS), 181–182
Surface hardening, 174, 174f
Suzuki effect, 166
Systems approach, to materials, 33–46, 34f, 45–46b
  flying cybersteel, 39–43, 40f, 41f, 42f
  by microstructural design, 35–43, 35f, 37f
  processing-microstructure-properties correlations, importance of, 43–46, 44f, 45f

T

Taylor factor, 154, 155f, 180
Taylor-Elam experiment, 153, 153f
Tearing or antiplane shear mode, of crack deformation, 224, 225f
Technology readiness levels (TRL), timeline of, 41–43, 41f
TEM. *see* Transmission electron microscopy
Temperature
  elastic modulus and, 95–96, 96b, 96f
  tempering, 222, 222f
  in tensile testing, 78, 79f
Tensile ductility, Brown-Embury analysis for, 191, 194f
Tensile fracture, 192f
Tensile opening mode, of crack deformation, 224, 225f
Tensile stress, 98–99
Tensile toughness, 214–215, 215f
Tension, test of, 71–78, 72f, 73f, 74f
  Hollomon's equation of, 74–77, 75f, 75t, 76f, 77t, 78f
  mechanical anisotropy in, 78–79, 80f
  temperature in, 78, 79f
Tensor of rank, 99
Tensorial shear strain, 101
Theoretical strength of crystal, 129–130, 129f
Thickness, transition temperature and, 222–224, 223f

Thompson tetrahedra, 138f
Threshold stress, 328
Titanium alloys, 55, 57f
Total dislocation density, 142
Total strain, 263
Toughness
    fracture, test of, 81–83, 82f, 83f, 84f, 85f
    versus strength, 243b
Toughness limiting design, 25–28, 27f, 28t, 213–250, 249–250b
    damage-tolerant design approach based on assumption of flaws in, 240–241, 241f
    definitions of, 213–227
        fracture toughness, 224–227, 226f, 227f
        impact toughness, 215–224
        tensile toughness, 214–215, 215f
    elastic-plastic fracture mechanics in, 235–240
        crack tip opening displacement in, 236–237, 236f
        J-integral in, 238–239, 238f
    fracture toughness testing in, 231–235
        for ductile materials, 233–235, 234f, 235f
        for less ductile materials, 231–233, 231f, 232f, 233t
    stress intensity and role of mechanics in, 227–230, 228f, 229f, 230b
    toughening of ceramics in, 244–247, 244t, 246f, 247f, 248f
    unintended consequence of constituent particles/inclusions in, 241–244, 242f, 243b, 243f
Transformation angle designations, 103t
Transformation toughening, 247
Transformation-induced plasticity (TRIP), 138, 139b
Transition temperature
    approach to fracture control, 215–218, 216f, 217f, 218f, 219f, 220f
    factors affecting, 219–224, 221f, 222f, 223f
Transmission electron microscopy (TEM)
    of aluminum alloys, 52–54, 53f, 54f
    for fatigue striations, 287
Transverse elastic modulus, 121
Tresca's yield criterion, 204
TRIP. *see* Transformation-induced plasticity
True strain, 92
True stress, 92, 186f
True stress-strain curve, 74, 74f
Twin boundaries, definition of, 49
Twinning-induced plasticity (TWIP), 138, 139b
TWIP. *see* Twinning-induced plasticity

U
Uniaxial tension test, 4
Unintended microstructural feature, 33–34, 34f

V
Vacancy, defined, 127–128
Vaulting pole, 114–115
Vickers hardness, test of, 66, 66f
Vickers indenter, microhardness testing with, 66, 67f
Viscoelasticity, 3
Viscoplasticity, 3
Void growth, 195–197
Void interlinkage for ductile fracture, 197f
Void nucleation
    Brown-Embury analysis of, 192–193, 196f
    strains at room temperature, 194t
Voigt average method, 113
von Mises criterion, 203–204

W
WC-Co cermet, microstructure of, 58–60, 59f
Wedge crack formation, 337
Weertman model, for dislocation creep, 310–312, 310f
Weertman pillbox model, 311
Welding, 35–36
Westergaard stress function, 227–228, 228f
Wilm, Alfred, 181
Work hardening, 150–157
    fine particle strengthening and, 186, 186f
    single crystal to polycrystal, 152–157, 153f, 155f, 156f
    stress-strain curve of single crystals, 150–152, 151f, 152f
Wrought condition, of metallic material, 52–54, 53f
Wrought products, deformation of, 43–46

Y
Yield locus, 204–206, 205f
Young's modulus, 22, 22f. *see also* Elastic modulus

Z
Zener pinning, in superplasticity, 324
Zener-Hollomon parameter, 341, 342f